유기화학의
어원과 용어

Etymology and Glossary of Organic Chemistry

유기화학의 어원과 용어

유기화학에서 사용하는 영어로 된 전문용어의 어원과 관련용어의 사전
유기화학의 어원을 알고나면 기본 개념을 쉽게 이해할 수 있다

Etymology and Glossary of Organic Chemistry

이왕근 편저

전남대학교출판부

어원(語源)이란, 순수한 말로 말밑이라고도 하며, 어떤 말이 생겨서 이루어진 역사적인 근원, 또는 어떤 말이 생겨난 근원이라고 정의할 수 있다.

인간은 언어를 사용하여 사회집단의 구성원으로서, 문화에 대한 참여자로서 의사전달과 개인 간의 정보전달을 한다.

장자는 말하기를 언어란 뜻을 전하면 사라지는 것이라 하였다. 그 뜻을 명확히 알고 전달하기 위하여 인류가 맨 처음 무슨 근거에서, 왜 이 말을 사용하였는가 알 필요가 있다. 그래야 그 당시의 문화를 유추할 수 있고, 그 당시 문화에서 그 언어의 참 뜻을 유추할 수 있기 때문이다. 낱말 하나 밑에 문화의 전 체계가 달려있다. 그래서 언어는 살아있는 생명체이다.

유기화학을 가르치면서 교과서에 소개된 영어로 된 전문(專門) 낱말의 어원을 알고 나면, 그 뜻을 이해하기가 한결 쉬워짐을 알 수 있었다. 이것이 동기가 되어 이 책을 쓰기 시작하였다. 사전을 찾아 그 낱말의 어원을 알아내고, 그 낱말과 관련된 전문 용어를 해설하는 것으로, 저서라기보다는 편집하였다는 것이 적절한 표현일지도 모른다.

유기화학의 용어를 올바르게 이해한다는 것은 유기화학에 대한 기본 개념을 정확히 이해하여, 기초지식을 튼튼히 쌓고 나아가 교과서나

해당 논문을 이해하는 데 큰 도움이 되는 것은 새삼 강조할 필요가 없다. 유기화학을 배우거나 가르치는 자라면 유기화학에서 사용되는 낱말의 어원과 관련 용어를 찾는 수고를 하지 않으면 안 되기에, 그 번거러움을 덜어주고 싶어서 이 책을 저술하였다. 정확히 말하면 이 책은 유기화학에서 사용하는 영어로 된 용어의 어원과 관련용어의 사전인 셈이다.

이 책은 유기화학을 공부하는 학부학생, 대학원생과 유기화학을 직접 가르치는 교수에게 참고서로 자리매김하리라 확신한다. 아울러 과학도들이 어원을 알아가면서, 서양문명의 발원지인 그리스나 로마의 문명을 조금이라도 이해하는 과정에서 인문학적 소양을 만난다면 창의적인 생각이 필요한 과학도들에게 많은 보탬이 될 것이라 생각한다.

이 편저가 유기화학을 공부하는 학문 후속세대에게 이어지는 드넓은 다리가 되기를 바란다.

이 책을 쓸 수 있도록 생각을 일깨워 준, 의학어원론을 쓴 정 상우 교수에게 감사함을 전하고 싶다.

2014년 7월

이 왕근

약어와 기호

Arabic. 아랍어에서 기원한 낱말이 영어에 들어 온 말

Aymara: 아이마라족어(볼리비아와 페루의 Titicaca 호수 주변 산악에 사는 인디오의 일족)

D. *Deutsch* (독일어), 독일어에서 기원한 낱말이 영어에 들어 온 말

E. English (영어)

F. *French* (프랑스어), 프랑스어에서 기원한 낱말이 영어에 들어 온 말

IE. *Indo-European root* (인도-유럽어 말 뿌리)

Italian. 이탈리아어

G. *Greek* (그리스어), 그리스어에서 기원한 낱말이 영어에 들어 온 말

L. *Latin* (라틴어), 라틴어에서 기원한 낱말이 영어에 들어 온 말

>, < 파생어나 복합어가 만들어진 방향

예: G > L: 그리스어에서 기원한 낱말이 라틴어화라는 통과의례를 거친 후 영어에 들어 온 말

G > L, end(o)- < endon: 그리스어원은 endon, 라틴어원은 end(o)-

ME. *middle English* (중세영어)

OF. *old French* (고전 프랑스어)

ON. *old norse* (고대 스칸디나비아어)

Portuguese. 포르투갈어

Quechua: 케추아어(잉카문명권의 공용어)

pl. *plural* (복수)

pp. past participle (과거분사)

Suffix, -ase, -ia, -ic, -ide, -ie, -in, -ine, -ium, -yl 등과 diminutives는 따로 명기를 하지 않았다.

두 단어가 명기된 경우, 뒤에 적은 단어로부터 앞의 단어가 유래된 것을 나타낸다.

예: *temperatura, temperare. temperare*로부터 *temperatura*가 유래되었다.

명사의 어원을 설명할 때 동사, 명사의 순으로 어원을 제시하였다.

예: *combinare*, *combination.*

어원과 용어를 설명할 때 맨 먼저 화학술어집(제 4개정증보판)에서 사용한 표기를 적어두고, 다른 표현을 따로 명기하였다.

absolute[ǽbsəlu:t]: L. *absolutus*, pp of *absolvere*, freed, unrestricted. 절대적인, 온전한, 실제의.

<u>absolute configuration</u>: 절대배열. The spatial arrangement of the atoms of a chiral molecular entity (or group) and its stereochemical description e.g. R or S. Absolute configurations for a chiral molecule (in pure form) are most often obtained by X-ray crystallography. Alternative techniques are Optical rotatory dispersion[광학 회전 분산(능)], vibrational circular dichroism (진동 원편광 이색성) and the use of chiral shift reagents in proton NMR. When the absolute configuration is obtained the assignment of R or S is based on the Cahn-Ingold-Prelog priority rules. Absolute configurations are also relevant to characterization of crystals.

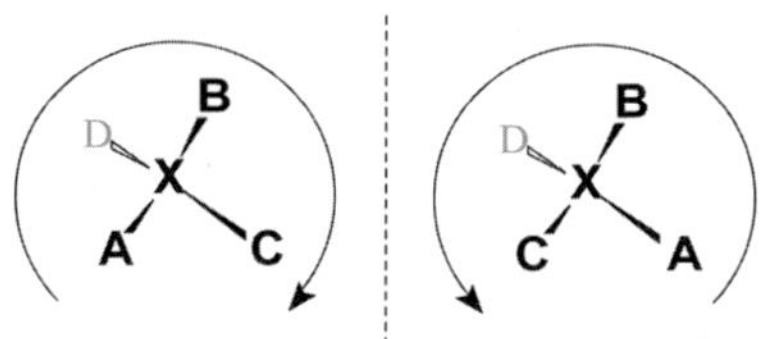

R S

Until 1951 it was not possible to obtain the absolute configuration of chiral compounds. It was at some time decided that (+)-glyceraldehyde was the (R)-enantiomer. The configuration of other chiral compounds was then related to that of (+)-glyceraldehyde by sequences of chemical reactions. For example (+)-glyceraldehyde (**1**) was related to (-)-glyceric acid **2** (oxidation by mercury oxide) which in turn was related to (+)-isoserine **3** (nitric acid oxidation) and bromide **4** and (-)-lactic acid **5** (zinc reduction). Because the chemical transformations did not affect the asymmetric carbon atom, this sequence demonstrated that (-)-lactic acid was also a (R)-enantiomer. In 1951 Bijvoet for the first time used X-ray crystallography and a new property called anomalous dispersion to determine absolute configuration. The compound investigated was (+)-sodium rubidium tartrate and from its configuration (R,R). It was deduced that the original guess for (+)-glyceraldehyde was correct.

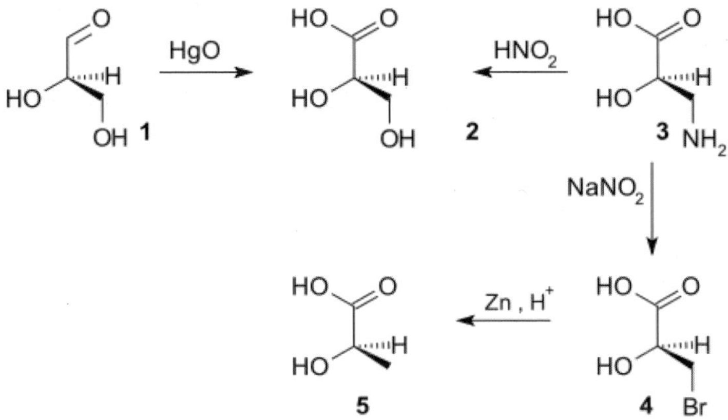

absorption[æbsɔ́:rpʃən, -zɔ́:rp- /əb-]: L. *absorbere*, *absorption*, swallow up. 흡수, 흡광. A physical or chemical phenomenon or a process in which atoms, molecules, or ions enter some bulk phase - gas, liquid or solid material. This is a different process from adsorption, since molecules undergoing absorption are taken up by the volume, not by the surface (as in the case for adsorption). A more general term is "sorption", which covers absorption, adsorption and ion exchange.

Absorption is basically where something takes in another substance.

acetal[ǽsətæl]: *acetic acid + alcohol*. 아세탈, 알데히드(케톤)와 알코올과의 화합물의 총칭. A molecule with two single bonded oxygens attached to the same carbon atom. Traditional usages distinguish ketal from acetal (whereas the ketal has two carbon-bonded R groups, the acetal has one carbon-bonded R group as H-). Current accepted terminology classifies ketals as a subset of acetals

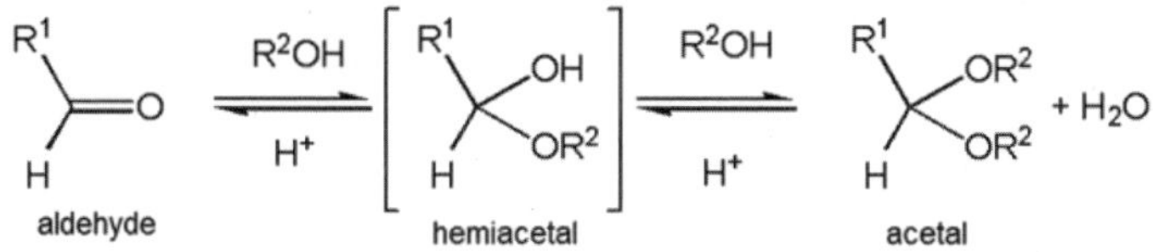

Aldehyde to acetal conversion

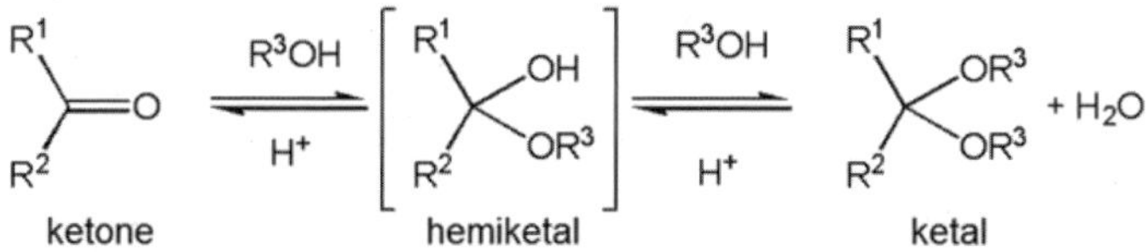

Ketone to ketal conversion

acetic[əsíːtik, əsét-]: L. *acetum*, vinegar. 식초의. 아세트산의.

acetic acid (= ethanoic acid): 식초산. An organic compound with the chemical formula CH_3CO_2H (also written as CH_3COOH). It is a colourless liquid that when undiluted is also called *glacial acetic acid*(빙초산). Acetic acid is the main component of vinegar (apart from water), and has a distinctive sour taste and pungent smell. It is mainly produced as a precursor to polyvinylacetate and cellulose acetate. Although it is classified as a weak acid, concentrated acetic acid is corrosive, and attacks the skin. Acetic acid is one of the simplest carboxylic acids. It is an important chemical reagent and industrial chemical, mainly used in the production of cellulose acetate mainly for photographic film and polyvinyl acetate for wood glue, as well as synthetic fibres and fabrics. In households, diluted acetic acid is often used in descaling agents. In the food industry, acetic acid is used under the food additive code E260 as an acidity regulator and as a condiment(조미료, 양념).

acetylcholine[əsi:tlkóulin, æsətl-]: 아세틸콜린(자율신경절의 주 신경전달체). A neurotransmitter in both the peripheral nervous system (PNS, 말초신경계) and central nervous system (CNS, 중추신경계) in many organisms including humans. Acetylcholine is one of many neurotransmitters in the autonomic nervous system (ANS, 자율신경계) and the only neurotransmitter used in the motor division of the somatic(체세포의) nervous system. (Sensory neurons use glutamate and various peptides at their synapses.) Acetylcholine is also the principal neurotransmitter in all autonomic ganglia(신경절).

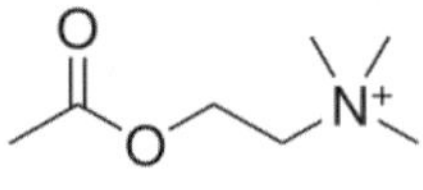

acetylcholinesterase (= cholinesterase): a family of enzymes that catalyze the hydrolysis of the neurotransmitter acetylcholine into choline and acetic acid, a reaction necessary to allow a cholinergic neuron to return to its resting state after activation.

acetyl CoA: an important molecule in metabolism, used in many biochemical reactions. Its main use is to convey the carbon atoms within the acetyl group to the citric acid cycle to be oxidized for energy production. In chemical structure, acetyl-CoA is the thioester between coenzyme A (a thiol) and acetic acid (an acyl group carrier). Acetyl-CoA is produced during the second step of aerobic cellular respiration, pyruvate decarboxylation, which occurs in the matrix of the mitochondria. Acetyl-CoA then enters the citric acid cycle. Acetyl-CoA is also an important component in the biogenic synthesis of the neurotransmitter acetylcholine. Choline, in combination with Acetyl-CoA, is catalyzed by the enzyme choline acetyltransferase to produce acetylcholine and a coenzyme as byproduct.

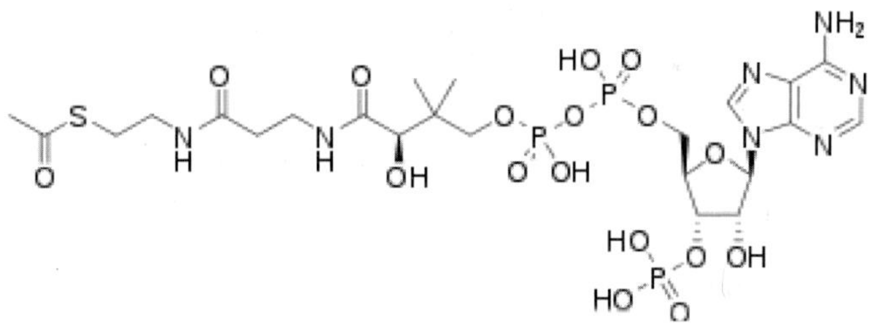

acetylene[əsétəlin]: mid 19th century. *acetum*, vinegar + *-yl* + *-enos,* one. 아세틸렌. Systematic name: ethyne. It is a hydrocarbon and the simplest alkyne. This colourless gas is widely used as a fuel and a chemical building block. It is unstable in pure form and thus is usually handled as a solution.

As an alkyne, acetylene is unsaturated because its two carbon atoms are bonded together in a triple bond.

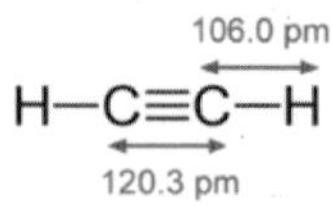

achiral[əkáiral]: G. *acheir* (*a*, not + *kheir*, hand). 손이 아닌. Describe an object that is superposable on its mirror image.

achiral (not chiral) objects: objects that are identical to their mirror image.

acid[æcid]: L. *acere, acidus, acida, acidum*, tasting sour or bitter, sharp-tongued. 신, 쓴, 혀에 예민한, 화학 (산, 산의 성질을 갖는). A substance which reacts with a base. Commonly, acids can be identified as tasting sour, reacting with metals such as calcium, and bases like sodium carbonate. Aqueous acids have a pH of less than 7, where an acid of lower pH is typically stronger. Chemicals or substances having the property of an acid are said to be acidic. Common examples of acids include acetic acid (in vinegar), sulfuric acid (used in car batteries), and tartaric acid (used in baking). As these three examples show, acids can be solutions, liquids, or solids. Gases such as hydrogen chloride can be acids as well. Strong acids and some concentrated weak acids are corrosive, but there are exceptions such as carboranes (a cluster composed of boron and carbon atoms) and boric acid. There are three common definitions for acids: the Arrhenius definition, the Brønsted-Lowry definition, and the Lewis definition.

Arrhenius acids

The Swedish chemist Svante Arrhenius attributed the properties of acidity to hydrogen in 1884. An Arrhenius acid is a substance that increases the concentration of the hydronium ion, H_3O^+, when dissolved in water. This definition stems from the equilibrium dissociation of water into hydronium and hydroxide (OH^-) ions:

$$H_2O(l) + H_2O(l) \rightleftharpoons H_3O^+(aq) + OH^-(aq)$$

In pure water the majority of molecules exist as H_2O, but a small number of molecules are constantly dissociating and re-associating. Pure water is neutral with respect to acidity or basicity because the concentration of hydroxide ions is always equal to the concentration of hydronium ions. An Arrhenius base is a molecule which increases the concentration of the hydroxide ion when dissolved in water. Note that chemists often write $H^+(aq)$ and refer to the hydrogen ion when describing acid-base reactions but the free hydrogen nucleus, a proton, does not exist alone in water, it exists as the hydronium ion, H_3O^+.

Brønsted-Lowry acids

While the Arrhenius concept is useful for describing many reactions, it is also quite limited in its scope. In 1923 chemists Johannes Nicolaus Brønsted and Thomas Martin Lowry independently recognized that acid-base reactions involve the transfer of a proton. A Brønsted-Lowry acid (or simply Brønsted acid) is a species that donates a proton to a Brønsted-Lowry base. Brønsted-Lowry acid-base theory has several advantages over Arrhenius theory. Consider the following reactions of acetic acid (CH_3COOH), the organic acid that gives vinegar its characteristic taste:

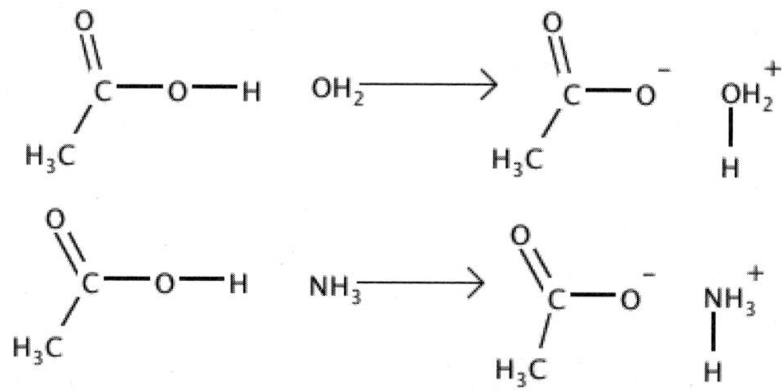

Both theories easily describe the first reaction: CH_3COOH acts as an Arrhenius acid because it acts as a source of H_3O^+ when dissolved in water, and it acts as a Brønsted acid by donating a proton to water. In the second example CH_3COOH undergoes the same transformation, in this case donating a proton to ammonia (NH_3), but cannot be described using the Arrhenius definition of an acid because the reaction does not produce hydronium. Brønsted-Lowry theory can also be used to describe molecular compounds, whereas Arrhenius acids must be ionic compounds. Hydrogen chloride (HCl) and ammonia combine under several different conditions to form ammonium chloride, NH_4Cl. In aqueous solution HCl behaves as hydrochloric acid and exists as hydronium and chloride ions. The following reactions illustrate the limitations of Arrhenius's definition:

1. $H_3O^+(aq) + Cl^-(aq) + NH_3 \rightarrow Cl^-(aq) + NH_4^+(aq)$
2. $HCl(benzene) + NH_3(benzene) \rightarrow NH_4Cl(s)$
3. $HCl(g) + NH_3(g) \rightarrow NH_4Cl(s)$

As with the acetic acid reactions, both definitions work for the first example, where water is the solvent and hydronium ion is formed. The next two reactions do not involve the formation of ions but are still proton transfer reactions. In the second reaction hydrogen chloride and ammonia (dissolved in benzene) react to form solid ammonium chloride in a benzene solvent and in the third gaseous HCl and NH_3 combine to form the solid.

Lewis acids

A third concept was proposed in 1923 by Gilbert N. Lewis which includes reactions with acid-base characteristics that do not involve a proton transfer. A Lewis acid is a species that accepts a pair of electrons from another species; in other words, it is an electron pair acceptor. Brønsted acid-base reactions are proton transfer reactions while Lewis acid-base reactions are electron pair transfers. All Brønsted acids are also Lewis acids, but not all Lewis acids are Brønsted acids. Contrast the following reactions which could be described in terms of acid-base chemistry.

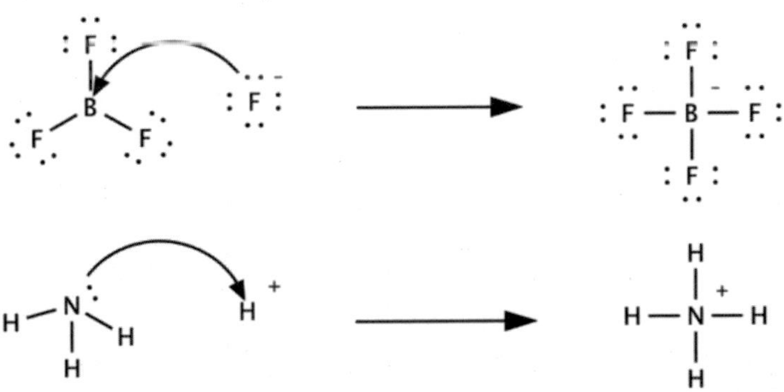

In the first reaction a fluoride ion, F^-, gives up an electron pair to boron trifluoride to form the product tetrafluoroborate. Fluoride "loses" a pair of valence electrons because the electrons shared in the B — F bond are located in the region of space between the two atomic nuclei and are therefore more distant from the fluoride nucleus than they are in the lone fluoride ion. BF_3 is a Lewis acid because it accepts the electron pair from fluoride. This reaction cannot be described in terms of Brønsted theory because there is no proton transfer. The second reaction can be described using either theory. A proton is transferred from an unspecified Brønsted acid to ammonia, a Brønsted base; alternatively, ammonia acts as a Lewis base and transfers a lone pair of electrons to form a bond with a hydrogen ion. The species that gains the electron pair is the Lewis acid; for example, the oxygen atom in H_3O^+ gains a pair of electrons when one of the H — O bonds is broken and the electrons shared in the bond become localized on oxygen. Depending on the context, a Lewis acid may also be described as an oxidizer or an electrophile.

acidity[əsídəti]: L. *aciditas, aciditatis*. 산성, 산성도.

acrylic[əkrílik]: L. *acrolein* (*acer, acri-,* pungent + *ol(eum)*, oil + *-in*, suffix) + *-yl*, suffix + *-ic, icus*, denoting an element in a higher valency. 아크릴 산의. 자극성 오일.

acrylic acid (IUPAC: prop-2-enoic acid): an organic compound with the formula $CH_2{=}CHCO_2H$. It is the simplest unsaturated carboxylic acid, consisting of a vinyl group connected directly to a carboxylic acid terminus. This colorless liquid has a characteristic acrid or tart(sour) smell. It is miscible with water, alcohols, ethers, and chloroform. More than one billion kilograms are produced annually.

addition[ədíʃən]: L. *addere, addition,* (*ad-*, to + *dare*, put). 첨가.

addition reaction: an organic reaction where two or more molecules combine to form a larger one. Addition reactions are limited to chemical compounds that have multiply-bonded atoms, such as molecules with carbon-carbon double bonds, i.e., alkenes, or with triple bonds, i.e., alkynes. Also included are molecules containing carbon - hetero double bonds like those with carbonyl (C=O) groups or those with imine (C=N) groups. There are two main types of polar addition reactions, electrophilic addition and nucleophilic addition. One non-polar addition reaction exists as well called free radical addition.

Fig. Addition reactions general overview. Top to bottom: electrophic addition to alkene, nucleophilic addition of nucleophile to carbonyl and free radical addition of halide to alkene

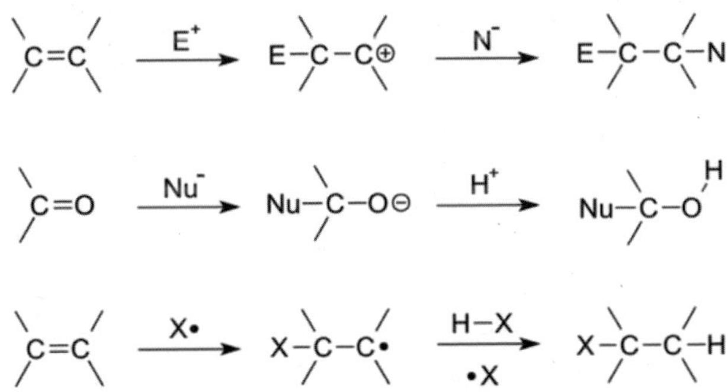

adenine[ǽdənin, -ni:n, -nain]: G. *adēn*, gland + *-ine, -ina*. 아데닌(핵산의 일부를 구성하는 염기). A nucleobase (a purine derivative) with a variety of roles in biochemistry including cellular respiration, in the form of both the energy-rich adenosine triphosphate (ATP) and the cofactors nicotinamide adenine dinucleotide (NAD) and flavin adenine dinucleotide (FAD), and protein synthesis, as a chemical component of DNA and RNA. The shape of adenine is complementary to either thymine in DNA or uracil in RNA.

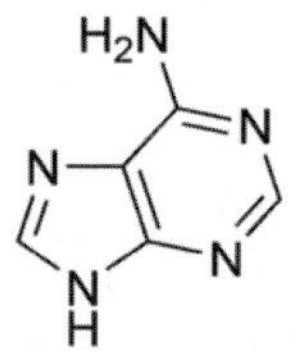

adenosine[ədénəsi(:)n]: 20th cent. blend of *adenine* (G. *adēn*, gland + *-ine*) and ribose (19th cent. arbitary alteration of *arabinose,* a related sugar). 아데노신. A nucleoside composed of a molecule of adenine attached to a ribose sugar molecule (ribofuranose) moiety via a β-N_9-glycosidic bond. Adenosine plays an important role in biochemical processes, such as energy transfer —as adenosine triphosphate (ATP) and adenosine diphosphate (ADP) —as well as in signal transduction as cyclic adenosine monophosphate, cAMP. It is also an inhibitory neurotransmitter, believed to play a role in promoting sleep and suppressing arousal(각성), with levels increasing with each hour an organism is awake. Adenosine is often abbreviated **Ado**.

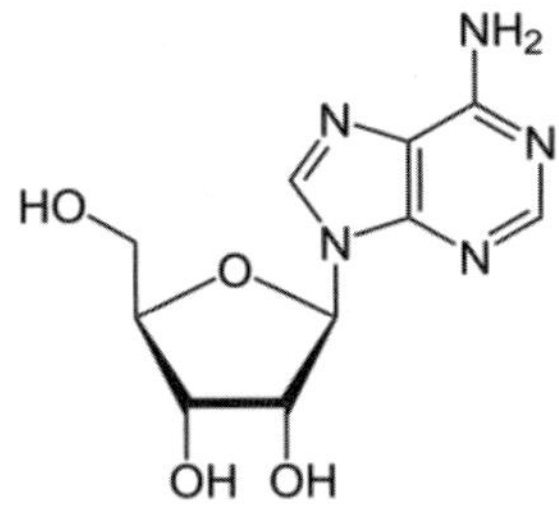

adrenaline[ədrénəlin, -li:n]: L. *ad-*, to + *renes*, *renal*, kidneys + *-in*. 아드레날린. Epinephrine (US only), widely called adrenaline (Rest of the world): hormone and neurotransmitter. When produced in the body it increases heart rate, blood vessels and dilates air passages and participates in the fight-or-flight [(생리) 공격-도피 반응(스트레스가 부가되는 자극에 대한 교감 신경의 반응)] recontracts sponse of the sympathetic nervous system. It is a catecholamine, a monoamine produced only by the adrenal glands from the amino acids phenylalanine and tyrosine.

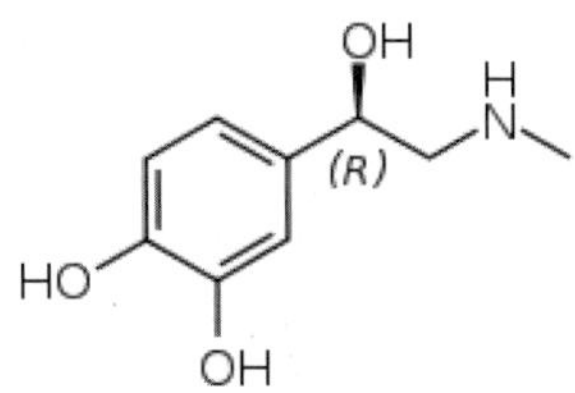

noradrenaline (abbreviated NA or NAd) or norepinephrine (abbreviated norepi or NE): a catecholamine with multiple roles including as a hormone and a neurotransmitter. As a stress hormone, norepinephrine affects parts of the brain where attention and responding actions are controlled. Along with epinephrine, norepinephrine also underlies the fight-or-flight response, directly increasing heart rate, triggering the release of glucose from energy stores, and increasing blood flow to skeletal muscle. Norepinephrine also has a neurotransmitter role when released diffusely in the brain as an antiinflammatory agent.

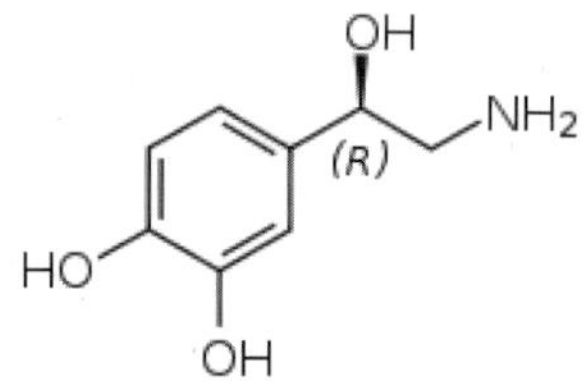

affinity[əfínəti]: L. *affinis*, *affinitas,* related (bordering on), (*ad-*, to + *finis*, border). 친화(親和)[성(性), 도(度)].

<u>electron affinity of a molecule or atom</u>: an energy change when an electron is added to the neutral atom to form a negative ion. This property can only be measured in an atom in gaseous state.

$X + e^{-} \rightarrow X^{-}$

Alternatively, electron affinity is often described as the amount of energy required to detach an electron from a singly charged negative ion, i.e. the energy change for the process

$X- \rightarrow X + e-$

A molecule or atom that has a positive electron affinity is often called an electron acceptor and may undergo charge-transfer reactions.

alanine[ǽləni:n]: D. *alanin, al*, aldehyde + *-an* (for ease of pronunciation), + *-ine*). 알라닌. An α-amino acid with the chemical formula $CH_3CH(NH_2)COOH$. The L-isomer is one of the 22 proteinogenic amino acids, i.e., the building blocks of proteins. It is classified as a nonpolar amino acid. L-Alanine is second only to leucine in rate of occurrence, accounting for 7.8% of the primary structure in a sample of 1,150 proteins. D-Alanine occurs in bacterial cell walls and in some peptide antibiotics.

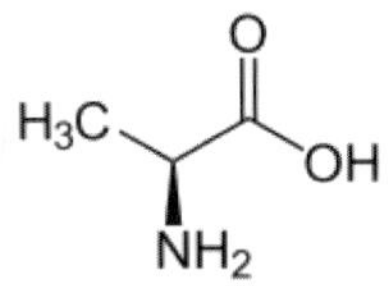

alchemy[ǽlkəmi]: Arabic, *al-*, the + *kimiyā,* G. *khēmia, khēmeia,* art of transmuting metals. 연금술(鍊金術), 연단술(鍊丹術). An influential philosophical tradition whose early practitioners' claims to profound powers were known from antiquity. The defining objectives of alchemy are varied; these include the creation of the fabled philosopher's stone possessing powers including the capability of turning base metals into the noble metals gold or silver, as well as an elixir of life conferring youth and immortality. Western alchemy is recognized as a protoscience that contributed to the development of modern chemistry and medicine. Alchemists developed a framework of theory, terminology, experimental process and basic laboratory techniques that is still recognizable today. But alchemy differs from modern science in the inclusion of Hermetic [비술(秘術)의, 연금술의] principles and practices related to mythology, religion, and spirituality.

alcohol[ǽlkəhɔ:l, -hɑl / -hɔl]: Arabic. *al-kuhl*, the kohl(콜먹, 회교국의 화장용 먹). A distilled or rectified(정류(精溜) spirit. Any organic compound in which a hydroxyl functional group (-OH) is bound to a carbon atom, usually connected to other carbon or hydrogen atoms. An important class are the simple acyclic alcohols, the general formula for which is $C_nH_{2n+1}OH$. Of those, ethanol (C_2H_5OH) is the type of alcohol found in alcoholic beverages, and in common speech the word alcohol refers specifically to ethanol. Other alcohols are usually described with a clarifying adjective, as in isopropyl alcohol (propan-2-ol) or wood alcohol (methyl alcohol, or methanol). The suffix -ol appears in the IUPAC chemical name of all alcohols.

aldehyde[ǽldəhaid]: L. shortened from *alcohol dehydrogenatum,* alcohol deprived of hydrogen. 알데히드. *A*n organic compound containing a formyl group. This functional group, with the structure RC(=O)H, consists of a carbonyl center bonded to hydrogen and an R group. The group without R is called the aldehyde group or formyl group. Aldehydes differ from ketones in that the carbonyl is placed at the end of a carbon skeleton rather than between two carbon atoms. Aldehydes are common in organic chemistry. Many fragrances are aldehydes.

aldehyde

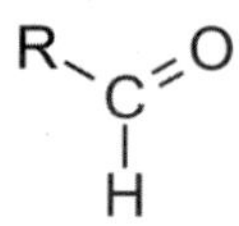

Formaldehyde

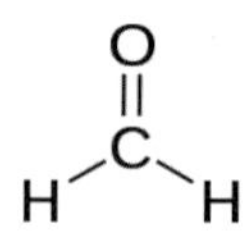

aldosterone[ældoustiróun /ældɔ́stəroun]: 1950s. blend of aldehyde and steroid + *-one*, ketone. 알도스테론(부신 피질 호르몬의 일종). A hormone that increases the reabsorption of sodium ions and water and the release (secretion) of potassium ions in the distal convoluted tubules[말초 회선형관(回旋形管)] of the kidneys. This increases blood volume and, therefore, increases blood pressure. Drugs that interfere with the secretion or action of aldosterone are in use as antihypertensives. One example is spironolactone, which lowers blood pressure by blocking the aldosterone receptor.

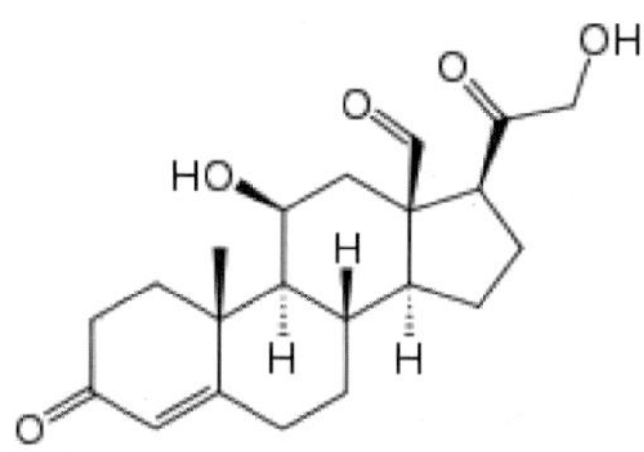

alicyclic[æləsáiklik, -sík-]: late 19th century. aliphatic + cyclic. 지방족 고리 화합물의.

<u>alicyclic compound</u>: an organic compound that is both aliphatic and cyclic. They contain one or more all-carbon rings which may be either saturated or unsaturated, but do not have aromatic character. Alicyclic compounds may or may not have aliphatic side chains attached. The simplest alicyclic compounds are the monocyclic cycloalkanes: cyclopropane, cyclobutane, cyclopentane, cyclohexane, cycloheptane, cyclooctane, cyclononane, cyclodecane, cycloundecane,

cyclododecane, and so on. Bicyclic alkanes include bicycloundecane and decalin. Polycyclic alkanes include cubane, basketane, and housane. Spiro compounds have two or more rings that are connected through only one carbon atom.

aliphatic[æləfǽtik]: G. *aleiphar*, *aleiphat-*, fat, oil + *-ic*. 지방족 화합물의, 사슬모양 유기화합물의. Compounds composed of carbon and hydrogen are divided into two classes: aromatic compounds, which contain benzene rings or similar rings of atoms, and aliphatic compounds which do not contain aromatic rings. Aliphatic compounds can be cyclic, like cyclohexane, or acyclic, like hexane. They also can be saturated, like hexane, or unsaturated, like hexene. In aliphatic compounds, carbon atoms can be joined together in straight chains, branched chains, or non-aromatic rings (in which case they are called alicyclic). They can be joined by single bonds (alkanes), double bonds (alkenes), or triple bonds (alkynes). Most aliphatic compounds are flammable, allowing the use of hydrocarbons as fuel, such as methane in Bunsen burners, and acetylene in welding.

alkaloid[ælkəlɔ́id]: D. *alkali,* ashes of plants. 알칼로이드, 식물염기(니코틴, 모르핀, 코카인 따위). A group of naturally occurring chemical compounds which mostly contain basic nitrogen atoms. This group also includes some related compounds with neutral and even weakly acidic properties. In addition to carbon, hydrogen and nitrogen, alkaloids may also contain oxygen, sulfur and more rarely other elements such as chlorine, bromine and phosphorus. Many alkaloids are toxic to other organisms. They often have pharmacological effects and are used as medications, as recreational drugs, or in entheogenic ("God inside us") rituals. Examples are the local anesthetic and stimulant cocaine, the stimulant caffeine, nicotine, the analgesic morphine, or the antimalarial drug quinine.

alkane[ǽlkein]: late 19th cent. alkyl, (D. *alkohol,* alcohol + *-yl*) + -ane. 엘게인, 알칸. Also known as paraffins or saturated hydrocarbons, chemical compounds that consist only of the elements carbon (C) and hydrogen (H) (i.e., hydrocarbons), wherein these atoms are linked together exclusively by single bonds (i.e., they are saturated compounds) without any cyclic structure. Alkanes belong to a homologous series of organic compounds with the general formula C_nH_{2n+2} in which the members differ by a constant relative molecular mass of 14.

alkene[ǽlki:n]: late 19th cent. alkyl, (D. *alkohol,* alcohol + *-yl*) + -ene. 엘킨, 알켄. Olefin, or olefine, an unsaturated chemical compound containing at least one carbon-to-carbon double bond. The simplest acyclic alkenes, with only one double bond and no other functional groups, form an homologous series of hydrocarbons with the general formula C_nH_{2n}. The simplest alkene is ethylene (C_2H_4), which has the International Union of Pure and Applied Chemistry (IUPAC) name ethene. For bridged alkenes, the Bredt's rule states that a double bond cannot be placed at the bridgehead of a bridged ring system, unless the rings are large enough. Aromatic compounds are often drawn as cyclic alkenes, but their structure and properties are different and they are not considered to be alkenes.

alkyl[ǽlkil]: D. *alkohol,* alcohol + *-yl.* A paraffin hydrocarbon(s) with the general formula C_nH_{2n+1} by removal of a hydroge atom. 알킬 (지방족의 포화탄화수소기).

alkyne[ǽlkain]: 20th cent. alkyl, (D. *alkohol,* alcohol + *-yl*) + -yne. 엘카인, 알킨. Hydrocarbons that have a triple bond between two carbon atoms, with the formula C_nH_{2n-2}. Alkynes are traditionally known as acetylenes, although the name acetylene also refers specifically to C_2H_2, known formally as ethyne using IUPAC nomenclature. Like other hydrocarbons, alkynes are generally hydrophobic but tend to be more reactive.

Fig. Illustrative alkynes: a, two depictions of propyne, b, the naturally-occurring 1-phenylhepta-1,3,5-triyne, and c, the strained cycloheptyne.

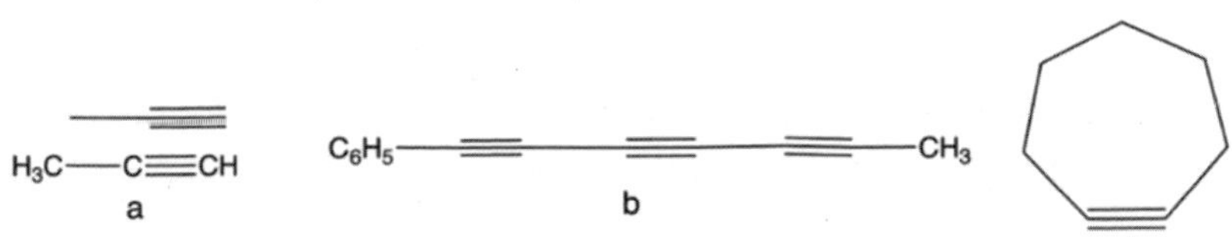

a b c

allergy[ǽlərdʒi]: G. *allos,* other + *energie,* energy. 알레르기, 이상민감증. A hypersensitivity disorder of the immune system. Allergic reactions occur when a person's immune system reacts to normally harmless substances in the environment. A substance that causes a reaction is called an allergen. These reactions are acquired, predictable, and rapid. Allergy is one of four forms of hypersensitivity and is formally called type I (or immediate) hypersensitivity. Allergic reactions are distinctive because of excessive activation of certain white blood cells called mast cells and basophils(호염기성 백혈구) by a type of antibody called Immunoglobulin E (IgE). This reaction results in an inflammatory response which can range from uncomfortable to dangerous. Mild allergies like hay fever are very common in the human population and cause symptoms such as red eyes, itchiness, and runny nose, eczema(습진), hives(발진, 두드러기), hay fever, or an asthma attack. Allergies can play a major role in conditions such as asthma. In some people, severe allergies to environmental or dietary allergens or to medication may result in life-threatening reactions called anaphylaxis(혈청주사를 맞은 뒤나 조개류를 먹은 뒤에 일어나는 이질 단백질에 대한 과민증). Food allergies, and reactions to the venom(독) of stinging insects such as wasps(말벌) and bees are often associated with these severe reactions. A variety of tests exist to diagnose allergic conditions. These include placing possible allergens on the skin and looking for a reaction such as swelling. Blood tests can also be done to look for an allergen-specific IgE. Treatments for allergies include avoiding known allergens, use of medications such as anti-histamines that specifically prevent allergic reactions, steroids that modify the immune system in general, and medications such as decongestants(소염제) that reduce the symptoms. Many of these medications are taken by mouth, though epinephrine, which is used to treat anaphylactic reactions, is injected. Immunotherapy uses injected allergens to desensitize the body's response.

allotropy[əlάtrəpi /əlɔ́t-]: G. *allotropos, allo-* other + *tropos*, manner, of another form. 동소체성, 동소체 현상. Allotropism, the property of some chemical elements to exist in two or more different forms, known as allotropes(동소체) of these elements. Allotropes are different structural modifications of an element; the atoms of the element are bonded together in a different manner. For example, the allotropes of carbon include diamond (where the carbon atoms are bonded together in a tetrahedral lattice arrangement), graphite (where the carbon atoms are bonded together in sheets of a hexagonal lattice), graphene (single sheets of graphite), and fullerenes (where the carbon atoms are bonded together in spherical, tubular, or ellipsoidal[타원체(면)의] formations). The term allotropy is used for elements only, not for compounds. The more general term, used for any crystalline material, is polymorphism. Allotropy refers only to different forms of an element within the same phase (i.e. different solid, liquid or gas forms); the changes of state between solid, liquid and gas in themselves are not considered allotropy. For some elements, allotropes have different molecular formulae which can persist in different phases – for example, two allotropes of oxygen (dioxygen, O_2 and ozone, O_3), can both exist in the solid, liquid and gaseous states. Conversely, some elements do not maintain distinct allotropes in different phases – for example phosphorus has numerous solid allotropes, which all revert to the same P_4 form when melted to the liquid state.

allyl[ǽlil]: L. *allium*, garlic + *-yl.* 알릴기(基). An allyl group is a hydrocarbon group with the structural formula $H_2C{=}CH{-}CH_2R$, where R is the connection to the rest of the molecule. It is made up of a methylene (-CH_2-), attached to a vinyl group (-CH=CH_2). Theodor Wertheim isolated an allyl derivative from garlic oil and named it Schwefelallyl. Allylic radicals, anions, and cations are often discussed as intermediates in reactions. All feature three contiguous sp^2-hybridized carbon centers.

Fig. MO diagram for pi-orbitals on allyl.

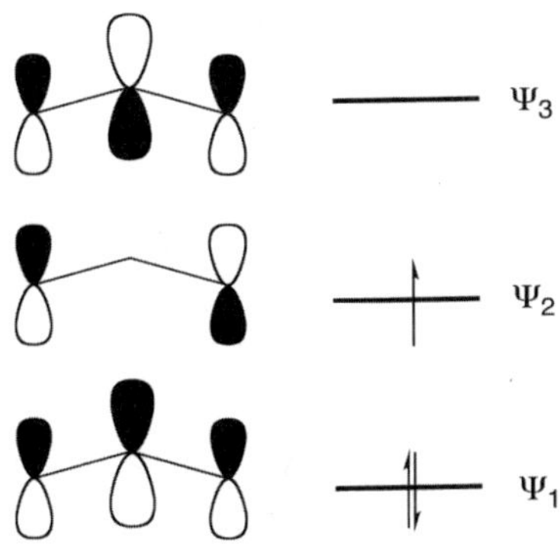

alpha[ǽlfə]: G. > L. the first letter of the Greek alphabet, A, α.

alpha cleavage (= α-cleavage): refers to the act of breaking the carbon-carbon bond, adjacent to the carbon bearing a specified functional group. Generally this topic is discussed when covering mass spectrometry and occurs generally by the same mechanisms. For example of a mechanism of alpha cleavage, an electron is knocked off an atom (usually by electron collision) to form a radical cation. Electron removal generally happens in the following order: 1) lone pair electrons, 2) pi bond electrons, 3) sigma bond electrons. One of the lone pair electrons moves down to form a pi bond with an electron from an adjacent (alpha) bond. The other electron from the bond moves to an adjacent atom (not one adjacent to the lone pair atom) creating a radical. This creates a double bond adjacent to the lone pair atom (oxygen is a good example) and breaks/cleaves the bond from which the two electrons were removed.

Fig. Example of alpha cleavage

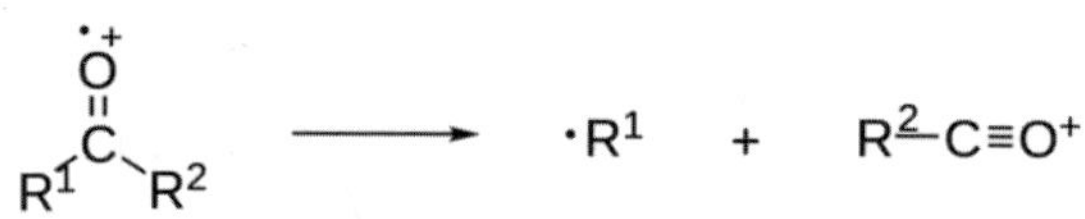

In molecules containing carbonyl groups, alpha cleavage often competes with McLafferty rearrangement.

aluminum[əlú:mənəm]: (E) aluminium. 19th cent. *aluminaium* (L. *alumen*, Al, + *-ium*). **Al**. 알루미늄. A silvery white member of the boron group of chemical elements. It has the symbol Al, and its atomic number is 13. It is not soluble in water under normal circumstances. Aluminium is the third most abundant element (after oxygen and silicon), and the most abundant metal, in the Earth's crust. It makes up about 8% by weight of the Earth's solid surface. Aluminium metal is too reactive chemically to occur natively. Instead, it is found combined in over 270 different minerals. The chief ore of aluminium is bauxite. Aluminium is remarkable for the metal's low density and for its ability to resist corrosion due to the phenomenon of passivation(보호막). Structural components made from aluminium and its alloys are vital to the aerospace industry and are important in other areas of transportation and structural materials. The most useful compounds of aluminium, at least on a weight basis, are the oxides and sulfates. Despite its prevalence in the environment, aluminium salts are not known to be used by any form of life. In keeping with its pervasiveness, it is well tolerated by plants and animals. Because of their prevalence, potential beneficial (or otherwise) biological roles of aluminium compounds are of continuing interest.

amalgam[əmǽlgəm]: L. < G. *amalgama* < *malagma*, alloy of mercury (esp. with gold or silver) < softening substance. 아말감(수은과 다른 금속과의 합금). A substance formed by the reaction of mercury with another metal. Almost all metals can form amalgams with mercury, notable exceptions being iron and platinum. Silver-mercury amalgams are important in dentistry, and gold-mercury amalgam is used in the extraction of gold from ore.

ambi-[ǽmbi, -bə]: L. *ambi-*. on both sides. both, around. 양쪽의, 주위의.

amines[əmí:n, ǽmin]: [*am(monia)*, L. *sal ammoniacus* (NH_4Cl, 암몬의 소금) + *ine*]. 아민. Organic compounds and functional groups that contain a basic nitrogen atom with a lone pair. Amines are derivatives of ammonia, wherein one or more hydrogen atoms have been replaced by a substituent such as an alkyl or aryl group. Important amines include amino acids, biogenic amines, trimethylamine, and aniline.

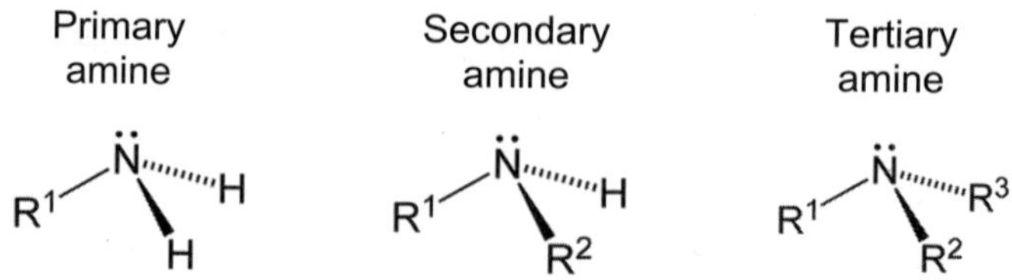

amide: usually an organic compound that contains the functional group consisting of an acyl group (R-C=O) linked to a nitrogen atom (N). 예: R-$CONH_2$, or R-$CONR_2$. The term amide also refers to deprotonated form of ammonia (NH_3) or an amine, often represented as anions R_2N^-. Example: "anionic amides," sodium amide and lithium diisopropylamide.

amino acid: molecules containing an amine group, a carboxylic acid group and a side chain that varies between different amino acids. These molecules are particularly important in biochemistry, where this term refers to alpha-amino acids with the general formula $H_2NCHRCOOH$, where R is an organic substituent. Amino acids serve as the building blocks of proteins, which are linear chains of amino acids. Amino acids can be linked together in varying sequences to form a vast variety of proteins. Twenty amino acids are naturally incorporated into polypeptides and are called proteinogenic or standard amino acids. These 20 are encoded by the universal genetic code. Nine standard amino acids are called "essential" for humans because they cannot be created from other compounds by the human body, and so must be taken

Amino acids with nonpolar side chains

Name	Structure	Abbreviation
Glycine	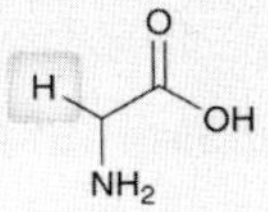	Gly G
Alanine	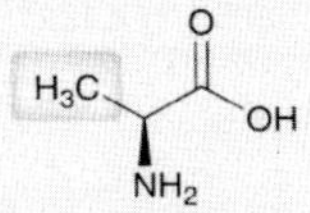	Ala A
Valine	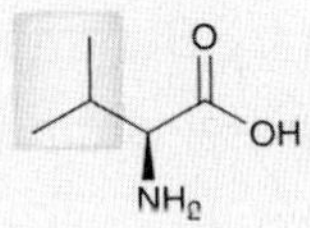	Val V
Leucine	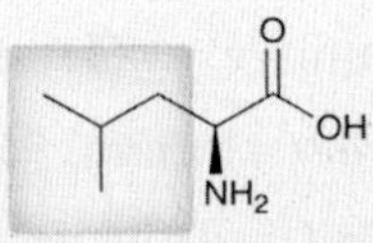	Leu L
Isoleucine	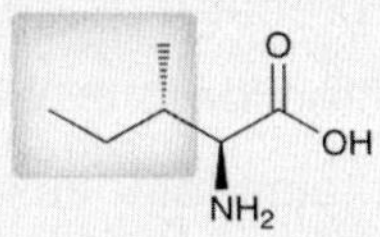	Ile I
Methionine	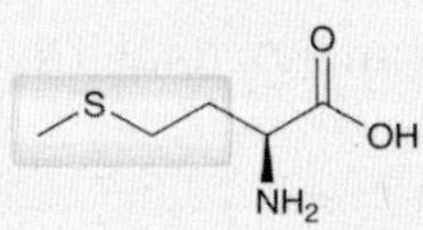	Met M
Proline	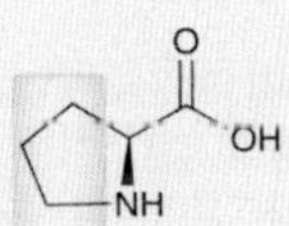	Pro P
Phenylalanine	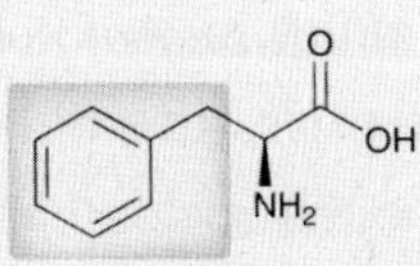	Phe F
Tryptophan	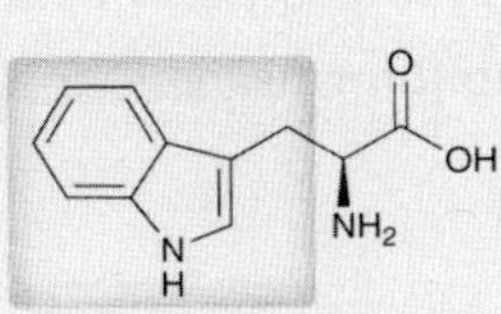	Trp W

Amino acids with polar side chains

Name	Structure	Abbreviation
Asparagine	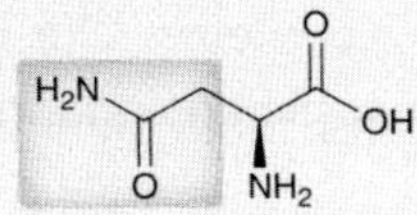	Asn N
Glutamine	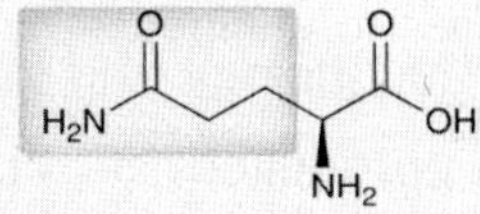	Gln Q
Serine	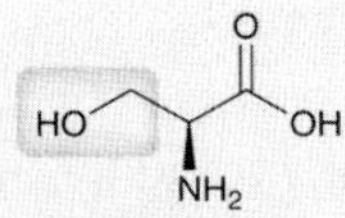	Ser S
Threonine	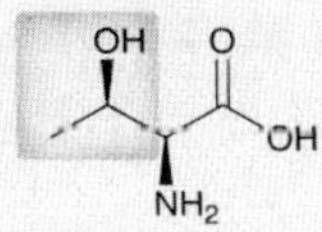	Thr T
Tyrosine	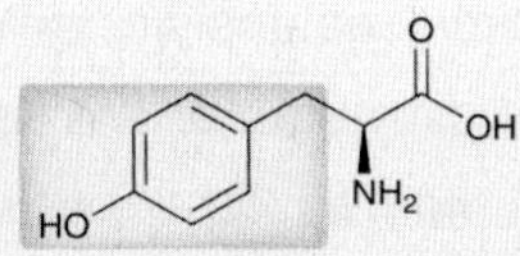	Tyr Y
Cysteine	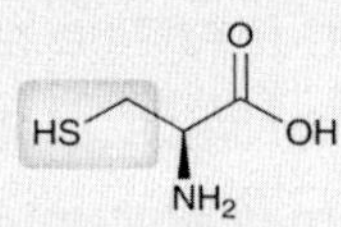	Cys C

Amino acids with acidic side chains

Name	Structure	Abbreviation
Aspartic acid	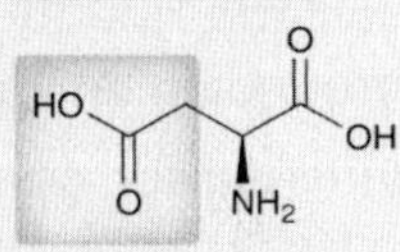	Asp D
Glutamic acid	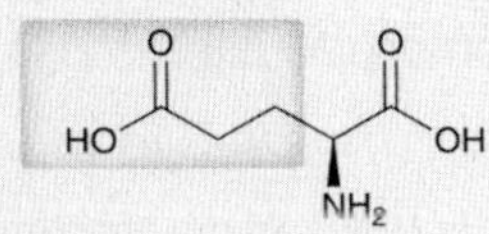	Glu E

Amino acids with basic side chains

Name	Structure	Abbreviation
Arginine	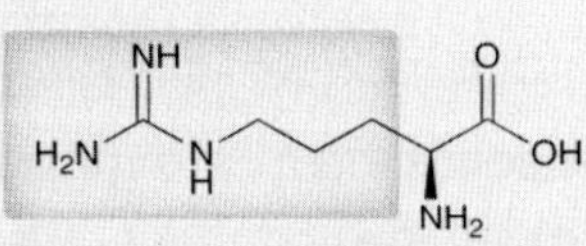	Arg R
Histidine	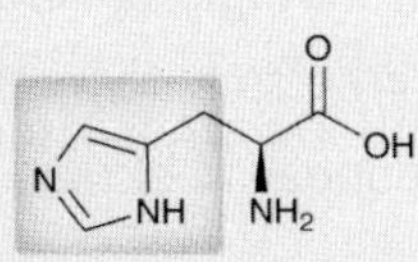	His H
Lysine	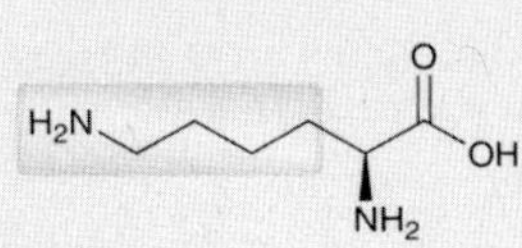	Lys K

in as food. Amino acids are important in nutrition and are commonly used in nutrition supplements, fertilizers, food technology and industry. In industry, applications include the production of biodegradable plastics, drugs, and chiral catalysts.

ammonia[əmóunjə, -niə]: L. *ammonia*, invented by Bergman, in 1782, as a name for the gas obtained from sal ammoniac (NH_4Cl, 암몬의 소금).

amniocentesis[æmniousentí:sis]: 1950s. amnion, G. *caul*, diminutive of *amnos*, lamb + G. *kentesis, ketein*, pricking. 양수(羊水)검사(태아의 성별, 염색체의 이상 여부 따위를 조사). Amniotic fluid test or AFT, a medical procedure used in prenatal diagnosis of chromosomal abnormalities and fetal infections, in which a small amount of amniotic fluid, which contains fetal tissues, is extracted from the amnion(태아를 싸는 양막) or amniotic sac surrounding a developing fetus(태아), and the fetal DNA is examined for genetic abnormalities.

amorphous[əmɔ́:rfəs]: G. > L. *amorphos* > *amorphus, (a,* without *+ morphē,* form, shape *+ -ous),* shapeless. 비결정질의, 일정한 모양이 없는. Non-crystalline solid in condensed matter physics, a solid that lacks the long-range order characteristic of a crystal that includes a ridgedly formed molecular structure. Amorphous solids have a similar molecular structure of crystals but they lack the rigid unified formation of the molecules, instead having an irregularly formed structure which allows for is visual structure to appear to slowly "flowing" or changing forms without applying a strong physical force such as heat to it.

amphetamine[æmfétəmi:n]: 1930s. abbreviation of its chemical name **a**lpha**m**ethyl**phe**nethyl**amine**.

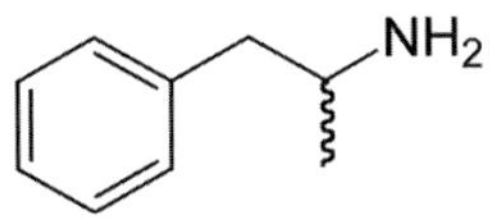

amphi-[ǽmfi, -fə]: G. *amphi-*, both, two, of both kinds, on both sides, about, around. 두 개의, 둘, 양쪽의, 주위의.

amphibia[æmfíbiə]: G. > L. *amphibium* < *amphbion, amphibios*, *amphi-*, both + *bios*, life, living both in water and on land. 양서류(兩棲類).

amphoteric[æmfətérik]: G. *amphoteros,* comparative of *amphō-*, both + -ic. 양성(兩性)의.

amylase[ǽməleis, -leiz]: L. *am(ylum),* starch + *-yl* + *-ase.* 아밀라아제, (전분 당화 효소). An enzyme that catalyses the breakdown of starch into sugars. Amylase is present in human saliva(침), where it begins the chemical process of digestion. Food that contains much starch but little sugar, such as rice and potato, taste slightly sweet as they are chewed because amylase turns some of their starch into sugar in the mouth. The pancreas also makes amylase (alpha amylase) to hydrolyse dietary starch into disaccharides and trisaccharides which are converted by other enzymes to glucose to supply the body with energy. Plants and some bacteria also produce amylase. As diastase, amylase was the first nzyme to be discovered and isolated (by Anselme Payen in 1833). Specific amylase proteins are designated by different Greek letters. All amylases are glycoside hydrolases and act on α-1,4-glycosidic bonds.

anabolism[ənǽbəlizm]: G. *anabolē*, *ana-*, up + *ballein*, to throw, ascent. 물질합성대사, 동화작용. The set of metabolic pathways that construct molecules from smaller units. These reactions require energy. One way of categorizing metabolic processes, whether at the cellular, organ or organism level is as 'anabolic' or as 'catabolic', which is the opposite. Anabolism is powered by catabolism, where large molecules are broken down into smaller parts and then used up in respiration. Many anabolic processes are powered by adenosine triphosphate (ATP). Anabolic processes tend toward "building up" organs and tissues. These processes produce growth and differentiation of cells and increase in body size, a process that involves synthesis of complex molecules. Examples of anabolic processes include the growth and mineralization of bone and increases in muscle mass.

anaerobic[ænəróubik, ænɛər-]: G. *anaērbios*, *an-*, not + *aēr*, air + *bios*, life, relating to or requiring an absence of free oxygen. 무산소성, 공기 (산소)가 없어도 자라는. 혐기성(嫌氣性) 생물(특히 균류).

analgesic[ænəldʒí:zik, -sik]: G. > L. *analgēsia* > *analgesia* (*an-*, not + *algesis*, pain, sorrow, grief + *-ia*), painlessness. 진통제.

analysis[ənǽləsis]: G. > L. *analusis, analuein*, *ana-*, up + *luein*, loosen, unloose. 분석, 분해, (수학)해석, 해석학(解析學). The field of chemistry uses analysis in at least three ways: to identify the components of a particular chemical compound (qualitative analysis), to identify the proportions of components in a mixture (quantitative analysis), and to break down chemical processes and examine chemical reactions between elements of matter.

anatomy[ənǽtəmi]: G. > L. *anatomia, ana,* up + *tomia, temnein,* cutting, to cut. 해부, 해부학. A branch of biology and medicine that is the consideration of the structure of living things. It is a general term that includes human anatomy, animal anatomy (zootomy), and plant anatomy (phytotomy). In some of its facets anatomy is closely related to embryology, comparative anatomy and comparative embryology, through common roots in evolution.

androgen[ǽndrədʒən]: G. *androgen, andro-*, 모음 앞에서는 *andr-*, man, male + *-gen,* to be born, become. 안드로겐(남성호르몬). Androgenic hormone, the generic term for any natural or synthetic compound, usually a steroid hormone, that stimulates or controls the development and maintenance of male characteristics in vertebrates by binding to androgen receptors. This includes the activity of the accessory male sex organs and development of male secondary sex characteristics. Androgens were first discovered in 1936. Androgens are also the original anabolic steroids and the precursor of all estrogens, the female sex hormones. The primary and most well-known androgen is testosterone, other less important androgens are dihydrotestosterone and androstenedione.

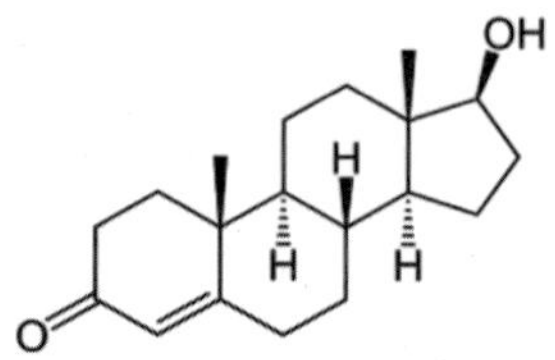

Testosterone

anaemia[əníːmiə]: (E) anemia. G. *anaimia*, *an-*, without + *haima*, blood. 빈혈.

anesthetize[ənésəətaiz]: G. *anaisthēsis*, *an-*, without + *aisthēsis*, sensation, insensitivity to pain. - 을 마취시키다. 마취상태.

angle[ǽŋgl]: L. *angulus*, *angli* (pl), corner. 각, 각도, 모퉁이. 귀퉁이.

anhydride[ænháidraid]: G. *anudros*, anhydrous + *-ide.* 무수물.

acid anhydride: an organic compound that has two acyl groups bound to the same oxygen atom. Most commonly, the acyl groups are derived from the same carboxylic acid, the formula of the anhydride being $(RC(O))_2O$. Symmetrical acid anhydrides of this type are named by replacing the word acid in the name of the parent carboxylic acid by the word anhydride. Thus, $(CH_3CO)_2O$ is called acetic anhydride. Mixed (or unsymmetrical) acid anhydrides, such as acetic formic anhydride are known.

Generic example of an acid anhydride:

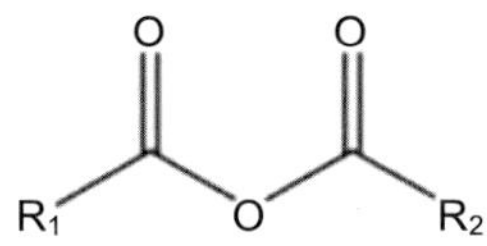

anhydrous[ænháidrəs]: G. *anudros*, *an-*, without + *hudōr*, water + *-ous*. 무수(無水)의, 결정수(結晶水)가 없는. As a general term, a substance is said to be anhydrous if it contains no water. The way of achieving the anhydrous form differs from one substance to another. That is to say, anhydrous matter is prepared or used as a form with no water, and specifically the form of a chemical without water of crystallization.

aniline[ǽnəlin, -lain]: mid 19th cent. *anil*, indigo (from which it was originally obtained) < Arabic. *an-nīl*, *ana-*, up + *nīla*, dark blue. 아닐린. Phenylamine or aminobenzene, an organic compound with the formula $C_6H_5NH_2$.

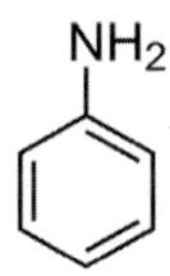

anion[ǽnaiən]: mid 19th cent. anode, (G. *anodos, ana-*, up + *hodos*, way, way up) + ion, G. *ienai* - go, going up. 음이온. An ion with more electrons than protons, giving it a net negative charge (since electrons are negatively charged and protons are positively charged). An ion carrying a negative charge which moves towards the anode (positive electrode) during electrolysis. Opposed to cation.

anisotropic[ænaisətrάpik / -trɔp-]: G. *anisos*, unequal + *tropos*, turn + *-ic*. 비등방성, 이방성(異方性)의. The property of being directionally dependent, as opposed to isotropy, which implies identical properties in all directions. It can be defined as a difference, when measured along different axes, in a material's physical or mechanical properties (absorbance, refractive index, conductivity, tensile strength, etc.) An example of anisotropy is the light coming through a polarizer. An example of an anisotropic material is wood, which is easier to split along its grain than against it. In NMR spectroscopy, the orientation of nuclei with respect to the applied magnetic field determines their chemical shift. In this context, anisotropic systems refer to the electron distribution of molecules with abnormally high electron density, like the pi system of benzene. This abnormal electron density affects the applied magnetic field and causes the observed chemical shift to change.

annular[ǽnjulər]: L. *anulus, annulus*, *annularis,* a ring. 고리 모양의, 환상(環狀)의.

anode[ǽnoud]: G. *anodos, ana-,* up + *hodos*, way way up. 양극, 산화전극. (전해조·전지의) 전류가 들어가는 전극, (볼타전지의, 축전지의) 음극, 전자가 나오는 전극, anion이 들어가는 전극. An electrode through which electric current flows into a polarized electrical device. The direction of electric current is, by convention, opposite to the direction of electron flow. In other words, the electrons flow from the anode into, for example, an electrical circuit. Mnemonic: ACID (Anode Current Into Device). A widespread misconception is that anode polarity is always positive (+). This is often incorrectly inferred from the correct fact that in all electrochemical devices negatively charged anions move towards the anode (hence their name) and/or positively charged cations move away from it. In fact anode polarity depends on the device type, and sometimes even in which mode it operates, as per the above electric current direction-based universal definition. Consequently, as can be seen from the following examples, in a device which consumes power the anode is positive, and in a device which provides power the anode is negative: In a discharging battery or galvanic cell (diagram at right) the anode is the negative terminal since that is where the current flows into the device (see drawing). This inward current is carried externally by electrons moving outwards, negative charge moving one way constituting positive current flowing the other way. It is continued internally by positive ions flowing into the electrolyte from the anode, i.e., away (surprisingly) from the more negative electrode and towards the more positive one (chemical energy is responsible for this "uphill" motion). In a recharging battery, or an electrolytic cell, the anode is the positive terminal, which receives current from an external generator. The current through a recharging battery is opposite to the direction of current during discharge; In other words, the electrode which was the cathode during battery discharge becomes the anode while the battery is recharging. In a diode, it is the positive terminal at the tail of the arrow symbol, where current flows into the device. Note electrode naming for diodes is always based on the direction of the forward current (that of the arrow, in which the current flows "most easily"), even for types such as Zener diodes or solar cells where the current of interest is the reverse current. In a cathode ray tube, it is the positive terminal where electrons flow out of the device, i.e., where positive electric current flows in.

Fig. Diagram of a zinc anode in a galvanic cell.

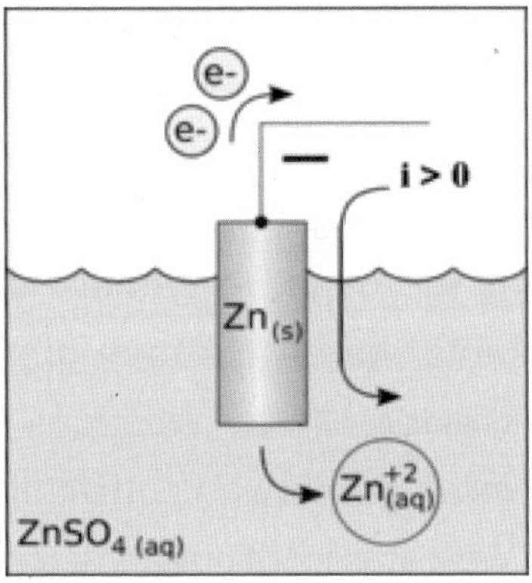

anomalous[ənάmələs / ənɔ́m]: G. *anōmalos*, *an-*, not + *homalos*, even. 변칙, 비정상. 이례적인, 외예적인.

<u>anomer</u>: a special type of epimer in carbohydrate chemistry. It is one of two stereoisomers of a cyclic saccharide that differs only in its configuration at the hemiacetal or hemiketal carbon, also called the anomeric carbon. Anomerization is the process of conversion of one anomer to the other. Anomerization is the anomeric analogue of epimerization.

Nomenclature: Two anomers are designated alpha (α) or beta (β), according to the configurational relationship between the *anomeric center* and the anomeric reference atom, hence they are relative stereodescriptors. The anomeric center is in hemiacetals the anomeric carbon C-1, which is attached to the hemiacetal oxygen (in the ring) and in hemiketals carbon C-2, attached to the hemiketal oxygen. In aldohexoses and smaller carbohydrates the anomeric reference atom is the furthest chiral centre in the ring (the configurational atom, defining the sugar as D or L). In α-D-glucopyranose the reference atom is C-5. If in the cyclic Fischer projection (see [1]) the exocyclic oxygen atom at the anomeric center is cis (on the same side) to the exocyclic oxygen attached to the anomeric reference atom (in the OH group) the anomer is α. If the two oxygens are trans (on different sides) the anomer is β. For cyclic compounds, however, use of the Fischer projection is complicated and not common. If the absolute configurations of the *anomeric carbon* and the *reference atom* are compared, then both are different (S,R or R,S) in the α anomer and the same (R,R or S,S) in the β anomer.

Fig. Different projections of α-D-Glucopyranose. 1. Fischer projection with at the top the anomeric centre C-1. C-5 is the anomeric reference atom. 4. absolute configurations.

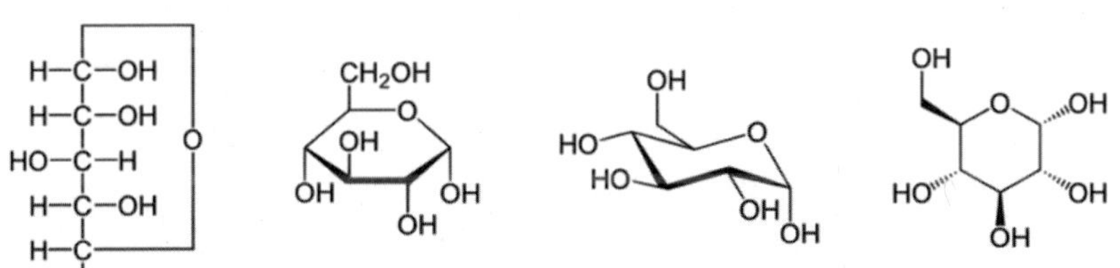

1 2 3 4

antenna[æntéनə]: G. > L. *antemna,* yard (of a ship) < *keraioi,* horns (of insects). 안테나, 촉각, 더듬이.

anthracene[ǽnθərəsi:n]: G. *anthrax, anthrak-*, coal + *-ēnos*, forming names of unsaturetd hydrocarbons containing a double bond. 안트라센(콜타르를 증류하여 얻는 알리자린 염료의 원료). A solid polycyclic aromatic hydrocarbon consisting of three fused benzene rings. It is a component of coal-tar. Anthracene is used in the production of the red dye alizarin and other dyes. Anthracene is colorless but exhibits a blue (400-500 nm peak) fluorescence under ultraviolet light.

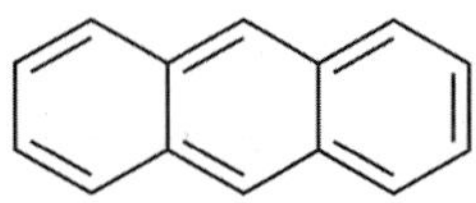

anti-[ǽntai / -ti]: G. *anti*, against, opposite. 반대의, - 에 반하여.

antibiotic[æntibaiɑ́tik, -tai- /æntibaiɔ́tik]: G. *anti-,* against, opposite + *biōtikos,* fit for life. A medicine that inhibits the growth of or destroys micro-organisms. 항생물질, 항생물질의.

antibody[ǽntibɑdi /-bɔdi]: D. *antikőrper, anti-*, against + *kőrper,* body. (혈청중의) 항체, 항독소. Also known as immunoglobulins, abbreviated Ig are gamma globulin proteins that are found in blood or other bodily fluids of vertebrates, and are used by the immune system to identify and neutralize foreign objects, such as bacteria and viruses.

antidote[ǽntidout]: G. *antidoton*, *antidotos*, *anti-*, against + *didonai*, give. given against. 해독제. A medicine taken or given to counteract a particular poison.

antigen[ǽntidʒən, -dʒen]: D. < F. *antigène*, *anti-*, against + *-gen*, be born. 항원(抗原)(생체 내에서 항체 형성을 촉진하는 독소, 세균). Originally the term came from antibody generator and was a molecule that binds specifically to an antibody, but the term now also refers to any molecule or molecular fragment that can be bound by a major histocompatibility(조직 적합성) complex (MHC) and presented to a T-cell receptor.

antiseptic[æntəséptik]: G. *ant-*, *anth-* opposite, against + *septikos*, putrefactive. antiputrescent. 살균의, 소독된, 소독제, 방부제.

apo-[ǽpou, ǽpə]: G. *apo*, from, away, off, (모음 앞에서는 ap-, 기음(h) 앞에서는 aph-). away from.

apoenzyme[æpouénzaim]: 결손 효소, 아포 효소. An enzyme without its cofactor.

aqua[ǽkwə, ɑ́:k-]: L. *aqua*, water, liquid, solution. 물, 액체, 용액.

aqua regia: 왕수(王水).

aqueous[éikwiəs, ǽk-]: L. *aqueus,* of water, watery, diluted with water. 물의, 물을 함유한, 수성(水性)의.

aramid[ǽrəmid]: 1970s. *ar(omatic) + (poly)amide*. 아라미드(합성 방향족 폴리아미드), 아라미드 섬유. Aramid fibers are a class of heat-resistant and strong synthetic fibers. They are used in aerospace and military applications, for ballistic(탄도의) rated body armor(갑옷과 투구) fabric and ballistic composites, in bicycle tires, and as an asbestos substitute. They are fibers in which the chain molecules are highly oriented along the fiber axis, so the strength of the chemical bond can be exploited.

Fig. Para-aramid structure.

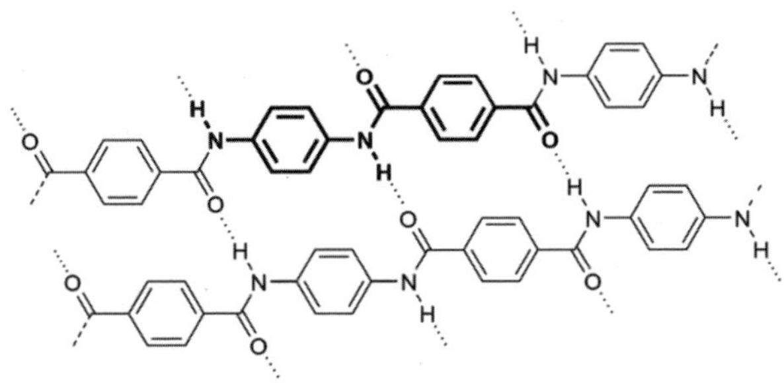

arene[əri:n]: aromatic hydrocarbon or aryl hydrocarbon. A hydrocarbon with alternating double and single bonds between carbon atoms forming rings. The term 'aromatic' was assigned before the physical mechanism determining aromaticity was discovered, and was derived from the fact that many of the compounds have a sweet scent. The configuration of six carbon atoms in aromatic compounds is known as a benzene ring. Aromatic hydrocarbons can be monocyclic (MAH) or polycyclic (PAH). Some non-benzene-based compounds called heteroarenes, which follows Hückel's rule, are also called as aromatic compounds. In these compounds, at least one carbon atom is replaced by one of the heteroatoms oxygen, nitrogen, or sulfur. Examples of non-benzene compounds with aromatic properties are furan, a heterocyclic compound with a five-membered ring that includes an oxygen atom, and pyridine, a heterocyclic compound with a six-membered ring containing one nitrogen atom.

arene substitution: arene substitution patterns are part of organic chemistry IUPAC nomenclature and pinpoint the position of substituents other than hydrogen in relation to each other on an aromatic hydrocarbon.

***Ortho*, *meta*, and *para* substitution**

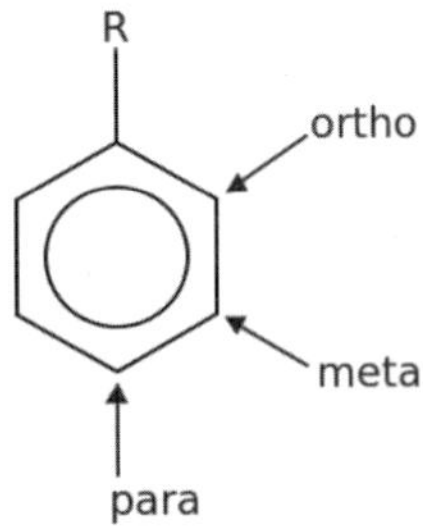

Main arene substitution patterns.

In ortho-substitution, two substituents occupy positions next to each other, which may be numbered 1 and 2. In the diagram, these positions are marked R and ortho.
In meta-substitution, the substituents occupy positions 1 and 3 (corresponding to R and meta in the diagram).
In para-substitution, the substituents occupy the opposite ends (positions 1 and 4, corresponding to R and para in the diagram).

***Ipso*, *meso*, and *peri* substitution**

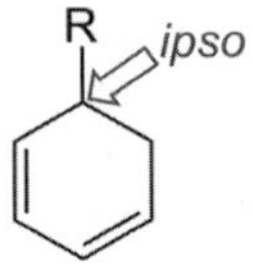

ipso- substitution.

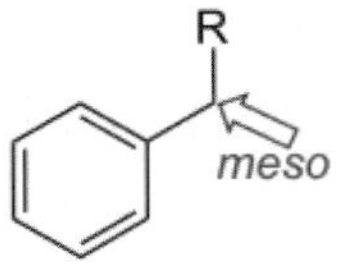

meso- substitution.

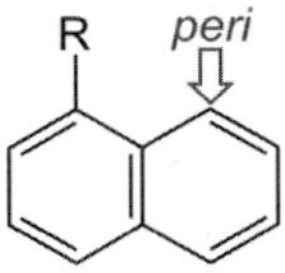

peri- substitution.

Ipso-substitution describes two substituents sharing the same ring position in an intermediate compound in an electrophilic aromatic substitution.

Meso-substitution refers to the substituents occupying a benzylic position. It is observed in compounds such as calixarenes and acridines.

Peri-substitution occurs in naphthalenes for substituents at the 1 and 8 positions.

Cine and *tele* substitution

In cine-substitution, the entering group takes up a position adjacent to that occupied by the leaving group. For example, cine-substitution is observed in aryne chemistry.

Tele-substitution occurs when the new position is more than one atom away on the ring.

Origins

The prefixes *ortho*, *meta*, and *para* are all derived from Greek, meaning *straight* or *correct*, *following* or *after*, and *akin to* or *similar*, respectively. The relationship to the current meaning is perhaps not obvious. The *ortho* description was historically used to designate the original compound, and an isomer was often called the *meta* compound. For instance, the trivial names orthophosphoric acid and trimetaphosphoric acid have nothing to do with aromatics at all. Likewise, the description *para* was reserved for just closely related compounds. Thus Berzelius originally called the racemic form of aspartic acid *paraaspartic acid* (another obsolete term: racemic acid) in 1830. The use of the descriptions ortho, meta and para for multiple substituted aromatic rings starts with Wilhelm Körner in the period 1866–1874 although he chose to reserve the *ortho* prefix for the 1,4 isomer and the *meta* prefix for the 1,2-isomer. The current nomenclature (different again from that of Körner) was introduced by the Chemical Society in 1879.

Examples

Examples of the use of this nomenclature are given for isomers of cresol:

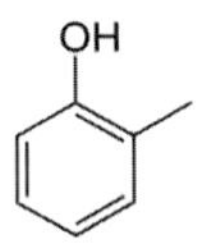

o-cresol

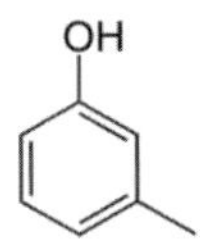

m-cresol

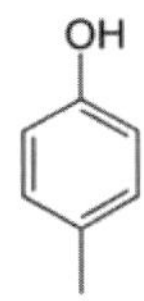

p-cresol

Catechol has two isomers, the meta isomer resorcinol and the para isomer hydroquinone:

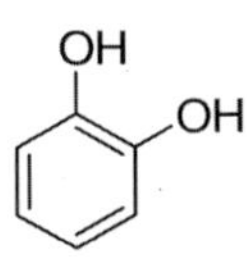

catechol

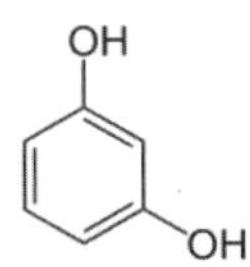

resorcinol

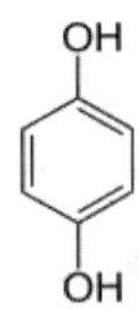

hydroquinone

Phthalic acid has two isomers, the meta isomer isophthalic acid and the para isomer terephthalic acid:

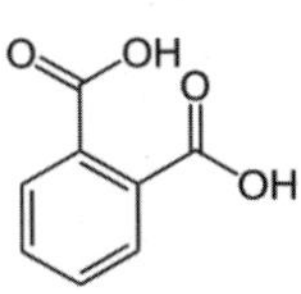

phthalic acid

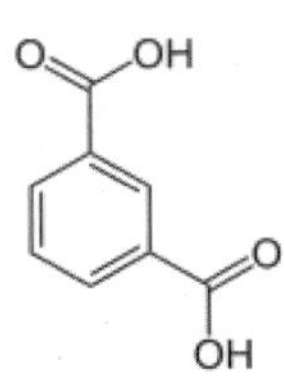

isophthalic acid

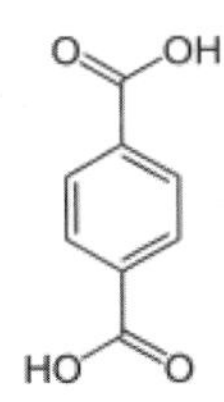

terephthalic acid

arginine[árdʒəni:n, -nain]: G. *arginoeis*, bright-shining, white. 아르기닌(아미노산의 일종). Abbreviated as Arg or R, an α-amino acid. The L-form is one of the 20 most common natural amino acids.

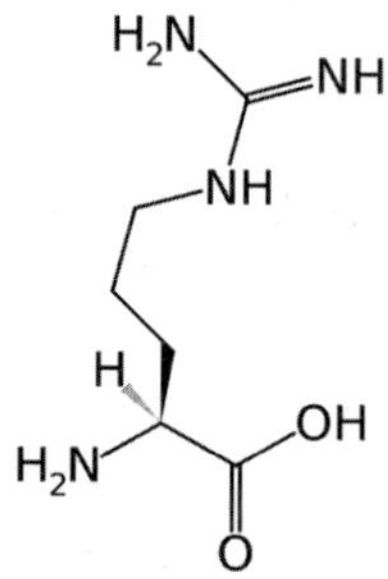

argon[á:rgɑn/-gɔn]: G. *argos, argon*, *a-,* without + *ergon,* work, idle, lazy, the inactive one, chemical inertness. 아르곤(Ar). A chemical element represented by the symbol **Ar**. Argon has atomic number 18 and is the third element in group 18 of the periodic table (noble gases). Argon is the third most common gas in the Earth's atmosphere, at 0.93%, making it more common than carbon dioxide. Nearly all of this argon is radiogenic argon-40 derived from the decay of potassium-40 in the Earth's crust. In the universe, argon-36 is by far the most common argon isotope, being the preferred argon isotope produced by stellar nucleosynthesis in supernovas[초신성(超新星)]. The complete octet (eight electrons) in the outer atomic shell makes argon stable and resistant to bonding with other elements. Argon is produced industrially by the fractional distillation of liquid air. Argon is mostly used as an inert shielding gas in welding and other high-temperature industrial processes where ordinarily non-reactive substances become reactive; for example, an argon atmosphere is used in graphite electric furnaces to prevent the graphite from burning. Argon gas also has uses in incandescent(백열의) and fluorescent lighting, and other types of gas discharge tubes. Argon makes a distinctive blue-green gas laser.

aromatic[ærəmǽtik]: L. < G. *aromaticus* < *arōmatikos,* from *arōma,* spice. 방향족(芳香族)의, 방향성의.

aromaticity: a chemical property in which a conjugated ring of unsaturated bonds, lone pairs, or empty orbitals exhibit a stabilization stronger than would be expected by the stabilization of conjugation alone. It can also be considered a manifestation of cyclic delocalization and of resonance. This is usually considered to be because electrons are free to cycle around circular arrangements of atoms, which are alternately single- and double-bonded to one another. These bonds may be seen as a hybrid of a single bond and a double bond, each bond in the ring identical to every other. This commonly-seen model of aromatic rings, namely the idea that benzene is a more stable molecule than would be expected without accounting for charge delocalization.

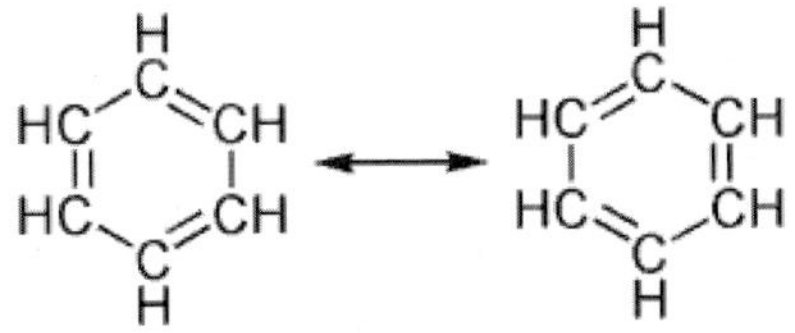

arthritis[ɑ:rəráitis]: L. < G. *arthron*, joint. 관절염.

artificial[ɑ:rtəfíʃəl]: L. *artificium*, *artificialis,* handcraft. 인공적인. 모조의, 거짓의, 가짜의, 부자연스러운.

aryl[ɛ́əril]: *G. arryl, aryl, ar(omatic) + -yl.* aromatic hydrocarbon(s). 아릴(방향족의 포화탄화수소기).

asbestos[æsbéstəs, æz-/-tɔs]: L. < G. *asbestos, a-*, not + *sbestos*, from *sbennumi*, quench, inextinguishable, unquenchable. 석면(石綿). The unquenchable stone, a fabulous stone. A set of six naturally occurring silicate minerals exploited commercially for their desirable physical properties. Asbestos in some forms can be toxic. The inhalation of asbestos fibres can cause serious illnesses, including malignant lung cancer, mesothelioma [a type of malignant neoplasm(종양) dependent mostly from exposure to asbestos].

ascorbic[əskɔ́:rbik]: L. *a*, without + *scorbutus*, scurvy + -ic, no scurvy. 괴혈병이 없는.

ascorbic acid: 아스코르브산(비타민 C). A sugar acid with antioxidant properties. Its appearance is white to light-yellow crystals or powder, and it is water-soluble. One form of ascorbic acid is commonly known as vitamin C. Scorbutus (scurvy); the disease caused by a deficiency of vitamin C.

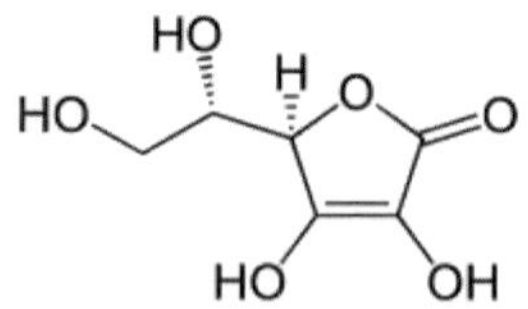

asparagine[əspǽrədʒi:n]: L. < G. *asparagos* + *-ine*. 아스파라긴(식물에 많은 아미노산의 일종). One of the 20 most common natural amino acids on Earth. It has carboxamide as the side chain's functional group. It is not an essential amino acid.

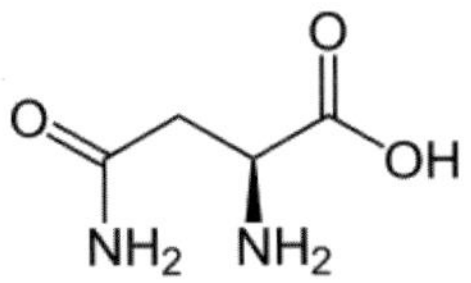

Aspirin[ǽspərin]: D. *acetylierte Spirsäure*, acetylsalicylic acid (ASA). *spir-,* being from the plant genus name *Spiraea.* 아세틸살리실산, 아스피린(해열 진통제, 독일 Bayer사 제품의 상표명). A salicylate drug, often used as an analgesic to relieve minor aches and pains, as an antipyretic to reduce fever, and as an anti-inflammatory medication. Aspirin was first isolated by Felix Hoffmann, a chemist with the German company Bayer in 1897. Salicylic acid, the main metabolite of aspirin, is an integral part of human and animal metabolism. While in humans much of it is attributable to diet, a substantial part is synthesized endogenously. Aspirin also has an antiplatelet effect by inhibiting the production of thromboxane, which under normal circumstances binds platelet molecules together to create a patch over damaged walls of blood vessels. Because the platelet patch can become too large and also block blood flow, locally and downstream, aspirin is also used long-term, at low doses, to help prevent heart attacks, strokes, and blood clot formation in people at high risk of developing blood clots. It has also been established that low doses of aspirin may be given immediately after a heart attack to reduce the risk of another heart attack or of the death of cardiac tissue. Aspirin may be effective at preventing certain types of cancer, particularly colorectal cancer. The main undesirable side effects of aspirin taken by mouth are gastrointestinal ulcers, stomach bleeding, and tinnitus(귀울림), especially in higher doses. In children and adolescents, aspirin is no longer indicated to control flu-like symptoms or the symptoms of chickenpox or other viral illnesses, because of the risk of Reye's syndrome. Aspirin is part of a group of medications called nonsteroidal anti-inflammatory drugs (NSAIDs), but differs from most other NSAIDs in the mechanism of action. Though it, and others in its group called the salicylates, have similar effects (antipyretic, anti-inflammatory, analgesic) to the other NSAIDs and inhibit the same enzyme cyclooxygenase, aspirin (but not the other salicylates) does so in an irreversible manner and, unlike others, affects more the COX-1 variant than the COX-2 variant of the enzyme. Today, aspirin is one of the most widely used medications in the world, with an estimated 40,000 tonnes of it being consumed each year. In countries where Aspirin is a registered trademark owned by Bayer, the generic term is acetylsalicylic acid (ASA).

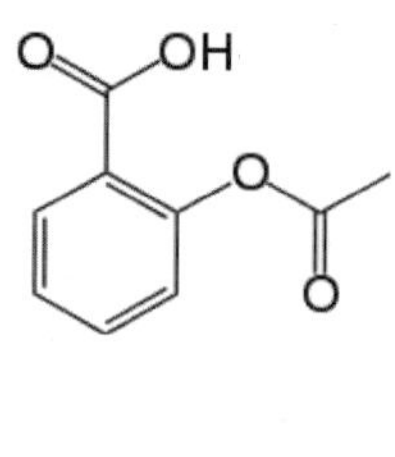

asymmetry[eisímətri, æ-]: G. asummetria, *a-*, without + *summetria*, symmetry, lacking symmetry. 비대칭, 불균형.

atherosclerosis[æθərousklərόusis]: D. < G. *athēre*, groats + *sklērosis*, hardening. 아테롬성(죽상, 粥狀) 동맥경화증.

atmosphere[ǽtməsfiər]: G. > L. *atmosphaera < atmossphaira,* G. *atmos,* vapor + *sphaira,* ball, globe, a unit of pressure equal to mean atmospheric pressure at sea level, 101,325 pascals. 기압(대기 압력의 단위로 1.013 헥토파스칼; atm), 기압.

atom[ǽtəm]: D. < G. *atomos*, *a-*, not + *temnein*, to cut. Indivisible. 원자. A basic unit of matter that consists of a dense central nucleus surrounded by a cloud of negatively charged electrons. The atomic nucleus contains a mix of positively charged protons and electrically neutral neutrons (except in the case of hydrogen-1, which is the only stable nuclide with no neutrons). The electrons of an atom are bound to the nucleus by the electromagnetic force. Likewise, a group of atoms can remain bound to each other, forming a molecule. An atom containing an equal number of protons and electrons is electrically neutral, otherwise it has a positive charge if there are fewer electrons (electron deficiency) or negative charge if there are more electrons (electron excess). A positively or negatively charged atom is known as an ion. An atom is classified according to the number of protons and neutrons in its nucleus: the number of protons determines the chemical element, and the number of neutrons determines the isotope of the element. The concept of an atom as an indivisible component of matter was first proposed by early Indian and Greek philosophers. In the 17th and 18th centuries, chemists provided a physical basis for this idea by showing that certain substances could not be further broken down by chemical methods. During the late 19th and early 20th centuries, physicists discovered subatomic components and structure inside the atom, thereby demonstrating that the 'atom' was divisible. The principles of quantum mechanics were used to successfully model the atom. Atoms are minuscule(대단히 작은) objects with proportionately tiny masses. Atoms can only be observed individually using special instruments such as the scanning tunneling microscope. Over 99.94% of an atom's mass is concentrated in the nucleus, with protons and neutrons having roughly equal mass. Each element has at least one isotope with unstable nuclei that can undergo radioactive decay. This can result in a transmutation that changes the number of protons or neutrons in a nucleus. Electrons that are bound to atoms possess a set of stable energy levels, or orbitals, and can undergo transitions between them by absorbing or emitting photons that match the energy differences between the levels. The electrons determine the chemical properties of an element, and strongly influence an atom's magnetic properties.

authentic[ɔ:θéntik]: G. > L. *authenticus, < authentikos,* principal, genuine. not a copy, original. 진품의, 믿을만한.

automatic[ɔ:təmǽtik]: G. *automatos*, acting of itself. 자동의. 무의식적인, 반사적인.

autonomy[ɔ:tánəmi / -tɔ́n-]: G. *autonomos, autonomia, autos*, self + *nomos*, law, having its own laws. 자율, 자발성.

axis[ǽksis]: L. *axle*, pivot(회전축). 축(軸).

<u>axial position</u>: in the lowest-energy chair conformation of cyclohexane, 6 of the 12 hydrogens are in **axial positions** (colored red) — their C-H bonds are parallel to each other and appear to stick up and down from the ring structure, the other 6 are in **equatorial positions** (colored blue) — they are splayed out around the perimeter of the ring. Note that in addition, one hydrogen at each position is "up" relative to the other being "down" at that position.

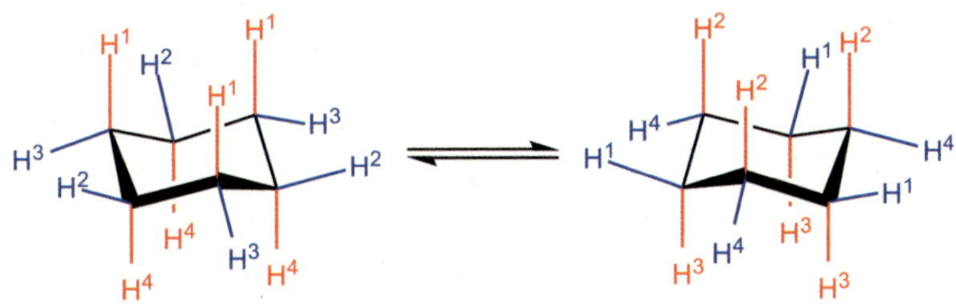

azeotrope[ə'zíətroup]: G. *a*, no + *zeein*, boil + *tropos*, change, no change on boiling. 공비점(共沸點). Mixture of two or more liquids in such a ratio that its composition cannot be changed by simple distillation. This occurs because, when an azeotrope is boiled, the resulting vapor has the same ratio of constituents as the original mixture. Because their composition is unchanged by distillation, azeotropes are also called (especially in older texts) constant boiling mixtures.

Fig. Vapor-liquid equilibrium of 2-propanol/water showing azeotropic behavior

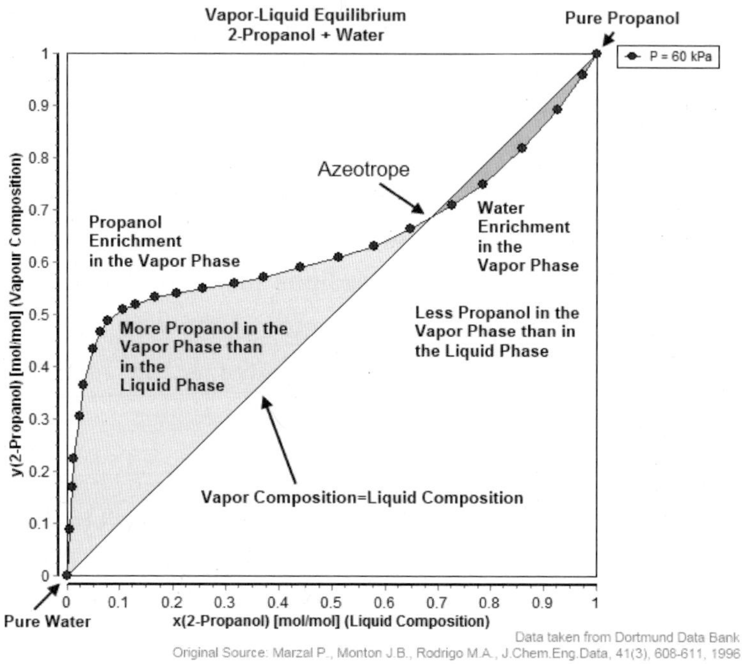

azo-[éizou, ǽzə-, éiz-]: G. > F. *azote, a-,* without + *zōē,* life, nitrogen. (화합물이)아조기를 함유한, 아조의.

azo compounds: a compounds bearing the functional group R-N=N-R', in which R and R' can be either aryl or alkyl. IUPAC defines azo compounds as: "Derivatives of diazene (diimide), HN=NH, wherein both hydrogens are substituted by hydrocarbyl groups, e.g. PhN=NPh azobenzene or diphenyldiazene. The more stable derivatives contain two aryl groups. The N=N group is called an azo group.

Fig. General chemical formula of azo compounds

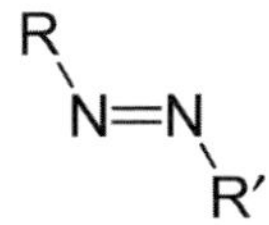

bacterium[bæktíəriəm]: G. *baktērion*, *baktēria* (pl). staff, cane (because the first ones to be discovered were rod-shaped). bacteria의 단수형.

barbituric[bɑ:rbət∫úərik] acid: D. *barbitursäure*. 바르비투르산의. Malonylurea or 6-hydroxyuracil, an organic compound based on a pyrimidine heterocyclic skeleton. It is an odorless powder soluble in water. Barbituric acid is the parent compound of barbiturate drugs, although barbituric acid itself is not pharmacologically active. The compound was discovered by the German chemist Adolf von Baeyer on December 4, 1864, the feast of Saint Barbara (who gave the compound its namesake), by combining urea and malonic acid in a condensation reaction. Malonic acid has since been replaced by diethyl malonate, as using the ester avoids the problem of having to deal with the acidity of the carboxylic acid and its unreactive carboxylate.

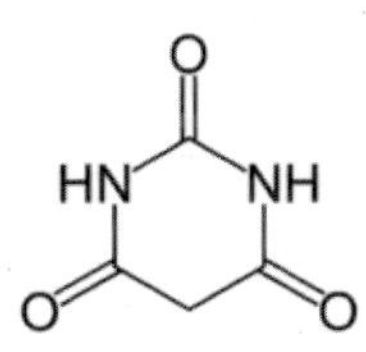

base[beis]: G. > L. *basis*, base, pedestal. 염기, 양성자를 받아들이는 분자. A substance that can accept hydrogen ions (protons) or more generally, donate electron pairs. A soluble base is referred to as an alkali if it contains and releases hydroxide ions (OH^-) quantitatively. The Brønsted-Lowry theory defines bases as proton (hydrogen ion) acceptors, while the more general Lewis theory defines bases as electron pair donors, allowing other Lewis acids than protons to be included. The oldest Arrhenius theory defines bases as hydroxide anions, which is strictly applicable only to alkali. In water, by altering the autoionization equilibrium, bases give solutions with a hydrogen ion activity lower than that of pure water, i.e. a pH higher than 7.0 at standard conditions. Examples of common bases are sodium hydroxide and ammonia. Metal oxides, hydroxides and especially alkoxides are basic, and counteranions of weak acids are weak bases. Bases can be thought of as the chemical opposite of acids. A reaction between an acid and base is called neutralization. Bases and acids are seen as opposites because the effect of an acid is to increase the hydronium ion (H_3O^+) concentration in water, whereas bases reduce this concentration. Bases and acids are typically found in aqueous solution forms. Aqueous solutions of bases react with aqueous solutions of acids to produce water and salts in aqueous solutions in which the salts separate into their component ions. If the aqueous solution is a saturated solution with respect to a given salt solute any additional such salt present in the solution will result in formation of a precipitate of the salt.

benign[bináin]: F. < L. *benigne* < *benignus*, *bene*, well + *-genus*, -born. not hamful in effect. 양성(良性)의.

benzene[bénzi:n, benzí:n]: 19th cent. benzoic acid + -ene. 벤젠. An organic chemical compound with the molecular formula C_6H_6. It is sometimes abbreviated Ph–H. Benzene is a colorless and highly flammable liquid with a sweet smell. Because it is a known carcinogen, its use as an additive in gasoline is now limited, but it is an important industrial solvent and precursor in the production of drugs, plastics, synthetic rubber, and dyes. Benzene is a natural constituent of crude oil, and may be synthesized from other compounds present in petroleum. Benzene is an aromatic hydrocarbon and the second [*n*]-annulene ([6]-annulene), a cyclic hydrocarbon with a continuous pi bond. It is also related to the functional group arene which is a generalized structure of benzene.

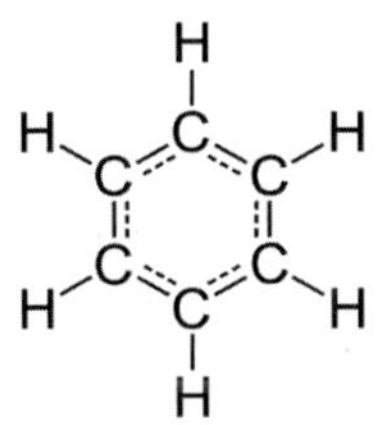
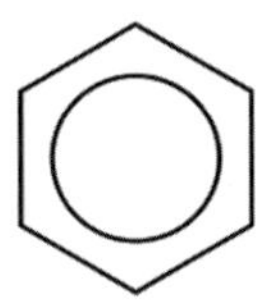

benzocaine[bénzoukein]: 1920s. benzoic acid + *-caine* (from cocaine). 벤조카인(국소 마취제용). A local anesthetic commonly used as a topical pain reliever. It is the active ingredient in many over-the-counter anesthetic ointments.

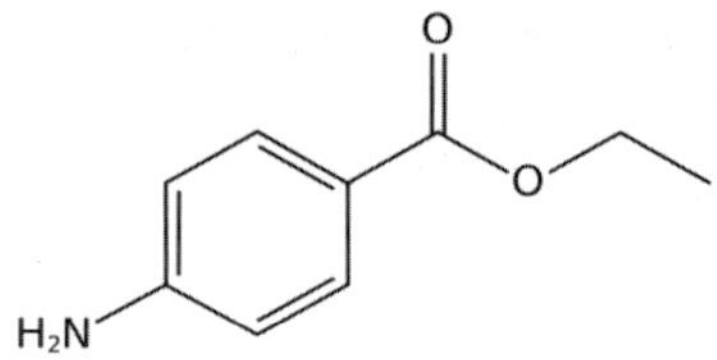

benzodiazepine[benzoudaiǽzəpi:n]: 1930s. benzene + *di-*, two, double + *azo-,* nitrogen, *epi-,* upon, near to, in addition + *-ine*. 벤조디아제핀(신경안정제). A psychoactive drug whose core chemical structure is the fusion of a benzene ring and a diazepine ring. The first benzodiazepine, chlordiazepoxide (Librium), was discovered accidentally by Leo Sternbach in 1955, and made available in 1960 by Hoffmann–La Roche, which has also marketed diazepam (Valium) since 1963.

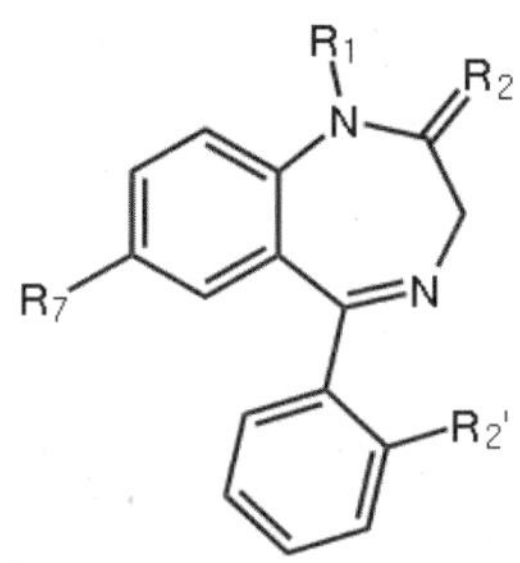

benzoin[bénzouin]: F. *benjoin*, incense of Java(자바산 커피향). An organic compound with the formula PhCH(OH)C(O)Ph. It is a hydroxy ketone attached to two phenyl groups. It appears as off-white crystals, with a light camphor-like odor. Benzoin is synthesized from benzaldehyde in the benzoin condensation. It is chiral and it exists as a pair of enantiomers: (*R*)-benzoin and (*S*)-benzoin.

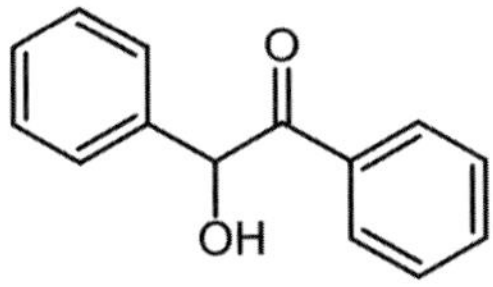

beryllium[bəríliəm/be-]: L. *beryllium, beryllii,* Be. 베릴륨(glucinum, beryllium의 옛 이름). The chemical element with the symbol **Be** and atomic number 4. It is a divalent element which occurs naturally only in combination with other elements in minerals. Notable gemstones which contain beryllium include beryl (aquamarine, emerald) and chrysoberyl. As a free element it is a steel-gray, strong, lightweight and brittle alkaline earth metal. Beryllium is used primarily as a hardening agent in alloys, notably beryllium copper. In structural applications, high flexural rigidity, thermal stability, thermal conductivity and low density (1.85 times that of water) make beryllium a quality aerospace material for high-speed aircraft, missiles, space vehicles and communication satellites. Because of its low density and atomic mass, beryllium is relatively transparent to X-rays and other forms of ionizing radiation; therefore, it is the most common window material for X-ray equipment and in particle physics experiments. The high thermal conductivity of beryllium and beryllium oxide have led to their use in heat transport and heat sinking applications.

bile[bail]: F. < L. *bilis*, gall. 담즙. A bitter-tasting, dark green to yellowish brown fluid, produced by the liver of most vertebrates, that aids the process of digestion of lipids in the small intestine. In many species, bile is stored in the gallbladder(쓸개) and upon eating is discharged into the duodenum(십이지장). Bile is a composition of the following materials: Water (85%), bile salts (10%), mucus and pigments (3%), fats(1%), inorganic salts (0.7%) and cholesterin (0.3%).

bilirubin[bilərú:bin]: D. < L. *bilis*, bile + *ruber*, red + *-in*. 빌리루빈(담즙 중의 적황색 색소). The yellow breakdown product of normal heme catabolism. Heme is found in hemoglobin, a principal component of red blood cells. Bilirubin is excreted in bile and urine, and elevated levels may indicate certain diseases. It is responsible for the yellow color of bruises(타박상, 멍), urine, and the yellow discoloration in jaundice(황달). It has also been found in plants.

biology[baiɑ́lədʒi/-ɔ́l-]: G. *bios*, life + *-logia*, speaking, study. 생물학. A natural science concerned with the study of life and living organisms, including their structure, function, growth, origin, evolution, distribution, and taxonomy(분류학).

biotic[baiɑ́tik / -ɔ́t-]: G. *biōtikos, bios*, life. relating to or resulting from living organisms. 생명(생물)의, 생명에 관한.

biopsy[baiɑpsi/-ɔp-]: G. *bios,* life + *-opsis,* seeing, sight. Examination of a tissue removed from a living body. 생체 조직 절편 검사(법), (검사를 위해) 생체 조직을 떼어내다.

biotin[báiə˞tin]: G. *bios*, life + *-in*. 비오틴(비타민 B복합체의 하나, 비타민 H와 동일). A water-soluble B-complex vitamin (vitamin B_7) that is composed of a ureido (tetrahydroimidizalone) ring fused with a tetrahydrothiophene ring. A valeric acid substituent is attached to one of the carbon atoms of the tetrahydrothiophene ring. Biotin is a coenzyme in the metabolism of fatty acids and leucine, and it plays a role in gluconeogenesis.

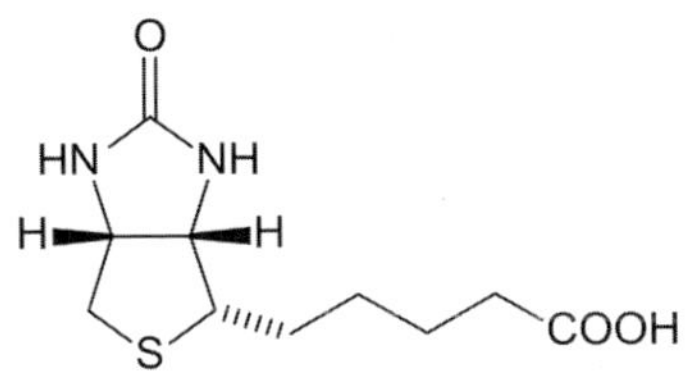

boil[bɔil]: F. < L. *boillir* < *bullire*, *bulla*, to bubble, bubble. 끓음, 비등(沸騰), 끓다, 비등하다.

<u>boiling point</u>: 비등점, 끓는점. The boiling point of an element or a substance is the temperature at which the vapor pressure of the liquid equals the environmental pressure surrounding the liquid. The boiling point of liquids varies with and depends upon the surrounding environmental pressure (which tends to vary with elevation). Different liquids (at a given pressure) boil at different temperatures.

bradykinin[brædikínin, -kái-]: G. *bradus*, slow + *kinēsis,* motion + *-in.* 브래디키닌(혈관확장제). A peptide that causes blood vessels to enlarge (dilate), and therefore causes blood pressure to lower. A class of drugs called ACE inhibitors, which are used to lower blood pressure, increase bradykinin (by inhibiting its degradation) further lowering blood pressure. Bradykinin works on blood vessels through the release of prostacyclin, nitric oxide, and Endothelium-Derived Hyperpolarizing Factor. Bradykinin is a physiologically and pharmacologically active peptide of the kinin group of proteins, consisting of nine amino acids.

bromine[bróumi:n]: F. < G. *brome*, *brōmos*, a stink, stench + -ine. 브롬(할로겐족 원소의 하나). A chemical element with the symbol **Br**, an atomic number of 35, and an atomic mass of 79.904. It is in the halogen element group. The element was isolated independently by two chemists, Carl Jacob Löwig and Antoine Jerome Balard, in 1825–1826. Elemental bromine is a fuming red-brown liquid at room temperature, corrosive and toxic, with properties between those of chlorine and iodine. Free bromine does not occur in nature, but occurs as colorless soluble crystalline mineral halide salts, analogous to table salt. Bromine is rarer than about three-quarters of elements in the Earth's crust, however the high solubility of bromide ion has caused its accumulation in the oceans, and commercially the element is easily extracted from brine(소금물) pools, mostly in the United States, Israel and China.

butyric[bju:tírik]: L. *butyric*, *butyrum*, butter + *-ic*, butter. 낙산(酪酸)에서 얻은.

butyric acid: the systematic name butanoic acid, a carboxylic acid with the structural formula $CH_3CH_2CH_2$-COOH. Salts and esters of butyric acid are known as butyrates or butanoates. Butyric acid is found in butter, parmesan cheese, vomit, and as a product of anaerobic fermentation (including in the colon and as body odor). It has an unpleasant smell and acrid taste, with a sweetish aftertaste (similar to ether).

cadaver[kədǽvər]: L. *cadere*, to fall, a corpse. 해부용의 시체.

cadmium[kǽdmiəm]: L. *cadmia,* calamine, **Cd**. 카드뮴. A chemical element with atomic number 48. This soft, bluish-white metal is chemically similar to the two other stable metals in group 12, zinc and mercury. Like zinc, it prefers oxidation state +2 in most of its compounds and like mercury it shows a low melting point compared to transition metals. Cadmium and its congeners(동족원소) are not always considered transition metals, in that they do not have partly filled d or f electron shells in the elemental or common oxidation states. The average concentration of cadmium in the Earth's crust is between 0.1 and 0.5 parts per million (ppm). It was discovered in 1817 simultaneously by Stromeyer and Hermann, both in Germany, as an impurity in zinc carbonate. Cadmium occurs as a minor component in most zinc ores and therefore is a byproduct of zinc production. It was used for a long time as a pigment and for corrosion resistant plating on steel while cadmium compounds were used to stabilize plastic. With the exception of its use in nickel–cadmium batteries and cadmium telluride solar panels, the use of cadmium is generally decreasing. These declines have been due to competing technologies, cadmium's toxicity in certain forms and concentration and resulting regulations.

camphor[kǽmfər]: Arabic > L. *kāfūr* > *camphora.* 장뇌(樟腦). A waxy, white or transparent solid with a strong, aromatic odor. It is a terpenoid with the chemical formula $C_{10}H_{16}O$. It is found in wood of the camphor laurel (Cinnamomum camphora), a large evergreen tree found in Asia (particularly in Borneo and Taiwan). It can also be synthetically produced from oil of turpentine. It is used for its scent, as an ingredient in cooking (mainly in India), as an embalming fluid, for medicinal purposes, and in religious ceremonies. A major source of camphor in Asia is camphor basil.

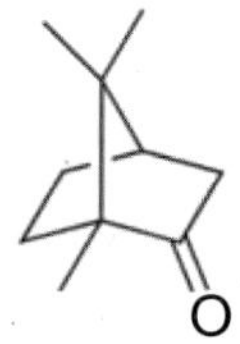

cancer[kǽnsər]: L. *cancer, cancri,* crab, cancer, the constellation (별자리). 암(癌), (천문)게자리(the Crab). Various diseases including apparently malignant tumours, cankers[궤양(潰瘍), 암종병(癌腫病)], gangrene[괴저(壞疽)].

capacity[kəpǽsəti]: F. < L. *capacitė* < *capacitas* (*capac-*, *capax*, that can contain from *capere*, take or hold), the maximum amount that something can contain. 용량(容量).

heat capacity (= thermal mass, thermal capacitance): the capacity of a body to store heat. It is typically measured in units of J/°C or J/K (which are equivalent). 열용량.

specific heat capacity (often shortened to specific heat): the measure of heat or thermal energy required to increase the temperature of a unit quantity of a substance by one unit. 비열(比熱). For example, at a temperature of 15°C, the heat required to raise the temperature of 1kg of water by 1K (equivalent. to 1°C) is 4186 joules, meaning that the specific heat of water is 4.186 kJ · kg^{-1} · K^{-1}.

caproick[kǽproic]: L. *caper*, goat. 염소.

Three such acids are named after goats: caproic (C6), caprylic (C8), capric (C10), totaling 15% in goat milk fat.

caproic acid (= hexanoic acid): $C_5H_{11}COOH$

caprylic acid (= octanoic acid): $C_7H_{15}COOH$

capric acid (= decanoic acid): $C_9H_{19}COOH$

carbon[kɑ́:rbən]: F. < L. *carbone < carbo, carbon,* coal, charcoal. 숯, 석탄, 탄소 (C). The chemical element with symbol **C** and atomic number 6. As a member of group 14 on the periodic table, it is nonmetallic and tetravalent — making four electrons available to form covalent chemical bonds. There are three naturally occurring isotopes, with ^{12}C and ^{13}C being stable, while ^{14}C is radioactive, decaying with a half-life of about 5,730 years. Carbon is one of the few elements known since antiquity. There are several allotropes of carbon of which the best known are graphite, diamond, and amorphous carbon. The physical properties of carbon vary widely with the allotropic form. For example, diamond is highly transparent, while graphite is opaque and black. Diamond is among the hardest materials known, while graphite is soft enough to form a streak on paper (hence its name, from the Greek word "to write"). Diamond has a very low electrical conductivity, while graphite is a very good conductor. Under normal conditions, diamond has the highest thermal conductivity of all known materials. All carbon allotropes are solids under normal conditions with graphite being the most thermodynamically stable form. They are chemically resistant and require high temperature to react even with oxygen. The most common oxidation state of carbon in inorganic compounds is +4, while +2 is found in carbon monoxide and other transition metal carbonyl complexes. The largest sources of inorganic carbon are limestones, dolomites (백운석) and carbon dioxide, but significant quantities occur in organic deposits of coal, peat(이탄), oil and methane clathrates (격자모양). Carbon forms more compounds than any other element, with almost ten million pure organic compounds described to date, which in turn are a tiny fraction of such compounds that are theoretically possible under standard conditions. Carbon is the 15th most abundant element in the Earth's crust, and the fourth most abundant element in the universe by mass after hydrogen, helium, and oxygen. It is present in all known life forms, and in the human body carbon is the second most abundant element by mass (about 18.5%) after oxygen. This abundance, together with the unique diversity of organic compounds and their unusual polymer-forming ability at the temperatures commonly encountered on Earth, make this element the chemical basis of all known life.

carbonate: a salt of carbonic acid, characterized by the presence of the carbonate ion, CO_3^{2-}. The name may also mean an ester of carbonic acid, an organic compound containing the carbonate group $C(=O)(O^-)_2$.

bicarbonate: (IUPAC-recommended nomenclature: hydrogen-carbonate), an intermediate form in the deprotonation of carbonic acid. Its chemical formula is HCO_3^-.

carbonic acid: the inorganic compound with the formula H_2CO_3 [equivalently $OC(OH)_2$]. It is also a name sometimes given to solutions of carbon dioxide in water, because such solutions contain small amounts of H_2CO_3. Carbonic acid salts forms two kinds of salts, the carbonates and the bicarbonates. It is a weak acid.

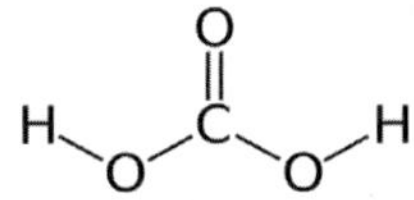

carbohydrate[kɑ:rbouháidreit]: L. *carbo, carbon*, coal, charcoal + G. *hydor, hydoratos*, water. 탄수화물. An organic compound with the general formula $C_m(H_2O)_n$, that is, consists only of carbon, hydrogen and oxygen, with the last two in the 2:1 atom ratio. Carbohydrates can be viewed as hydrates of carbon, hence their name. Structurally however, it is more accurate to view them as polyhydroxy aldehydes and ketones. The term is most common in biochemistry, where it is a synonym of saccharide. The carbohydrates (saccharides) are divided into four chemical groupings: monosaccharides, disaccharides, oligosaccharides, and polysaccharides. In general, the monosaccharides and disaccharides, which are smaller (lower molecular weight) carbohydrates, are commonly referred to as sugars.

Fig. Lactose is a disaccharide found in milk. It consists of a molecule of D-galactose and a molecule of D-glucose bonded by a-1-4 glycosidic linkage.

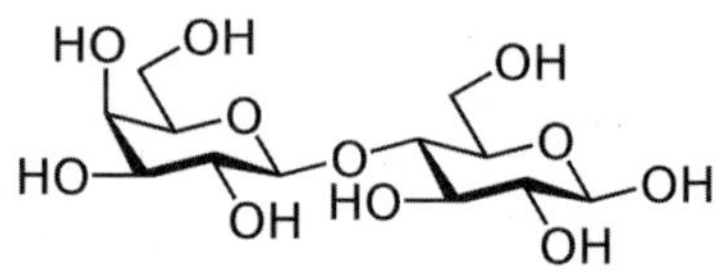

carboxyl[kɑ:rbɑ́ksil /-bɔ́k-]: mid 19th century. *carbo-*, carbon + *ox-*, oxygen + *-yl*). 카르복시기를 함유하는.

carboxylic acids: organic acids characterized by the presence of at least one carboxyl group. The general formula of a carboxylic acid is R-COOH, where R is some monovalent functional group. A carboxyl group (or carboxy) is a functional group consisting of a carbonyl (RR'C=O) and a hydroxyl (R-O-H), which has the formula -C(=O)OH, usually written as -COOH or -CO_2H. Carboxylic acids are Brønsted-Lowry acids, they are proton donors. They are the most common type of organic acid. Among the simplest examples are formic acid H-COOH, that occurs in ants, and acetic acid CH_3-COOH, that gives vinegar its sour taste. Acids with two or more carboxyl groups are called dicarboxylic, tricarboxylic, etc. The simplest dicarboxylic example is oxalic acid $(COOH)_2$, which is just two connected carboxyls. Mellitic acid (benzene-hexacarboxylic acid) is an example of a hexacarboxylic acid. Other important natural examples are citric acid (in lemons) and tartaric acid (in tamarinds). Salts and esters of carboxylic acids are called carboxylates. When a carboxyl group is deprotonated, its conjugate base, a carboxylate anion is formed. Carboxylate ions are resonance stabilized and this increased stability makes carboxylic acids more acidic than alcohols. Carboxylic acids can be seen as reduced or alkylated forms of the Lewis acid carbon dioxide; under some circumstances they can be decarboxylated to yield carbon dioxide.

carcinogen(s)[kɑ:rsínədʒən]: L. *carcinoma, carcinomatis,* tumor, + -gen. A substance capable of causing cancer in living tissue. 발암 물질, 발암(성)의.

cardiovascular[kɑ:rdiouvǽskjulər]: G. *kardia*, heart + *vascularis*, vessel, 심장혈관의.

carotene[kǽrətin]: L. *Carota*, carrot. 카로틴 (당근 고추 등에 함유 되어있는 적황색 색소. Hydrocarbon substances having the formula $C_{40}Hx$, which are synthesized by plants but cannot be made by animals. Carotene is an orange photosynthetic pigment important for photosynthesis. They are responsible for the orange colour of the carrot, for which this class of chemicals is named, and for the colours of many other fruits and vegetables (for example, sweet potatoes and orange cantaloupe melon). The two primary isomers of carotene, α-carotene and β-carotene, differ in the position of double bonds in the cyclic group at the end. β-Carotene is the more common form and can be found in yellow, orange, and green leafy fruits and vegetables. As a rule of thumb, the greater the intensity of the orange colour of the fruit or vegetable, the more β-carotene it contains. Carotene protects plant cells against the destructive effects of ultraviolet light. β-Carotene is an anti-oxidant.

Fig. α-Carotene

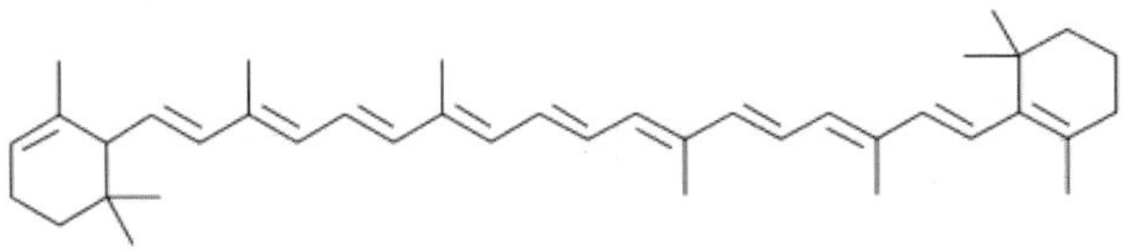

Fig. β-Carotene

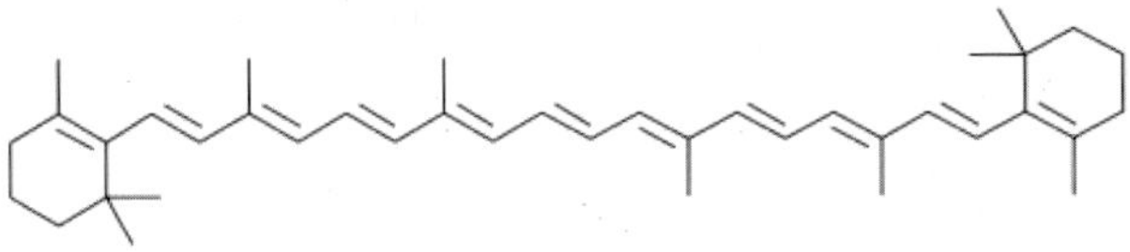

Cartesian[kɑ:rtíʒən /-zjən]: L. *Cartesius, Cartesianus,* Latinized form of the name of Descartes. 데카르트의(프랑스의 철학자·수학자).

<u>Cartesian coordinates</u>: specifies each point uniquely in a plane by a pair of numerical coordinates, which are the signed distances from the point to two fixed perpendicular directed lines, measured in the same unit of length. 데카르트 좌표, 평행 좌표.

Fig. Illustration of a Cartesian coordinate plane. Four points are marked and labeled with their coordinates: (2,3), (−3,1), (−1.5,−2.5), and the origin (0,0).

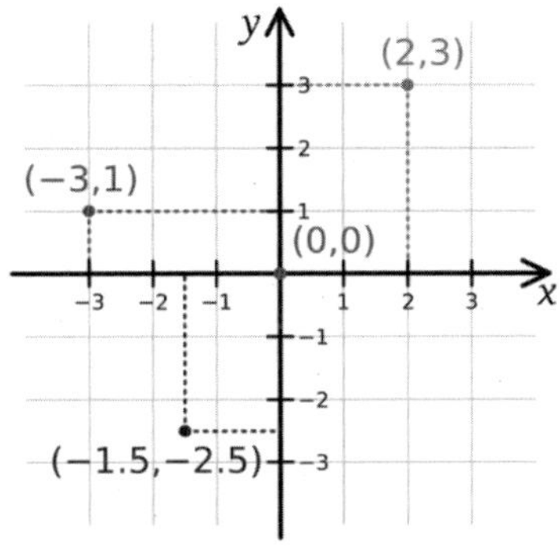

catabolism[kətǽbəlizm]: G. *katabolē, kata-,* down *+ ballein,* to throw, throwing down. 이화작용. The set of pathways that break down molecules into smaller units and release energy. In catabolism, large molecules such as polysaccharides, lipids, nucleic acids and proteins are broken down into smaller units such as monosaccharides, fatty acids, nucleotides, and amino acids, respectively.

catalysis[kətǽləsis]: G. *katalusis*, *kataluein, kata-*, down + *luein*, loosen, dissolve. 촉매작용. The change in rate of a chemical reaction due to the participation of a substance called a catalyst. Unlike other reagents that participate in the chemical reaction, a catalyst is not consumed by the reaction itself. A catalyst may participate in multiple chemical transformations. Catalysts that speed the reaction are called positive catalysts.

Substances that slow a catalyst's effect in a chemical reaction are called inhibitors. Substances that increase the activity of catalysts are called promoters, and substances that deactivate catalysts are called catalytic poisons. Catalytic reactions have a lower rate-limiting free energy of activation than the corresponding uncatalyzed reaction, resulting in higher reaction rate at the same temperature. However, the mechanistic explanation of catalysis is complex. Catalysts may affect the reaction environment favorably, or bind to the reagents to polarize bonds, e.g. acid catalysts for reactions of carbonyl compounds, or form specific intermediates that are not produced naturally, such as osmate esters in osmium tetroxide-catalyzed dihydroxylation of alkenes, or cause lysis of reagents to reactive forms, such as atomic hydrogen in catalytic hydrogenation. Kinetically, catalytic reactions are typical chemical reactions; i.e. the reaction rate depends on the frequency of contact of the reactants in the rate-determining step. Usually, the catalyst participates in this slowest step, and rates are limited by amount of catalyst and its "activity". In heterogeneous catalysis, the diffusion of reagents to the surface and diffusion of products from the surface can be rate determining. Analogous events associated with substrate binding and product dissociation apply to homogeneous catalysts. Although catalysts are not consumed by the reaction itself, they may be inhibited, deactivated, or destroyed by secondary processes. In heterogeneous catalysis, typical secondary processes include coking where the catalyst becomes covered by polymeric side products.

Additionally, heterogeneous catalysts can dissolve into the solution in a solid–liquid system or evaporate in a solid–gas system.

cathode[kǽθoud]: G. *kathodos* (*kata-,* down + *hodos,* way), a going down, way down. (전해조·전지의) 전류가 나오는 전극, 음극, (볼타전지의, 축전지의) 양극. 전자가 들어가는 전극, cation이 들어가는 전극. An electrode through which electric current flows out of a polarized electrical device. The direction of electric current is, by convention, opposite to the direction of electron flow. Therefore the electrons flow into the polarized electrical device and out of, for example, the connected electrical circuit. Mnemonic: CCD (Cathode Current Departs). Cathode polarity is not always negative. Although positively charged cations always move towards the cathode (hence their name) and negatively charged anions move away from it, cathode polarity depends on the device type, and can even vary according to the operating mode. In a device which consumes power, the cathode is negative, and in a device which provides power, the cathode is positive: In a discharging battery or a galvanic cell the cathode is the positive terminal since that is where the current flows out of the device (see drawing). This outward current is carried internally by positive ions moving from the electrolyte to the positive cathode (chemical energy is responsible for this "uphill" motion). It is continued externally by electrons moving inwards, negative charge moving one way constituting positive current flowing the other way. For example, the Daniell galvanic cell's copper electrode is the positive terminal and the cathode. In a recharging battery, or an electrolytic cell, the cathode is the negative terminal, which sends current back to the external generator. For example, reversing the current direction in a Daniell galvanic cell would produce an electrolytic cell, where the copper electrode is the positive terminal and the anode. In a diode, it is the negative terminal at the pointed end of the arrow symbol, where current flows out of the device. Note: electrode naming for diodes is always based on the direction of the forward current (that of the arrow, in which the current flows "most easily"), even for types such as Zener diodes or solar cells where the current of interest is the reverse current. In vacuum tubes (including cathode ray tubes) it is the negative terminal where electrons flow in from the wiring and through the tube's near vacuum, constituting a positive current flowing out of the device.

Fig. Diagram of a copper cathode in a galvanic (= voltaic) cell.

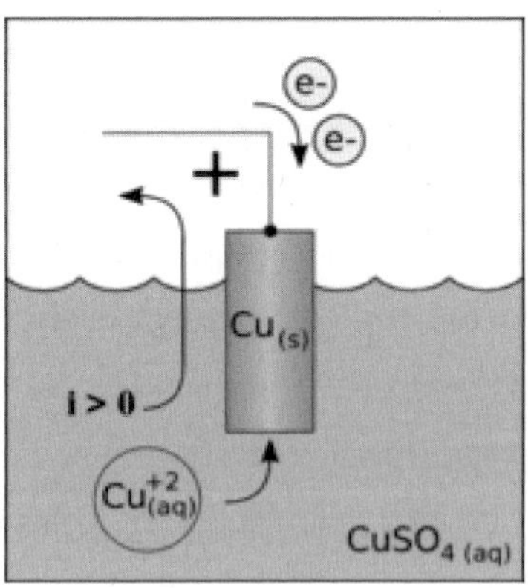

cation[kǽtaiən, -aiɑn]: G. (*kat(a),* alongside, down + *ion* - go), going down. 양이온. An ion with fewer electrons than protons, giving it a positive charge. Since the charge on a proton is equal in magnitude to the charge on an electron, the net charge on an ion is equal to the number of protons in the ion minus the number of electrons. An ion carrying a positive charge which moves towards the cathode (negative electrode) during electrolysis. Opposed to anion.

cerebroside[sərí:brəsaid, sérə-]: L. *cerebrum*, brain + *-oside*, the pattern of (gluc)ose + *-ide*. 세레브로시드[뇌, 신경 조직의 당지질(糖脂質)]. The common name for a group of glycosphingolipids called monoglycosylceramides which are important components in animal muscle and nerve cell membranes. They consist of a ceramide with a single sugar residue at the 1-hydroxyl moiety. The sugar residue can be either glucose or galactose; the two major types are therefore called glucocerebrosides and galactocerebrosides. Galactocerebrosides are typically found in neural tissue, while glucocerebrosides are found in other tissues.

chemical[kémikəl]: F. < L. *chimique* < *chimicus, chymicus*. 화학의, 화학적인, 화학작용에 의한.

chemotherapy[ki:mouθérəpi /kem-]: L. *chem- chemo-*, chemical, chemistry, + G. *therapeia*, healing, treat medically. 화학요법. The treatment of disease by the use of chemical substances, especially the treatment of cancer by cytotoxic (=killing) and other drugs.

chiral[káiral]: G. *cheir*, hand. 손. Describe an object that is non-superposable on its mirror image.

chiral molecule: a type of molecule that lacks an internal plane of symmetry and has a non-superimposable mirror image. The feature that is most often the cause of chirality in molecules is the presence of an asymmetric carbon atom.

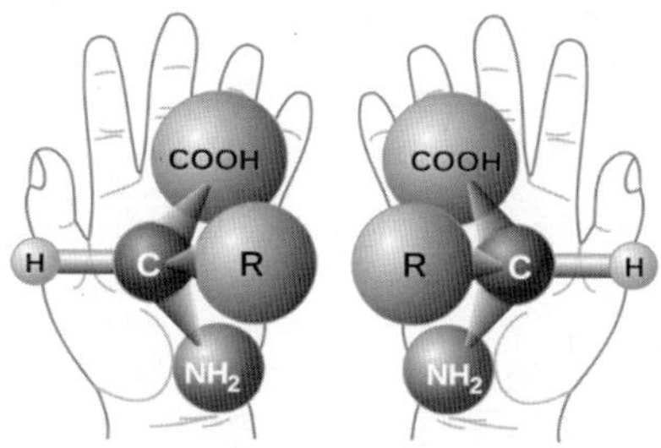

(S)-Alanine (left) and (R)-alanine (right) in zwitterionic form at neutral pH.

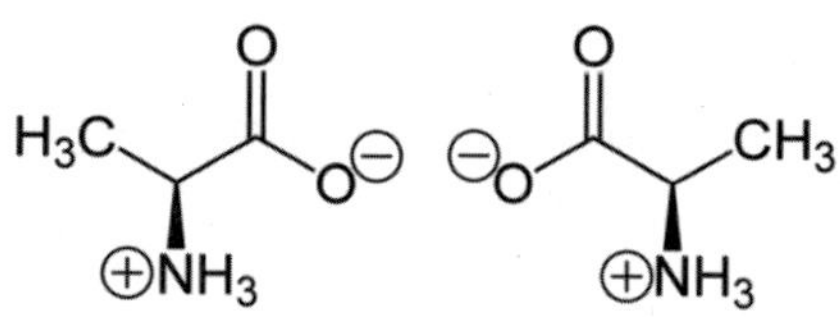

chitin[káitin]: F. < G. *chitine* < *khitōn*, chiton, tunic(피막), 키틴질(갑각류 껍질의 주성분). $(C_8H_{13}O_5N)_n$. A long-chain polymer of a N-acetylglucosamine, a derivative of glucose, and is found in many places throughout the natural world. It is the main component of the cell walls of fungi, the exoskeletons of arthropods(절지동물) such as crustaceans(갑각류) (e.g., crabs, lobsters and shrimps) and insects, the radulas(치설) of mollusks (연체동물), and the beaks of cephalopods(두족류) including squid (오징어) and octopuses. In terms of structure, chitin may be compared to the polysaccharide cellulose and, in terms of function, to the protein keratin. Chitin has also proven useful for several medical and industrial purposes.

Fig. Structure of the chitin molecule, showing two of the N-acetylglucosamine units that repeat to form long chains in β-1,4 linkage.

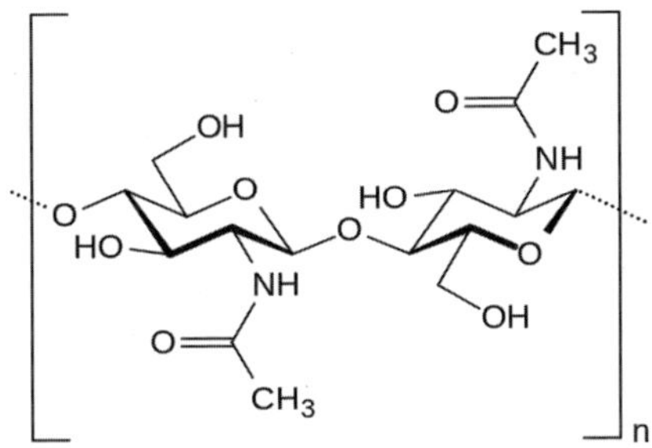

chlorine[klɔ́:ri(:)n]: G. *khlōros*, pale green + *-ine.* 염소(鹽素). The chemical element with atomic number 17 and symbol **Cl**. It is a halogen, found in the periodic table in group 17. As the chloride ion, which is part of common salt and other compounds, it is abundant in nature and necessary to most forms of life, including humans. In its elemental form (Cl_2 or "dichlorine") under standard conditions, chlorine is a powerful oxidant and is used in bleaching and disinfectants, as well as an essential reagent in the chemical industry. As a common disinfectant, chlorine compounds are used in swimming pools to keep them clean and sanitary. In the upper atmosphere, chlorine-containing molecules such as chlorofluorocarbons have been implicated in ozone depletion.

chlorofluorocarbons(CFC): an organic compound that contains carbon, chlorine, and fluorine, produced as a volatile derivative of methane and ethane. A common subclass is the hydrochlorofluorocarbons (HCFCs), which contain hydrogen, as well. They are also commonly known by the DuPont trade name Freon. The most common representative is dichlorodifluoromethane (R-12 or Freon-12). Many CFCs have been widely used as refrigerants, propellants (in aerosol applications), and solvents. The manufacture of such compounds is being phased out by the Montreal Protocol because they contribute to ozone depletion.

chlorophyll[klɔ́:rəfil]: G. *khlōros*, green + *phullon*, leaf. 엽록소(葉綠素), 클로로필. A green pigment found in almost all plants, algae, and cyanobacteria. Chlorophyll is an extremely important biomolecule, critical in photosynthesis, which allows plants to obtain energy from light. Chlorophyll absorbs light most strongly in the blue portion of the electromagnetic spectrum, followed by the red portion. However, it is a poor absorber of green and near-green portions of the spectrum; hence the green color of chlorophyll-containing tissues.

chlorpromazine[klɔ:prɑ́məzi:n]: 1950s. chloro- + *prom(eth)azine*, 클로르프로마진(鎮靜劑). Marketed in the United States as Thorazine and elsewhere as Largactil is a typical antipsychotic

(항정신병성의). First synthesized on December 11, 1950, chlorpromazine was the first drug developed with specific antipsychotic action, and would serve as the prototype for the phenothiazine class of drugs, which later grew to comprise several other agents. The introduction of chlorpromazine into clinical use has been described as the single greatest advance in psychiatric care, dramatically improving the prognosis(예후) of patients in psychiatric hospitals worldwide; the availability of antipsychotic drugs curtailed indiscriminate use of electroconvulsive therapy(전기 충격 요법) and psychosurgery, and was one of the driving forces behind the deinstitutionalization (탈시설화) movement.

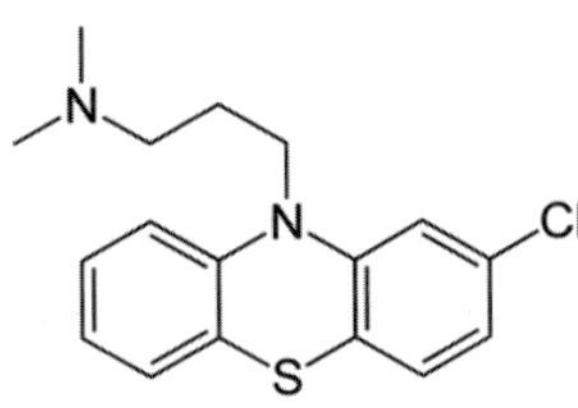

chocolate[tʃɔ́:kələt, tʃɑ́k- / tʃɔ́k-]: F. *chocolat*, Spanish. *chocolate*, food made from cacao seeds. 초콜릿. A raw or processed food produced from the seed of the tropical Theobroma cacao tree. The seeds of the cacao tree have an intense bitter taste, and must be fermented to develop the flavor. The possible health effects of chocolate include both positive and negative effects. While chocolate is regularly eaten for pleasure, there are potential beneficial health effects of eating chocolate. Cocoa or dark chocolate may positively affect the circulatory system.

Other possible effects under basic research include anticancer, brain stimulator, cough preventor and antidiarrhoeal activities. On the other hand, the unconstrained consumption of large quantities of any energy-rich food such as chocolate is thought to increase the risk of obesity(비만) without a corresponding increase in activity. Raw chocolate is high in cocoa butter, a fat which is removed during chocolate refining, then added back in varying proportions during the manufacturing process. Manufacturers may add other fats, sugars, and milk as well, all of which increase the caloric content of chocolate.

cholesterol[kəléstəroul, -rɔːl /-rɔl]: G. *khole*, bile(담즙) + *stereos*, solid, stiff + *-ol,* alcohol. 콜레스테롤. A waxy steroid metabolite found in the cell membranes and transported in the blood plasma of all animals. It is an essential structural component of mammalian cell membranes, where it is required to establish proper membrane permeability and fluidity. In addition, cholesterol is an important component for the manufacture of bile acids, steroid hormones, and fat-soluble vitamins including Vitamin A, Vitamin D, Vitamin E, and Vitamin K. Cholesterol is the principal sterol synthesized by animals, but small quantities are synthesized in other eukaryotes(진핵생물), such as plants and fungi. It is almost completely absent among prokaryotes(원핵동물), which include bacteria. Although cholesterol is an important and necessary molecule for animals, a high level of serum cholesterol is an indicator for diseases such as heart disease.

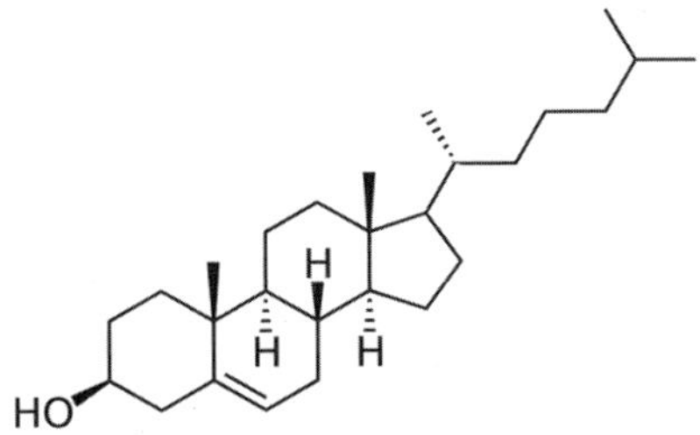

chromatography[kroumətάgrəfi/-tɔ́g-]: G. *Khrōma, Khrōmato-,* colour + *graphien,* to write. 크로마토그래피(색층 분석). The collective term for a set of laboratory techniques for the separation of mixtures. It involves passing a mixture dissolved in a "mobile phase" through a stationary phase, which separates the analyte to be measured from other molecules in the mixture based on differential partitioning between the mobile and stationary phases. Subtle differences in compounds partition coefficient results in differential retention on the stationary phase and thus changing the separation.

chromium[króumiəm]: G. *Khrōma,* colour (some chromium compounds having brilliant colours) + *-ium.* 크롬. A chemical element which has the symbol **Cr** and atomic number 24, first element in Group 6. It is a steely-gray, lustrous, hard metal that takes a high polish and has a high melting point. It is also odorless, tasteless, and malleable. Chromium was regarded with great interest because of its high corrosion resistance and hardness. A major development was the discovery that steel could be made highly resistant to corrosion and discoloration by adding chromium to form stainless steel.

chromic acid: usually used for a mixture made by adding concentrated sulfuric acid to a dichromate, which may contain a variety of compounds, including solid chromium trioxide. This kind of chromic acid may used as a cleaning mixture for glass. Chromic acid may also refer to the molecular species, H_2CrO_4 of which the trioxide is the anhydride. Chromic acid features chromium in an oxidation state of +6 (or VI). It is a strong and corrosive oxidising agent.

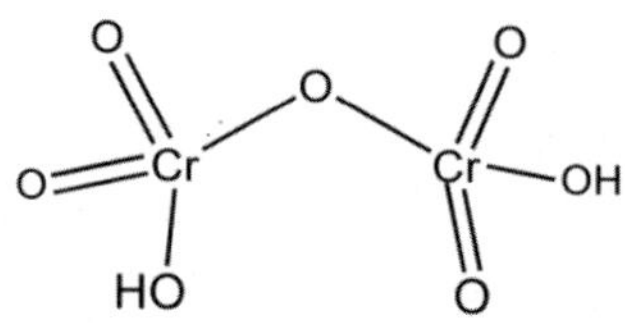

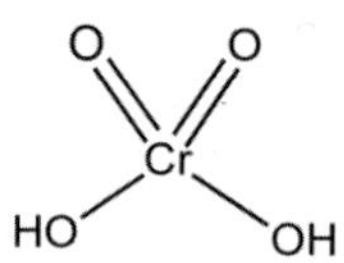

chromium trioxide: the inorganic compound with the formula CrO_3. It is the acidic anhydride of chromic acid, and is sometimes marketed under the same name. This compound is a dark red/orange brown solid, which dissolves in water concomitant with hydrolysis. Millions of kilograms are produced annually, mainly for electroplating.

chromophore[króumoufɔ:r]: G. *Khrōma, Khrōmat-,* colour + *phoros, -phoron,* bearing, bearer. color bearer. 발색단. The part of a molecule responsible for its color. The color arises when a molecule absorbs certain wavelengths of visible light and transmits or reflects others. The chromophore is a region in the molecule where the energy difference between two different molecular orbitals falls within the range of the visible spectrum. Visible light that hits the chromophore can thus be absorbed by exciting an electron from its ground state into an excited state. In biological molecules that serve to capture or detect light energy, the chromophore is the moiety that causes a conformational change of the molecule when hit by light.

Fig. Chemical structure of beta-carotene. The eleven conjugated double bonds that form the chromophore of the molecule are highlighted in red.

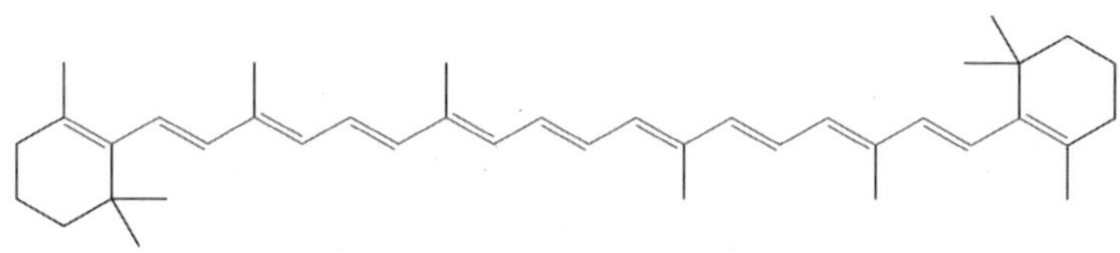

cimetidine[səmétidi:n]: 1970s. *ci-*, *cy-* in cyano, + *met(hyl)* + *-ide* + *-ine.* 십이장궤양 치료제, 제산제. A histamine H2-receptor antagonist(길항제) that inhibits the production of acid in the stomach. It is largely used in the treatment of heartburn and peptic ulcers. It is marketed by GlaxoSmithKline under the trade name Tagamet (sometimes Tagamet HB or Tagamet HB200).

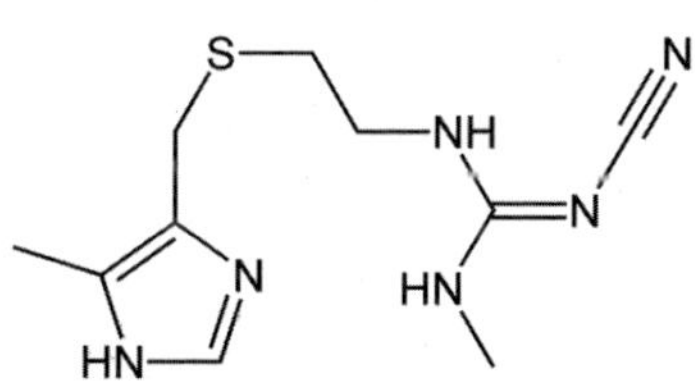

cis-[sis]: L. *cis*-, on this side of, on the same side of. 시스, 같은 쪽에.

cis-trans isomerism (= geometric isomerism or configuration isomerism or E-Z isomerism): a form of stereoisomerism describing the orientation of functional groups within a molecule. In general, such isomers contain double bonds, which cannot rotate, but they can also arise from ring structures, wherein the rotation of bonds is greatly restricted. *Cis* and *trans* isomers occur both in organic molecules and in inorganic coordination complexes.

Fig. *Cis*-2-butene and *Trans*-2-butene

Fig. *Trans*-1,2-dichlorocyclohexane and *cis*-1,2-dichlorocyclohexane

citric[sítrik]: L. *citrus*, citron(감귤류 식물) tree + *-ic*. 시트르산의, 감귤류의.

citric acid: a weak organic acid. It is a natural preservative and is also used to add an acidic, or sour, taste to foods and soft drinks. In biochemistry, it is important as an intermediate in the citric acid cycle and therefore occurs in the metabolism of virtually all living things. It can also be used as an environmentally benign cleaning agent.

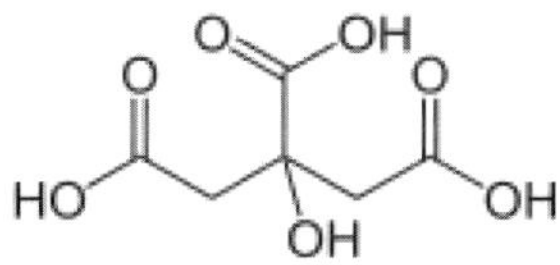

clinical[klínikəl]: G. *klinike*, bedside + *-al.* 임상(치료, 진찰)의, 병상의, 병실에서 사용되는.

coagulant[kouǽgjulənt]: L. *coagulare*, *coagulant*, curdling. 응고제, 지혈제.

cocain(e)[koukéin, kóukein]: Spain. < Aymara = Quechua. *coca* < *kuka* = *koka*, the name of a tropical american shrub + *-ine.* 코카인(코카 잎에서 채취하는 마약, 마취제). Or benzoylmethylecgonine: a crystalline tropane alkaloid that is obtained from the leaves of the coca plant. It is a stimulant of the central nervous system and an appetite suppressant. Specifically, it is a serotonin-norepinephrine-dopamine reuptake inhibitor, which mediates functionality of such as an exogenous catecholamine transporter ligand. Because of the way it affects the mesolimbic reward pathway, cocaine is addictive.

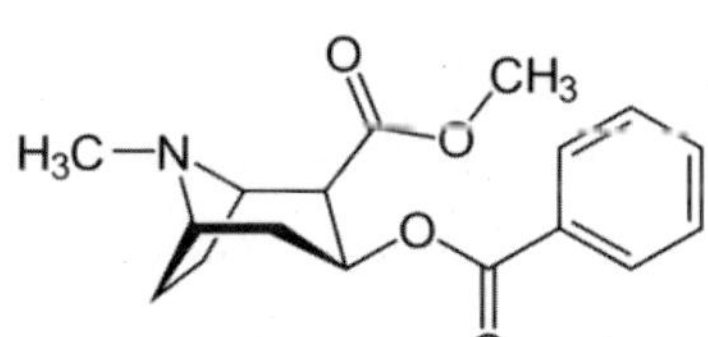

code[koud]: L. *codex*, block of wood. 유전 암호.

genetic code: biological organisms contain genetic material that is used to control their function and development. This is the

DNA, which contains units named genes that can produce proteins through a code (genetic code) in which a series of triplets of four possible nucleotides are translated into one of twenty possible amino acids.

codon[cóudɑn / -dɔn]: 1960s. *code,* block of wood *+ -on.* 코돈(특정 아미노산을 지정하는 유전정보의 최소단위). A sequence of three bases in mRNA that specifies the amino acid to be incorporated into a protein.

coenzyme[kouénzaim]: G. *co-,* with, joint + *enzumos*, *en-*, within + *zume*, leaven, leavened. 보조효소. 조효소. A loosely-bound cofactors that is an organic molecule.

cofactor[kóufæktər]: G. *co-,* with, joint + L. *facere, factor.* 보조요인, 공동인자. A non-protein chemical compound that is bound to a protein and is required for the protein's biological activity. These proteins are commonly enzymes, and cofactors can be considered "helper molecules" that assist in biochemical transformations. Cofactors can also be classified depending on how tightly they bind to an enzyme, with loosely-bound cofactors termed coenzymes and tightly-bound cofactors termed prosthetic groups. Some sources also limit the use of the term "cofactor" to inorganic substances. An inactive enzyme, without the cofactor is called an apoenzyme, while the complete enzyme with cofactor is the holoenzyme.

collagen[kɑ́lədʒən / kɔ́l-]: F. *collagēne*, G. *kolla*, glue + F. *-gēne*, born. 콜라겐. A group of naturally occurring proteins. In nature, it is found exclusively in animals, especially in the flesh and connective tissues of mammals. It is the main component of connective tissue, and is the most abundant protein in mammals, making up about 25% to 35% of the whole-body protein content. Collagen, in the form of elongated fibrils(근육·신경의 원섬유) is mostly found in fibrous tissues such as tendon(힘줄), ligament (인대) and skin, and is also abundant in cornea(각막), cartilage(연골), bone, blood vessels, the gut(장, 창자), and intervertebral disc.

collision[kəlíʒən]: L. *collidere*, *collision,* strike together. 충돌, 대립.

collision theory: a theory proposed by Max Trautz and William Lewis in 1916 and 1918, that qualitatively explains how chemical reactions occur and why reaction rates differ for different reactions. For a reaction to occur the reactant particles must collide. Only a certain fraction of the total collisions cause chemical change; these are called successful collisions. The successful collisions have sufficient energy (activation energy) at the moment of impact to break the existing bonds and form new bonds, resulting in the products of the reaction. Increasing the concentration of the reactants and raising the temperature bring about more collisions and therefore more successful collisions, increasing the rate of reaction. When a catalyst is involved in the collision between the reactant molecules, less energy is required for the chemical change to take place, and hence more collisions have sufficient energy for reaction to occur. The reaction rate therefore increases. Collision theory is closely related to chemical kinetics. The rate constant for a bimolecular gas phase reaction, as predicted by collision theory is:

$$k(T) = Z\rho \exp\left(\frac{-E_a}{RT}\right).$$

where:

- Z is the collision frequency.
- ρ is the steric factor.
- E_a is the activation energy of the reaction.
- T is the temperature.
- R is gas constant.

The collision frequency is:

$$Z = N_A \sigma_{AB} \sqrt{\frac{8k_B T}{\pi \mu_{AB}}}$$

where:

- N_A is the Number of Molecules per unit volume
- σ_{AB} is the reaction cross section
- k_B is Boltzmann's constant
- μ_{AB} is the reduced mass of the reactants.

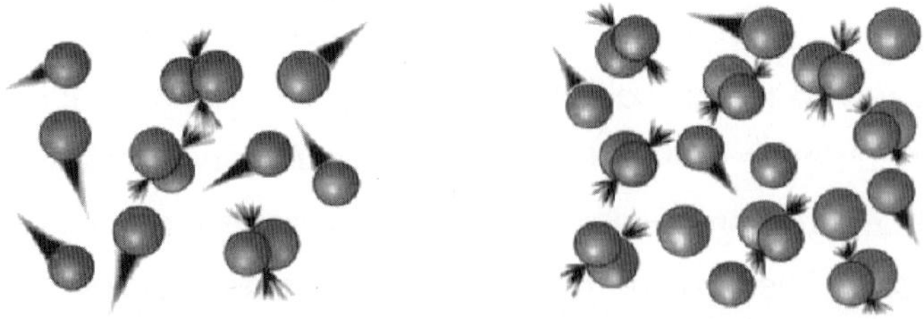

Low concentration = Few collisions High concentration = More collisions

combination[kɑmbənéiʃən]: L. *combinare*, *combination,* join two by two. 결합, 화합.

combustion[kəmbʌ́stʃən]: L. *comburere*, *combustion,* burn up. 연소, 산화.

common[kámən / kɔ́m-]: F. < L. *comun* < *communis*, occurring, found, or done often. 공통의, 일반적인, 속명의.

common name: a taxon(분류군) or organism [also known as a vernacular name(고유명), English name, colloquial name(구어명), trivial name(통속명), trivial epithet(별칭), country name, popular name, or farmer's name] is a name in general use within a community; it is often contrasted with the scientific name for the same organism. A so-called "common name" is not always one that is commonly used. Sometimes common names are created by authorities on one particular subject, in an attempt to make it possible for members of the general public (including interested parties such as fishermen, farmers etc.) to refer to a species of organism without needing to be able to memorise or pronounce the Latinized scientific name. Creating common names can also be an attempt to standardize the use of common names which can sometimes vary a great deal between one part of a country and another as well as between one country and another, even where the same language is spoken in both places.

complex[kəmpléks, kámpleks / kɔ́mpleks]: L. *complectere*, *complexus(*p.p*)*, embrace, comprise. 착물(錯物)의.

coordination complex (= coordination compound, or metal complex): a structure consisting of a central atom or ion (usually metallic), bonded to a surrounding array of molecules or anions (ligands, complexing agents). The atom within a ligand that is directly bonded to the central atom or ion is called the donor atom. Polydentate (multiple bonded) ligands can form a chelate complex. A ligand donates at least one pair of electrons to the central atom/ion.

chelate complex: the formation or presence of two or more separate bindings between a polydentate (multiple bonded) ligand and a single central atom. Usually these ligands are organic compounds, and are called chelants, chelators, chelating agents, or sequestering agents. The ligand forms a chelate complex with the substrate. Chelate complexes are contrasted with coordination complexes with monodentate legands, which form only one bond with the central atom.

Fig. Metal-EDTA chelate

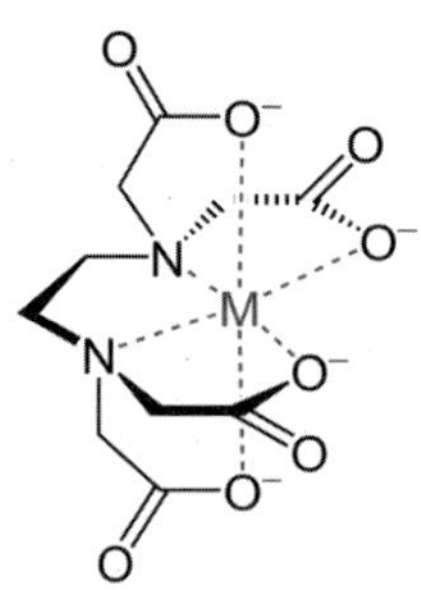

composite[cəmpɑsit / kɔ́mpəzit]: L. *componere*, *compositus*(p.p), put together. 복합물

compound[kάmpaund, kɔ́mpound]: F. < L. compound < *componere,* put together. 화합물.

con-[kən, kɑn / kɔn]: L. *com-,* concord, with, together, jointly, altogether. com-의 변형(b, h, l, m, p, r, w를 제외한 자음 앞에서).

<u>conrotatory mode</u>: in an electrocyclic reaction (a class of organic chemical reactions) the substituents located at the termini of a conjugated double bond system move in the same (clockwise or counter clockwise) direction during ring opening or ring closure. In a disrotatory mode, they move in opposite directions. A simple arrow-pushing analysis fails to explain the quantitative conversion of *trans-cis-trans*-2,4,6-octatriene to *cis*-dimethylcyclohexadiene (top of figure). However, a detailed look at the orbital mechanics behind the process reveals that the orbital symmetry of the octatriene's HOMO requires for the end pi orbitals to move in opposite directions to form the correct symmetry found in the sigma bond. This stereospecificity is found in thermal rearrangements of all conjugated systems containing 4n + 2 pi electron and is based on preservation of orbital symmetry in the highest occupied molecular orbital. Systems containing 4n pi electrons show the opposite conrotatory mode, as do photoinduced rearrangements of 4n + 2 pi electrons. Photoinduced rearrangements of 4n electron systems follow the disrotatory rule.

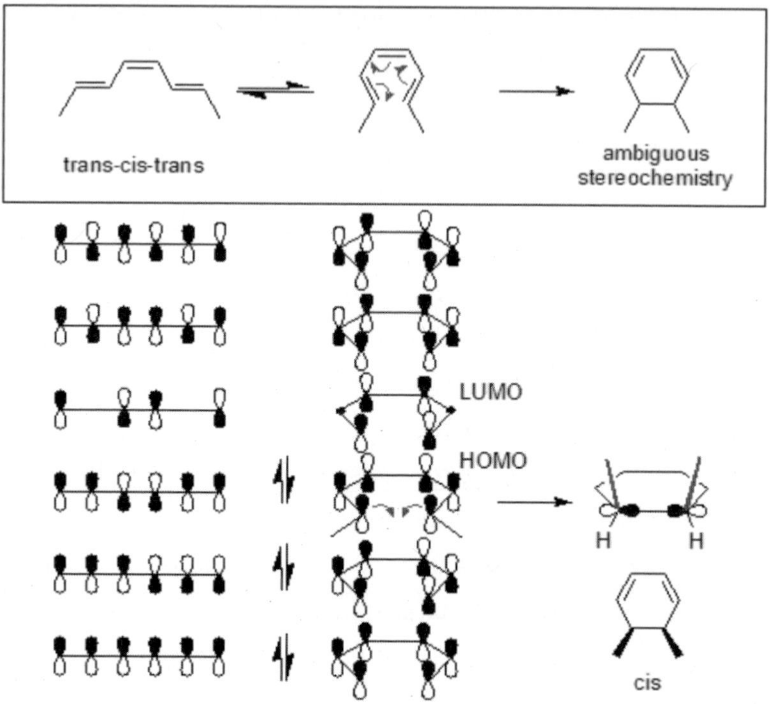

concentration[kάnsəntreiʃən/kɔ́n-]: L. *concentratio, concentrationis, con-,* with, together, + *centro,* center + *-atus,* -ation. 농도(濃度). Defined as the abundance of a constituent divided by the total volume of a mixture. Furthermore, in chemistry, four types of mathematical description can be distinguished: mass concentration, molar concentration, number concentration, and volume concentration. The term concentration can be applied to any kind of chemical mixture, but most frequently it refers to solutes in solutions.

Table of concentrations and related quantities

Concentration type	Symbol	Definition	SI-unit	Other unit(s)
mass concentration	ρ_i or γ_i	m_i / V	kg/m^3	g/100mL(=g/dL)
molar concentration	c_i	n_i / V	mol/m^3	M(=mol/L)
number concentration	C_i	N_i / V	$1/m^3$	$1/cm^3$
volume concentration	ϕ_i	V_i / V	m^3/m^3	

Related quantities	Symbol	Definition	SI-unit	Other unit(s)
normality		c_i / f_{eq}	mol/m^3	N(=mol/L)
molality	b_i	n_i / $m_{solvent}$	mol/kg	
mole fraction	x_i	n_i / n_{tot}	mol/mol	ppm, ppb, ppt
mole ratio	r_i	n_i / (n_{tot} − n_i)	mol/mol	ppm, ppb, ppt
mass fraction	w_i	m_i / m_{tot}	kg/kg	ppm, ppb, ppt
mass ratio	ζ_i	m_i / (m_{tot} − m_i)	kg/kg	ppm, ppb, ppt

concert[kánsə(:)rt / kɔ́nsət]: F. *concerter*, Italian. *concertare*, harmonize. 협동, 제휴.

concerted reaction: a chemical reaction in which all bond breaking and bond making occurs in a single step. Reactive intermediates or other unstable high energy intermediates are not involved. Concerted reaction rates tend not to depend on solvent polarity ruling out large buildup of charge in the transition state. Pericyclic reactions are concerted reactions. In the S_N2 reaction a concerted reaction also takes place. Because the reaction rate is bimolecular there are two molecular species controlling the rate of the reaction. Because the rate is dependent on two molecules the reaction does not have any intermediate steps, only a transition state. This means that all the bond making and bond breaking takes place in a single step. In order for the reaction to occur both molecules must be situated correctly. Therefore, the reaction is said to be concerted because it occurs in one step.

condensation[kɑndenséiʃən / kɔn-]: L. *condensare, condensation,* press close together. 축합(縮合).

condensation reaction: 축합 반응. A chemical reaction in which two molecules or moieties (functional groups) combine to form one single molecule, together with the loss of a small molecule. When this small molecule is water, it is known as a dehydration reaction; other possible small molecules lost are hydrogen chloride, methanol, or acetic acid. The word "condensation" suggests a process in which two or more things are brought "together" to form something "dense", like in condensation from gaseous to liquid state of matter; this does not imply, however, that condensation reaction products have greater density than reactants.

Ex) The condensation of two amino acids to form a peptide bond with expulsion of water

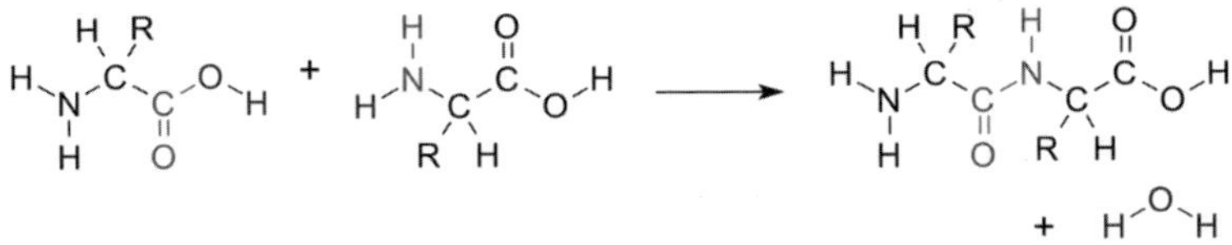

conduct[kɑndʌ́kt / kɔn-]: L. *conducere, conduct*, brought together, lead, guide, transmit. (열. 전기, 소리 따위를)전도(傳導)하다.

configuration[kənfigjuréiʃən]: L. *configurare*, *configuration,* shape after a pattern. 배치 (전자의), 배열 (원자의), 배위 (공간).

<u>configuration of a molecule</u>: 분자의 배치, 분자의 배열. The permanent geometry that results from the spatial arrangement of its bonds. The ability of the same set of atoms to form two or more molecules with different configurations is stereoisomerism. Configuration is distinct from chemical conformation, a shape attainable by bond rotations.

conformer[kɑnfɔ́:rmər/kɔn-]: L. *conformare*, to shape, fashion, form. (= conformational isomerism, 이형태체): a form of stereoisomerism in which the isomers can be interconverted by rotations about formally single bonds. Such conformational isomers or conformers can rotate about one or more σ bonds. Rotamers are conformers that differ by rotation about only one single σ bond.

Fig. Conformers of butane. The gauche and anti forms are staggered conformations. The eclipsed conformation of butane is unstable because it is at an energy maximum.

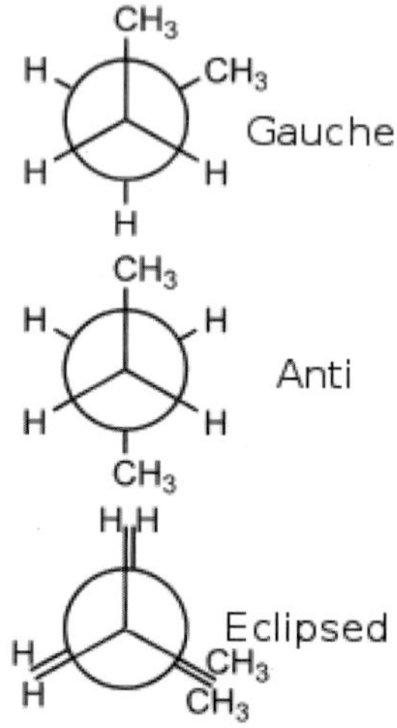

conjugate[kάndʒugeit / kɔ́n-]: L. *conjugare*, *conjugat*, *con*, together + *jugum*, yoke, yoked together. 접합(체), 짝, 켤레, 공액(共軛). (of an acid or base) Related to the corresponding base or acid by loss or gain of a proton.

<u>conjugated dienes</u>: conjugated double bonds separated by one single bond.

<u>unconjugated dienes</u>: the double bonds separated by two or more single bonds. They are usually less stable than isomeric conjugated dienes.

constitution[kɑnstətjú:ʃən / kɔnstitjú:-]: L. *constituere*, *constitution*, establishment, appointment. 구조, 조직.

constitutional isomer: 구조 이성질체. A form of isomerism in which molecules with the same molecular formula have atoms bonded together in different orders, as opposed to stereoisomerism. Three categories of constitutional isomers are skeletal, positional, and functional isomers. In skeletal isomerism, or chain isomerism, components of the (usually carbon) skeleton are distinctly re-ordered to create different structures. Pentane exists as three isomers: *(n-)pentane*, isopentane (or methylbutane) and neopentane (or dimethylpropane). In position isomerism a functional group changes position on the chain. As for example, the hydroxyl group can occupy three different positions on an n-pentane chain forming three different compounds.

Ex) 1-pentanol, 2-pentanol, 3-pentanol

Functional isomers are structural isomers that have the same molecular formula (that is, the same number of atoms of the same elements), but the atoms are connected together in different ways so that the groupings are dissimilar.

Ex) dimethyl ether, ethanol

contraception[kɑntrəsépʃən]: L. *contra-*, against + *conception*, conceiving, prevent conception. 피임(법).

contraceptive[kɑntrəséptiv/kɔn-]: L. *contra,* contrary to law + *ceptive,* conceiving (in the womb). (of method or device) Serving to prevent pregnancy. 피임(용)의, 피임약(기구).

convert[kənvə́:rt]: F. < L. *convertir* < *convertere*, *con-*, altogether + *vertere*, turn, turn about. - 을 변환(전환, 변화)시키다, 화학변화를 일으키다.

convulsant[kənvʌ́lsənt]: F. < L. *convulser* < *convellere*, *convulsant*, pulled violently, wrenched. 경련을 일으키는, 경련성의, 경련 유발제.

coordinate[kouɔ́:rdənət, -neit]: L. *coordinare, co-,* with, together + *ordinare, ordo,* order, of the same rank, place in the same rank. 배위, 좌표.

1) Denoting a type of covalent bond in which one atom provides both the shared electrons. 배위.
2) Each of a group of numbers used to indicate the position of a point, line, or plane. 좌표.

copolymer[koupɑ́limə/ -pɔl-]: L. *copolymer*, *co-*, with, together + *polumeros*, *polu-*, many + *meros*, a share. 혼성중합체(= heteropolymer), 공중합체 : A polymer derived from two (or more) monomeric species, as opposed to a homopolymer where only one monomer is used. Commercially relevant copolymers include ABS plastic, SBR, Nitrile rubber, styrene-acrylonitrile, styrene-isoprene-styrene (SIS) and ethylene-vinyl acetate.

copolymerization: Methods used to chemically synthesize a copolymer. 혼성중합 (반응), 공중합 (반응)

coronary[kɔ́:rəneri, kɑ́r- / kɔ́rənəri]: L. *corona*, *coronarius,* wreath

(화환, 화관), crown. heart attack. 동맥(관상)의, 심장의, 동맥(관상)질환에 걸린.

couple[kʌpl]: F. < L. *copler, cople < copulare, copula*, *co-*, together, + *apere*, fasten. 짝, 짝지우다.

coupled protons: protons that split each other in NMR.

coupling constant: 짝지움 상수. The distance (in hertz) between two adjacent peaks of a split NMR signal.

coupling reaction: 짝지움 반응. A catch-all term for a variety of reactions where two hydrocarbon fragments are coupled with the aid of a metal catalyst. In one important reaction type a main group organometallic compound of the type RM (R = organic fragment, M = main group centre) reacts with an organic halide of the type R'X with formation of a new carbon-carbon bond in the product R-R'. Contributions to coupling reactions by Ei-ichi Negishi and Akira Suzuki were recognized with the 2010 Nobel Prize in Chemistry, which was shared with Richard F. Heck. Broadly speaking, two types of coupling reactions are recognized:

- cross couplings involve reactions between two different partners, for example bromobenzene (PhBr) and vinyl chloride to give styrene ($PhCH=CH_2$).
- homocouplings couple two identical partners, for example, the conversion of iodobenzene (PhI) to biphenyl (Ph-Ph).

crown[kraun]: F. < L. *coroner, corone < corona*, wreath(화환, 화관), chaplet(화관). 왕관.

crown ether: cyclic chemical compounds that consist of a ring containing several ether groups. The most common crown ethers are oligomers of ethylene oxide, the repeating unit being ethyleneoxy, i.e., $-CH_2CH_2O-$. Important members of this series are the tetramer (n = 4), the pentamer (n = 5), and the hexamer (n = 6). The term "crown" refers to the resemblance between the structure of a crown ether bound to a cation, and a crown sitting on a person's head. The first number in a crown ether's name refers to the number of atoms in the cycle, and the second number refers to the number of those atoms that are oxygen. Crown ethers are much broader than the oligomers of ethylene oxide; an important group are derived from catechol. Crown ethers strongly bind certain cations, forming complexes. The oxygen atoms are well situated to coordinate with a cation located at the interior of the ring, whereas the exterior of the ring is hydrophobic. The resulting cations often form salts that are soluble in nonpolar solvents, and for this reason crown ethers are useful in phase transfer catalysis. The denticity(ligand) of the polyether influences the affinity of the crown ether for various cations. For example, 18-crown-6 has high affinity for potassium cation, 15-crown-5 for sodium cation, and 12-crown-4 for lithium cation. The high affinity of 18-crown-6 for potassium ions contributes towards its toxicity.

Fig. Structures of common crown ethers: 12-crown-4, 15-crown-5, 18-crown-6, dibenzo-18-crown-6, and diaza-18-crown-6

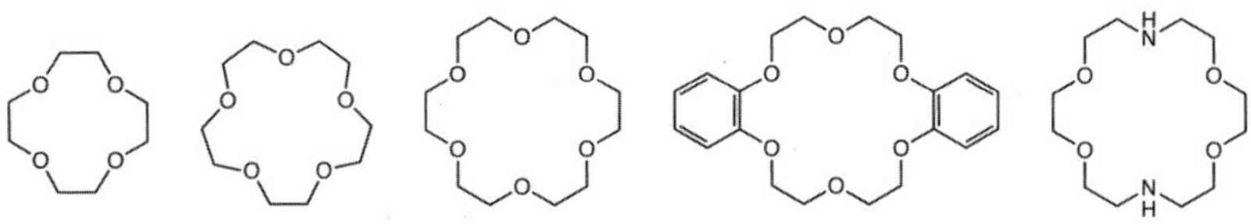

cryptate[krípteit]: G. crypto-, *kruptos*, hidden, the complex formed when a cryptand binds a substrate. 크립데이트(원자가 결합에 관여하고 있지 않은 전자와 함께 금속이온을 둘러싸고 있는 화합물).

cryptand: a family of synthetic bi- and polycyclic multidentate ligands for a variety of cations. The term cryptand implies that this ligand binds substrates in a crypt, interring(매장하다, 묻다) the guest as in a burial. These molecules are three dimensional analogues of crown ethers but are more selective and complex the guest ions more strongly. The resulting complexes are lipophilic.

Fig. 2,2,2-Cryptand

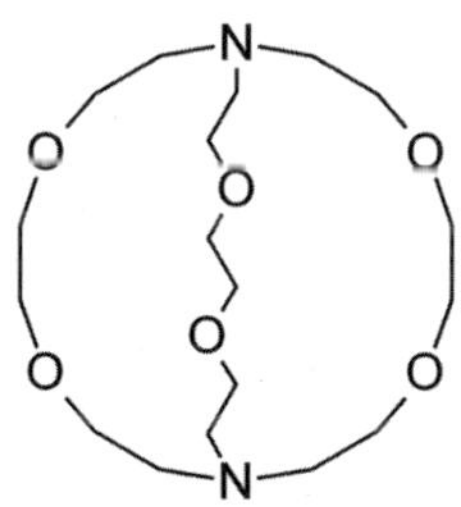

crystal[krístl]: G. > L. *krustallos* > *crystallum*, ice, crystal. 결정(체).

cumulate[kjú:mjuleit]: L. *cumulare*, *cumulate*, heaped. 쌓아올리다. 집적하다.

<u>cumulated dienes</u>: the double bonds sharing a common atom as in a group of compounds called allenes (a hydrocarbon in which one atom of carbon is connected by double bonds with two other atoms of carbon. Allene also is the common name for the parent compound of this series, propadiene. Such pair of bonds make allenes much more reactive than other alkenes. For example, their reactivity with gaseous chlorine is more like the reactivity of alkynes. Allenes can be chiral.

cyano-[sáiənou, -nə, saiǽ-]: 모음 앞에서는 cyan- G. *kuan(e)os*, dark blue. cyanogen, cyanogen group.

<u>cyanide</u>: a chemical compound that contains the cyano group (C≡N), which consists of a carbon atom triple-bonded to a nitrogen atom. Most commonly, cyanides refers to salts of the anion (:C≡N:$^-$). These compounds are highly toxic.

cyanogen[saiǽnədʒə˞n, -dʒen]: F. < G. *cyanogène* < *kuanos*, dark blue mineral + *-gène*, so named because it is a constituent of Prussian blue. 시아노겐, 청소(靑素, 무색의 유독가스). The chemical compound with the formula $(CN)_2$. It is a colorless, toxic gas with a pungent odor. The molecule is a pseudohalogen. Cyanogen molecules consist of two CN groups — analogous to diatomic halogen molecules, such as Cl_2, but far less oxidizing. The two cyano groups are bonded together at their carbon atoms: N≡C−C≡N, although other isomers have been detected. Certain derivatives of cyanogen are also called "cyanogen" even though they contain only one CN group. For example cyanogen bromide has the formula NCBr. Cyanogen is the anhydride of oxamide:

$H_2NC(O)C(O)NH_2 \rightarrow NCCN + 2\ H_2O$

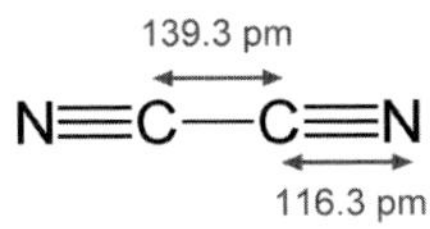

cyclo-[sáiklou, -lə, sík-]: G. > L. *cyclicus* < *kuklos*, 모음 앞에서는 cycl-, circle, cycle. 고리, 순환, 한바퀴.

cyclic compound: a mostly organic, compound in which a series of atoms is connected to form a loop or ring. Cyclic compounds may or may not be aromatic. Benzene is a well known example. The term "polycyclic" is used when more than one ring is formed in a single molecule for instance in naphthalene, and the term macrocycle is used for a ring containing more than a dozen atoms.

Cyclic compounds can be categorized:

- Alicyclic compound
- Cycloalkane
- Cycloalkene
- Aromatic hydrocarbon
- Polycyclic aromatic hydrocarbon
- Heterocyclic compound
- Macrocycle

dec(a)-[dék(ə)]: G. *deka*, ten, tenfold. 10, 10배의.

<u>decane</u>: an alkane hydrocarbon with the chemical formula $CH_3(CH_2)_8CH_3$. 75 structural isomers of decane exist, all of which are flammable liquids. Decane is one of the components of gasoline (petrol). Like other alkanes, it is nonpolar and therefore will not dissolve in polar liquids such as water. It has a surface tension of 0.0238 $N \cdot m^{-1}$.

degenerate[didʒénəreit]: L. *degencrare*, *degereratus, degener de-*, away from + *genus*, *gener*, race, kind, no longer of its kind, debased. 축퇴. 퇴화.

degradation[degrədéiʃən]: L. *degradare, degradation*, *de-*, down, away from + *gradus*, step or grade. 분해.

dehydrate[di:háidreit]: Late 19th century. G. *de-*, removal + G. *hudros, hudr-*, water), remove water. 탈수(脫水)하다, - 에서 수분을 빼다.

deleterious[delitíəriəs]: G. > L. *dēlētērios > noxius*, *noxa*, harm + *-ous*, harmful, poisonous, or very unpleasant. 해로운, 유독한.

delocalize[di:lóukəlaiz]: L. *de-*, away from, removal + *localis, locus*, place. - 의 지방성(地方性)을 제거하다. (of electrons) Be shared among more than two atoms in a molecule. 둘 이상의 원자가 전자를 공유하다.

<u>delocalized electrons</u>: electrons in a molecule or solid metal that are not associated with a single atom or one covalent bond. Delocalized electrons are contained within an orbital that

extends over several adjacent atoms. Classically, delocalized electrons can be found in conjugated systems and mesoionic compounds. It is increasingly appreciated that electrons in sigma bonding levels are also delocalized. For example, in methane, the bonding electrons are shared by all five atoms equally. Pervasive existence of delocalization is implicit in molecular orbital theory. In the simple aromatic ring of benzene the delocalization of six π electrons over the C_6 ring is often graphically indicated by a circle. The fact that the six C-C bonds are equidistant is one indication of this delocalization. In valence bond theory, delocalization in benzene is represented by resonance structures. Another example of delocalized electrons can be found in a carboxylate group, wherein the negative charge is centered equally on the two oxygen atoms.

Fig. Benzene, with the delocalization of the electrons indicated by the circle.

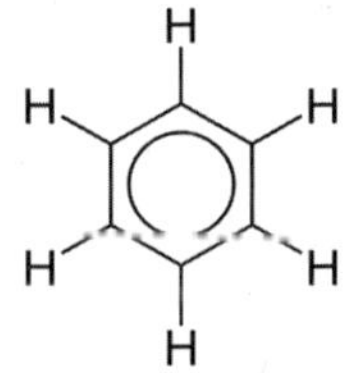

deoxyribose[di:ɑksiráibous / -ɔk-]: 1930s: *de-*, reduction + *oxy-, ox-*, oxygen + ribose. 디옥시리보오스(DNA의 구성 요소). More precisely 2-deoxyribose, is a monosaccharide with formula $C_5H_{10}O_4$. Its name indicates that it is a deoxy sugar, meaning that it is derived from the sugar ribose by loss of an oxygen atom.

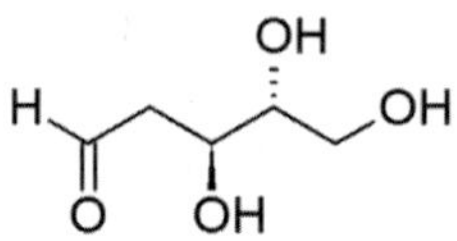

Fig. Chemical equilibrium of deoxyribose in solution.

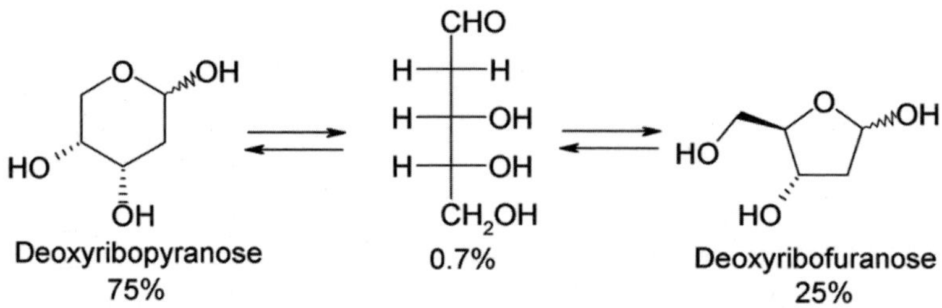

ribose: an organic compound with formula $C_5H_{10}O_5$; specifically, a monosaccharide (simple sugar) with linear form H-(C=O)-$(CHOH)_4$-H, which has all the hydroxyl groups on the same side in the Fischer projection. The term may refer to any of two enantiomers: almost always to D-ribose, which occurs widely in nature and is of limited interest. D-ribose was first reported in 1891 by Emil Fischer. It is a C'-2 carbon enantiomer of the sugar D-arabinose (both isomers of which are named for their source, *gum arabic*) and ribose itself is named as a transposition of the name of arabinose. Ribose comprises the backbone of RNA, a biopolymer that is the basis of genetic transcription. It is related to deoxyribose, as found in DNA. Once phosphorylated, ribose can become a subunit of ATP, NADH, and several other compounds that are critical to metabolism.

Fig. D-Ribose

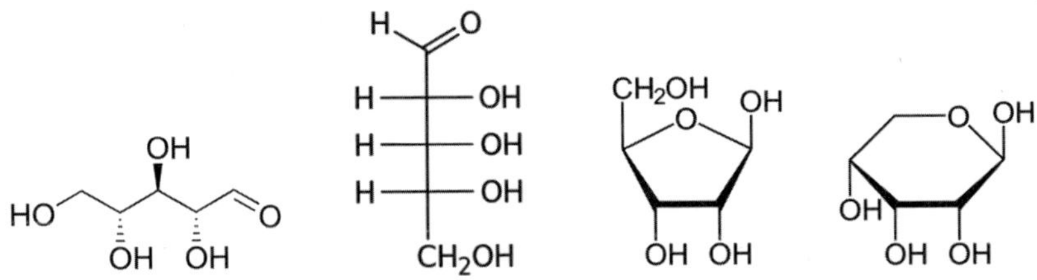

derivative[dirívətiv]: F. < L. *dérivatif* < *derivare, derivatives, de-,* down, away + *rivus,* book, stream. 유도체. A compound that is derived from a similar compound by some chemical or physical process. In the past it was also used to mean a compound that can be imagined to arise from another compound, if one atom is replaced with another atom or group of atoms, but modern chemical language now uses the term structural analogue for this meaning - thus eliminating ambiguity of both terms. The term "structural analogue" is common in organic chemistry. In biochemistry, the word is used for compounds that at least theoretically can be formed from the precursor compound. Chemical derivatives may be used to facilitate analysis. For example, melting point analysis can assist in identification of many organic compounds. A crystalline derivative may be prepared, such as a semicarbazone or 2,4-dinitrophenylhydrazone (derived from aldehydes/ketones), as a simple way of verifying the identity of the original compound, assuming that a table of derivative melting point values is available. Prior to the advent of spectroscopic analysis, such methods were widely used.

detergent[ditə́:rdʒənt]: L. *detergere*, *detergent, de-,* away from + *tergere*, to wipe, wiping away, to wipe off. 세탁제.

deuterium[dju:tíəriəm]: G. *deuteros*, second. heavy hydrogen. 중수소. ^{2}H, D. A stable isotope of hydrogen with a natural abundance in the oceans of Earth of approximately one atom in 6,420 of hydrogen (~156.25 ppm on an atom basis). Deuterium thus accounts for approximately 0.0156% (alternately, on a mass basis: 0.0312%) of all naturally occurring hydrogen in the oceans on Earth. While the most common isotope (hydrogen-1 or protium) accounts for more than 99.98%. The abundance of deuterium changes slightly from one kind of natural water to another. The nucleus of deuterium, called a deuteron, contains one proton and one neutron, whereas the far more common hydrogen isotope, protium, has no neutron in the nucleus. The deuterium isotope's name is formed from the Greek deuteros meaning "second", to denote the two particles composing the nucleus. Deuterium was discovered and named in 1931 by Harold Urey, earning him a Nobel Prize in 1934 after the discovery of the neutron in 1932 made the structure of deuterium obvious. Soon after deuterium's discovery, Urey and others produced samples of water in which deuterium has been highly concentrated with respect to protium, a substance popularly known as heavy water. Because deuterium is destroyed in the interiors of stars faster than it is produced, and because other natural processes are thought to produce only an insignificant amount of deuterium, it is presently thought that nearly all deuterium found in nature was produced in the Big Bang 13.7 billion years ago, and that the basic or primordial ratio of hydrogen-1 (protium) to deuterium (about 26 atoms of deuterium per million hydrogen) has its origin from that time. This is the ratio found in the gas giant planets, such as Jupiter. However, different astronomical bodies are found to have different ratios of deuterium to hydrogen-1, and this is thought to be as a result of natural isotope separation processes that occur from solar heating of ices in comets. Like the water-cycle in Earth's weather, such heating processes may enrich deuterium with respect to protium. In fact, the discovery of deuterium/protium ratios in a number of comets very similar to the mean ratio in Earth's oceans has led to theories that much of Earth's ocean water has a cometary origin. Deuterium/protium ratios thus continue to be an active topic of research in both astronomy and climatology.

dextro-[dékstrou, -trə]: L. *dexter, dextra, dextrum,* 모음 앞에서는 *dextr-*, right, turning clockwise. 오른쪽의, 우(右)의, 시계방향의 회전의.

dexstrorotatory: 우선성(右旋性). Used in the sense '(turning or turned) to the right', in physical and chemical terms, chiefly having reference to the property possessed by certain substances of causing the plane of a ray of polarized light to rotate to the right.

dextrose: dextrorotatory glucose, D-glucose.

di-[dai]: G. *dis*, twice.

dia-[dáiə]: 모음 앞에서는 di-, G. *dia*, through, across, apart.

<u>diastereoisomer (=diastereomer)</u>: 부분입체 이성질체. Stereoisomers that are not enantiomers. Diastereomerism occurs when two or more stereoisomers of a compound have different configurations at one or more (but not all) of the equivalent (related) stereocenters and are not mirror images of each other. When two diastereoisomers differ from each other at only one stereocenter they are epimers.

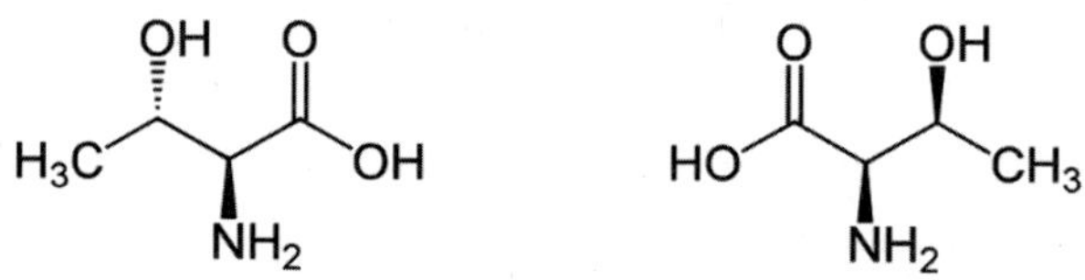

diabetes[daiəbí:tis, -ti:z]: G. > L. *diabetes, diabetae, di(a)-,* through, apart + *bainein,* to go, passer through, siphon. 당뇨병(糖尿病).

diamagnetic[daiəmægnétik]: G. *dia*, across, through + *magenetic, magneticus,* L. *magneta*, lodestone (magnetized mineral able to be used as a magnet, 천연자석). 반자성(反磁性)의. The property of an object which causes it to create a magnetic field in opposition to an externally applied magnetic field, thus causing a repulsive effect. Specifically, an external magnetic field alters the orbital velocity of electrons around their nuclei, thus changing the magnetic dipole moment.

diamagnetism: 반자성적 현상. The property of an object which causes it to create a magnetic field in opposition to an externally applied magnetic field, thus causing a repulsive effect.

diamond[dáiəmənd / daiə-]: L. *diamas, diamant, adamans, adamant, a-*, not + *daman*, to tame, untameable, invincible, later used to denote the hardest metal, or stone hence diamond. 다이아몬드, 금강석. An allotrope of carbon, where the carbon atoms are arranged in a variation of the face-centered cubic crystal structure called a diamond lattice. Diamond is less stable than graphite, but the conversion rate from diamond to graphite is negligible at ambient conditions. Diamond is renowned as a material with superlative physical qualities, most of which originate from the strong covalent bonding between its atoms. In particular, diamond has the highest hardness and thermal conductivity of any bulk material. Those properties determine the major industrial application of diamond in cutting and polishing tools. Diamond has remarkable optical characteristics. Because of its extremely rigid lattice, it can be contaminated by very few types of impurities, such as boron and nitrogen. Combined with wide transparency, this results in the clear, colorless appearance of most natural diamonds. Small amounts of defects or impurities (about one per million of lattice atoms) color diamond blue (boron), yellow (nitrogen), brown (lattice defects), green (radiation exposure), purple, pink, orange or red. Diamond also has relatively high optical dispersion (ability to disperse light of different colors), which results in its characteristic luster. Excellent optical and mechanical properties, combined with efficient marketing, make diamond the most popular gemstone. Most natural diamonds are formed at high-pressure high-temperature conditions existing at depths of 140 to 190 kilometers (87 to 120 mile) in the Earth mantle. Carbon-containing minerals provide the carbon source, and the growth occurs over periods from 1 billion to 3.3 billion years (25% to 75% of the age of the Earth). Diamonds are brought close to the Earth surface through deep volcanic eruptions by a magma, which cools into igneous rocks known as kimberlites and lamproites. Diamonds can also be produced synthetically in a high-pressure high-temperature process which approximately simulates the conditions in the Earth mantle. An alternative, and completely different growth technique is chemical vapor deposition (CVD). Several non-diamond materials, which include cubic zirconia and silicon carbide and are often called diamond simulants, resemble diamond in appearance and many properties. Special gemological techniques have been developed to distinguish natural and synthetic diamonds and diamond simulants.

diarrh(o)ea[daiəríːə/-ríə]: L. < G. *diarrhoia, diarrhein, di(a),* through, thoroughly, apart + *rhein,* to flow, flux, a flowing through, diarrhea. 설사(泄瀉).

dielectric[daiiléktrik]: 19th cent. *di-, dia,* through, across, apart + *electric*, L. < G. *ēlectron* > *electrum*, across which electricity is transmitted (without conduction). 절연체, 유전체(誘電體). An electrical insulator that may be polarized by the action of an applied electric field. When a dielectric is placed in an electric field, electric charges do not flow through the material, as in a conductor, but only slightly shift from their average equilibrium positions causing dielectric polarization: positive charges are displaced along the field and negative charges shift in the opposite direction. This creates an internal electric field which partly compensates the external field inside the dielectric. If a dielectric is composed of weakly bonded molecules, those molecules not only become polarized, but also reorient so that their symmetry axis aligns to the field.

dimer[dáimər]: 1930s. G. *dis-*, two, twice + *meros*, part, a share. 이량(二量)체, 이합(二合)체.

dipole[dáipoul]: G. *dis-*, two + D. *pole*, field. 이중극자, 쌍극자.

<u>dipole moment</u>:

electric dipole moment; the measure of the electrical polarity of a system of charges.

bond dipole moment; the measure of polarity of a chemical bond.

electron electric dipole moment; the measure of the charge distribution within an electron.

molecular dipole moment; the electric dipole moment of a molecule.

disproportion[disprəpɔ́:rʃən]: L. *dis-*, absence + *proportion, proportione*, in respect of share. 불균화 반응.

<u>disproportionation or dismutation</u>: used to describe two particular

types of chemical reaction. A chemical reaction of the type: 2A → A' + A" where A, A' and A" are different chemical species. Most but not all are redox reactions. For example: The UV photolysis of mercury(I) chloride $Hg_2Cl_2 \rightarrow Hg + HgCl_2$ is a redox disproportionation while the self-ionization of water $2H_2O \rightarrow H_3O^+ + OH^-$ is a disproportionation but not a redox reaction. A chemical reaction (reversible or irreversible) in which a species is simultaneously reduced and oxidized so as to form two different products. The reverse of disproportionation is called comproportionation.

dissociation[disousiéiʃən, -ʃi-]: L. *dissociare*, *dis-*, reversal + *sociare*, join together, *dissociation,* separation. 해리(解離), 분리, 분열. A general process in which ionic compounds (complexes, or salts) separate or split into smaller particles, ions, or radicals, usually in a reversible manner. For instance, when a Brønsted-Lowry acid is put in water, a covalent bond between an electronegative atom and a hydrogen atom is broken by heterolytic fission, which gives a proton and a negative ion. Dissociation is the opposite of association and recombination. The process is frequently confused with ionization.

dissociation constant: for reversible dissociations in a chemical equilibrium AB ⇌ A + B, the dissociation constant K_d is the ratio of dissociated to undissociated compound

$$K_d = \frac{[A][B]}{[AB]}$$

where the brackets denote the equilibrium concentrations of the species.

dissolve[dizálv /-zɔ́lv]: L. *dissolvere*, *dis-,* apart + *solvere*, loosen or solve. 녹(이)다, 용해하다.

<u>dissolution</u>: the process by which a solid, liquid or gas forms a solution in a solvent. For the dissolution of solids, the process of dissolution can be explained as the breakdown of the crystal lattice into individual ions, atoms or molecules and their transport into the solvent. For liquids and gases, the molecules must be compatible with those of the solvent for a solution to form. The outcome of the process of dissolution (the amount dissolved at equilibrium, i.e., the solubility) is governed by the thermodynamic energies involved, such as the heat of solution and entropy of solution, but the dissolution itself (a kinetic process) is not. Overall the free energy must be negative for net dissolution to occur. In turn, those energies are controlled by the way in which different chemical bond types interact with those in the solvent. Solid solutions occur in metal alloys and their formation and description is governed by the relevant phase diagram. Dissolution process is of fundamental importance to the description of numerous natural processes on earth, and it is commonly utilized by humans. Dissolution testing is widely used in the pharmaceutical industry for optimization of formulation and quality control.

distillation[distəléiʃən]: L. *distillare*, *distillatus,* fallen in drops. 증류. A method of separating mixtures based on differences in volatilities of components in a boiling liquid mixture. Distillation is a unit operation, or a physical separation process, and not a chemical reaction. Commercially, distillation has a number of applications. It is used to separate crude oil into more fractions for specific uses such as transport, power generation and heating. Water is distilled to remove impurities, such as salt

Fig. Laboratory display of distillation: **1**: A heating device **2**: Still pot **3**: Still head **4**: Thermometer/Boiling point temperature **5**: Condenser **6**: Cooling water in **7**: Cooling water out **8**: Distillate/receiving flask **9**: Vacuum/gas inlet **10**: Still receiver **11**: Heat control **12**: Stirrer speed control **13**: Stirrer/heat plate **14**: Heating (Oil/sand) bath **15**: Stirring means e.g.(shown), boiling chips or mechanical stirrer **16**: Cooling bath.

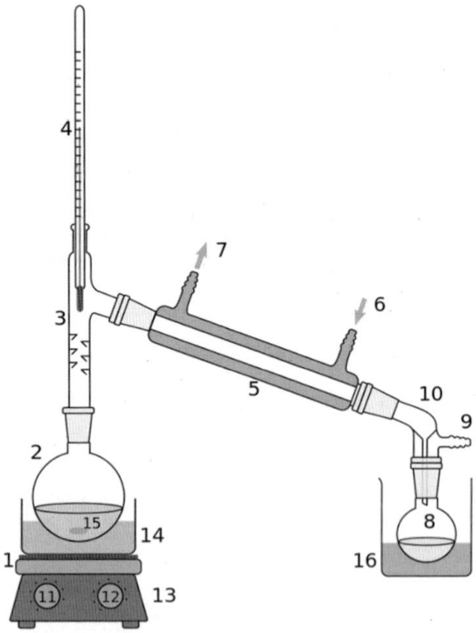

from seawater. Air is distilled to separate its components — notably oxygen, nitrogen, and argon — for industrial use. Distillation of fermented solutions has been used since ancient times to produce distilled beverages with a higher alcohol content. The premises where distillation is carried out, especially distillation of alcohol, are known as a distillery.

distort[distɔ́:rt]: L. *distorquere*, *dis-*, apart + *torquere*, to twist, twisted apart. - (의 형상)을 일그러지게 하다, 뒤틀다, 비틀다.

drug[drʌg]: F. *drogue*, dry goods. 약(藥), 약품. Broadly speaking, is any substance that, when absorbed into the body of a living organism, alters normal bodily function. There is no single, precise definition, as there are different meanings in drug control law, government regulations, medicine, and colloquial usage. In pharmacology, a drug is "a chemical substance used in the treatment, cure, prevention, or diagnosis of disease or used to otherwise enhance physical or mental well-being." Drugs may be prescribed for a limited duration, or on a regular basis for chronic disorders. Recreational drugs are chemical substances that affect the central nervous system, such as opioids(마취제) or hallucinogens(환각제). They may be used for perceived beneficial effects on perception, consciousness, personality, and behavior. Some drugs can cause addiction and/or habituation. Drugs are usually distinguished from endogenous biochemicals by being introduced from outside the organism. For example, insulin is a hormone that is synthesized in the body; it is called a hormone when it is synthesized by the pancreas inside the body, but if it is introduced into the body from outside, it is called a drug. Many natural substances, such as beers, wines, and psychoactive mushrooms, blur the line between food and recreational drugs, as when ingested they affect the functioning of both mind and body and some substances normally considered drugs such as DMT (Dimethyltryptamine) are actually produced by the human body in trace amounts.

dynamic[dainǽmik]: F. < G. *dynamique < dunamis, dunamikos,* power. 힘 있는, 동적인, 동력의. The study of the causes of motion and changes in motion is dynamics. In other words the study of forces and why objects are in motion. Dynamics includes the study of the effect of torques on motion. These are in contrast to kinematics, the branch of classical mechanics that describes the motion of objects without consideration of the causes leading to the motion. Generally speaking, researchers involved in dynamics study how a physical system might develop or alter over time and study the causes of those changes. In addition, Isaac Newton established the undergirding physical laws which govern dynamics in physics. By studying his system of mechanics, dynamics can be understood. In particular dynamics is mostly related to Newton's second law of motion. However, all three laws of motion are taken into consideration, because these are interrelated in any given observation or experiment. For classical electromagnetism, it is Maxwell's equations that describe the dynamics. And the dynamics of classical systems involving both mechanics and electromagnetism are described by the combination of

Newton's laws, Maxwell's equations, and the Lorentz force.

echolocation[ekouloukéiʃən]: L. < G. *ēkhō, ēkhē*, a sound + L. *locare*, *location*, place, position. 음파 탐지, 반향 위치 탐지법.

eclipse[iklíps]: G. > F. *eclipser, eclipse*, < *ekleipein*, *ekleipsis, ek,* out + *leipein*, to leave, fail to appear, to cease to exsist, eclipsed. (해, 달의) 식(蝕), - 을 가리다.

ecology(=oecology)[ikάlədʒi/-kɔ́l-]: G. *oikos,* house + *-logia,* study of, science, bionomics. 생태학, 자연환경, 환경보존. The scientific study of the relations that living organisms have with respect to each other and their natural environment. Variables of interest to ecologists include the composition, distribution, amount (biomass), number, and changing states of organisms within and among ecosystems. Ecosystems are hierarchical systems that are organized into a graded series of regularly interacting and semi-independent parts (e.g., species) that aggregate into higher orders of complex integrated wholes (e.g., communities). Ecosystems are sustained by the biodiversity within them. Biodiversity is the full-scale of life and its processes, including genes, species and ecosystems forming lineages that integrate into a complex and regenerative spatial arrangement of types, forms, and interactions. Ecosystems create biophysical feedback mechanisms between living (biotic) and nonliving (abiotic) components of the planet. These feedback loops regulate and sustain local communities, continental climate systems, and global biogeochemical cycles. Ecology is a sub-discipline of biology, the study of life. The word "ecology"

("Ökologie") was coined in 1866 by the German scientist Ernst Haeckel (1834–1919). Ancient philosophers of Greece, including Hippocrates and Aristotle, were among the earliest to record observations and notes on the natural history of plants and animals. Modern ecology branched out of natural history and matured into a more rigorous science in the late 19th century. Charles Darwin's evolutionary treatise including the concept of adaptation, as it was introduced in 1859, is a pivotal cornerstone in modern ecological theory. Ecology is not synonymous with environment, environmentalism, natural history or environmental science. It is closely related to physiology, evolutionary biology, genetics and ethology[(동물) 행동학]. An understanding of how biodiversity affects ecological function is an important focus area in ecological studies. Ecologists seek to explain:

- Life processes and adaptations
- Distribution and abundance of organisms
- The movement of materials and energy through living communities
- The successional development of ecosystems, and
- The abundance and distribution of biodiversity in context of the environment.

Ecology is a human science as well. There are many practical applications of ecology in conservation biology, wetland management, natural resource management (agriculture, forestry, fisheries), city planning (urban ecology), community health, economics, basic and applied science and human social interaction (human ecology). Ecosystems sustain every life-supporting function on the planet, including climate regulation, water filtration, soil formation, food, fibers, medicines, erosion control, and many other natural features of scientific, historical or spiritual value.

effective[iféktiv]: L. *effectivus, efficere*, accomplish. 유효한, 유능한, 실제의.

effective magnetic field: the magnetic field that a proton "senses" through the surrounding cloud of electrons.

elastic[ilǽstik]: L. < G. *elasticus < elaunein*, *elastikos*, propulsive. 탄성 있는, 융통성 있는, 탄력적인.

elastomer[ilǽstəmər]: 1930s. elastic, G. > L. *elasticus < elaunein*, *elastikos,* propulsive, to drive + -mer, G. *meros*, part. 탄성 중합체 (실리콘, 고무 따위). A polymer with the property of viscoelasticity (colloquially "elasticity"), generally having notably low Young's modulus and high yield strain compared with other materials. The term, which is derived from elastic polymer, is often used interchangeably with the term rubber. Each of the monomers which link to form the polymer is usually made of carbon, hydrogen, oxygen and/or silicon. Elastomers are amorphous polymers existing above their glass transition temperature, so that considerable segmental motion is possible. At ambient temperatures rubbers are thus relatively soft and deformable. Their primary uses are for seals, adhesives and molded flexible parts.

electric[iléktrik]: G. > L. *electrum*, *electricus* < *ēlektron,* amber (because rubbing amber from which the word electricity is derived). 기전(起電)물체(호박(琥珀), 유리 따위의), 전기의, 전도하는.

electrode[iléktroud]: electric + G. *hodos,* way, path, tarvel. 전극. An electrical conductor used to make contact with a nonmetallic part of a circuit (e.g. a semiconductor, an electrolyte or a vacuum). The word was coined by the scientist Michael Faraday.

electrode potential, *E*: in electrochemistry, according to an IUPAC definition, is the electromotive force of a cell built of two electrodes:

- on the left-hand side is the standard hydrogen electrode, and
- on the right-hand side is the electrode the potential of which is being defined.

By convention:

$$E_{\text{Cell}} = E_{\text{Cathode}} - E_{\text{Anode}}$$

From the above, for the cell with the standard hydrogen electrode (potential of 0 by convention), one obtains:

$$E_{\text{Cell}} = E_{\text{Right}} - 0 = E_{\text{Electrode}}$$

The left-right convention is consistent with the international agreement that redox potentials be given for reactions written in the form of reduction half-reactions. Electrode potential is measured in volts (V).

electromagnet[ilektrəmǽgnit]: electro- + L. *magnes, magnet*, lodestone(천연자석). 전자석(電磁石).

electron[iléktrɑn / -trɔn]: G. > L. *electrum,* < *ēlektron,* amber [호박(琥珀)] (because rubbing amber causes electrostatic phenomena). 전자.

electrophile: (electron + G. *philos*, loving, lover). A reagent attracted to electrons that participates in a chemical reaction by accepting an electron pair in order to bond to a nucleophile. Because electrophiles accept electrons, they are Lewis acids (see acid-base reaction theories). Most electrophiles are positively charged, have an atom that carries a partial positive charge, or have an atom that does not have an octet of electrons. The electrophiles attack the most electron-populated part of a nucleophile. The electrophiles frequently seen in the organic syntheses are cations such as H^+ and NO^+, polarized neutral molecules such as HCl, alkyl halides, acyl halides, and carbonyl compounds, polarizable neutral molecules such as Cl_2 and Br_2, oxidizing agents such as organic peracids, chemical species that do not satisfy the octet rule such as carbenes and radicals, and some lewis acids such as BH_3 and DIBAL.

electronegative: electrically negative

electronegativity[ilektrənegətívəti]: 전기 음성도. A chemical property that describes the tendency of an atom or a functional group to attract electrons (or electron density) towards itself and thus the tendency to form negative ions. An atom's electronegativity is affected by both its atomic number and the distance that its valence electrons reside from the charged nucleus. The higher the associated electronegativity number, the more an element or compound attracts electrons towards it. First proposed by Linus Pauling in 1932 as a development of valence bond theory, it has been shown to correlate with a number of other chemical properties. Electronegativity cannot be directly measured and must be calculated from other atomic or molecular properties. Several methods of calculation have been proposed and, although there may be small differences in the numerical values of the electronegativity, all methods show the same periodic trends between elements. The most commonly used method of calculation is that originally proposed by Pauling. This gives a dimensionless quantity, commonly referred to as the Pauling scale, on a relative scale running from around 0.7(Fr) to 3.98(F) (hydrogen = 2.20). When other methods of calculation are used, it is conventional (although not obligatory) to quote the results on an electronegativity in Pauling units.

electrophoresis[ilektrəfərí:sis]: electro- + G. *phorēsis,* being carried. 전기이동, 전기영동(泳動). The motion of dispersed particles relative to a fluid under the influence of a spatially uniform electric field. This electrokinetic phenomenon was observed for the first time in 1807 by Reuss, who noticed that the application of a constant electric field caused clay particles dispersed in water to migrate. It is ultimately caused by the presence of a charged interface between the particle surface and the surrounding fluid.

electroposotivity: 전기 양성도. A measure of an element's ability to donate electrons, and therefore form positive ions; thus, it is opposed to electronegativity. Mainly, this is an attribute of metals, meaning that for the most part, the greater the metallic character of an element, the greater the electropositivity. Therefore the alkali metals are most electropositive of all. This is because they have a single electron in their outer shell and, as this is relatively far from the nucleus of the atom, it is easily lost; in other words, these metals have low ionization energies. While electronegativity increases along periods in the periodic table, and decreases down groups, electropositivity decreases along periods and increases down groups.

electrostatic[ilektroustǽtik]: mid 19th cent. electro- + G. > L. *statikos* > *staticus*, causing to stand. 정(靜)전기의.

element[éləmənt]: G. > L. *stoikheion* > *elementum*, principle, rudiment. 원소.

chemical element: a pure chemical substance consisting of one type of atom distinguished by its atomic number, which is the number of protons in its nucleus. Familiar examples of elements include carbon, oxygen, aluminum, iron, copper, gold, mercury, and lead. As of November 2011, 118 elements have been identified, the latest being ununseptium(Uus, artificial element, atomic number 117) in 2010. Of the 118 known elements, only the first 98 are known to occur naturally on Earth. Of these, 80 are stable or essentially so, while the others are radioactive, decaying into lighter elements over various timescales from fractions of a second to billions of years. Those elements that do not occur naturally on Earth have been produced artificially as the synthetic products of man-made nuclear reactions. Hydrogen and helium are by far the most abundant elements in the universe. However, iron is the most abundant element (by mass) making up the Earth, and oxygen is the most common element in the Earth's crust. Although all known chemical matter is composed of these elements, chemical matter itself constitutes only about 15% of the matter in the universe. The remainder is dark matter, a mysterious substance which is not composed of chemical elements since it lacks protons, neutrons or electrons. The chemical elements are thought to have been produced by various cosmic processes, including hydrogen, helium (and smaller amounts of lithium, beryllium and boron) created during the Big Bang and cosmic-ray spallation. Production of heavier elements, from carbon to the very heaviest elements, proceeds by stellar nucleosynthesis, and these were made available for later solar system and planetary formation by supernovae [초신성(超新星)], which blast these elements into space. The high abundance of oxygen, silicon, and iron on Earth reflect their common production in such stars, after the lighter gaseous elements and their compounds have been subtracted. While most elements are generally viewed as stable, a small amount of natural transformation of one element to another also occurs in the present time, through decay of radioactive elements as well as other natural nuclear processes. Relatively pure samples of isolated elements are uncommon in nature. While all of the 98 naturally occurring elements have been identified in mineral samples from the Earth's crust, only a small minority of elements are found as recognizable, relative pure minerals. Among the more common of such "native elements" are copper, silver, gold, carbon (as coal, graphite, or diamonds), sulfur, and mercury. All but a few of the most inert elements, such as noble gases and noble metals, are usually found on Earth in chemically combined form, as chemical compounds. While about 32 of the chemical elements occur on Earth in native uncombined form, most of these occur as mixtures. For example, atmospheric air is primarily a mixture of nitrogen, oxygen, and argon, and native solid elements occur in alloys, such as that of iron and nickel. When two distinct elements are chemically combined, with the atoms held together by chemical bonds, the result is termed a chemical compound. Two thirds of the chemical elements occur on Earth only as compounds, and in the remaining third, often the compound forms of the element are most common. Chemical compounds may be composed of elements combined in exact whole-number ratios of atoms, as in water, table salt, and minerals as quartz, calcite, and some ores. However, chemical bonding of many types of elements results in crystalline solids and metallic alloys for which exact chemical formulas do not exist. The properties of the chemical elements are often summarized using the periodic table that organizes the elements by increasing atomic number into rows ("periods") in which the columns ("groups") share recurring ("periodic") physical and chemical properties. Either in its pure forms, or in various chemical compounds or mixtures, almost every element has at least one important human use. Around two dozen of the elements are essential to various kinds of biological life. Most rare elements on Earth are not needed by life (exceptions being selenium and iodine), while a few quite common ones (aluminum and titanium) are not used. Most organisms share element needs, with a few differences. For example, ocean algae use bromine but land plants and animals seem to need none, and all animals require sodium, but some plants do not. Just six elements — carbon, hydrogen, nitrogen, oxygen, calcium, and phosphorus — make up almost 99% of the mass of a human body (see composition of the human body for a complete list). In addition to the six major elements that compose most of the human body, humans require consumption of at least a dozen more elements in the form of certain chemical compounds.

elimination[ilimənéiʃən]: L. *eliminare*, *elimination*, *e-, ex,* out + *limen, limin,* threshold, turned out of doors. 제거(반응). Generate a simple substance as a product in the course of a reaction involving larger molecules.

<u>elimination reaction</u>: 제거반응. A type of organic reaction in which two substituents are removed from a molecule in either a one or two-step mechanism. Either the unsaturation of the molecule increases (as in most organic elimination reactions) or the valence of an atom in the molecule decreases by two, a process known as reductive elimination. An important class

of elimination reactions are those involving alkyl halides, or alkanes in general, with good leaving groups, reacting with a Lewis base to form an alkene in the reverse of an addition reaction. When the substrate is asymmetric, regioselectivity is determined by Zaitsev's rule. The one and two-step mechanisms are named and known as E2 reaction and E1 reaction, respectively.

Ex) Elimination reaction of cyclohexanol to cyclohexene with sulfuric acid and heat

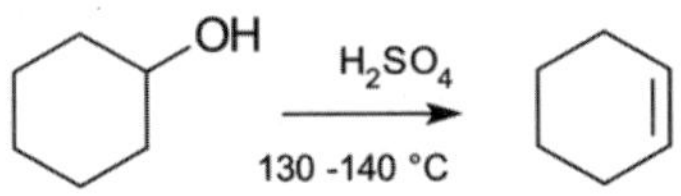

E1cB elimination reaction (Elimination Unimolecular conjugate Base): A special type of elimination reaction in organic chemistry. This reaction mechanism explains the formation of alkenes from (mostly) alkyl halides through a carbanion intermediate given specified reaction conditions and specified substrates. The reaction takes place around a sp^3 - sp^3 carbon to carbon covalent bond with an α-acidic hydrogen atom substituent and a β-leaving group. This leaving group can be a halide or a sulfonic acid ester such as a tosyl group. A strong base abstracts the α proton generating a carbanion. The electron pair then expels the leaving group and the double bond is formed. When the first step to the carbanion is slow and the second step fast the reaction is irreversible and named $(E1cB)_i$. When the first step is fast and the deprotonation reversible then the reaction mechanism is $(E1cB)_r$. In the $(E1cB)_{anion}$ variation the carbanion is especially stable with a rapid first step and a slow second step. Reaction of 1-methyl-2-(2-fluoroethyl)-pyridinium iodide) with hydroxide. A named reaction displaying E1cB elimination mechanism is the Boord olefin synthesis. Being able to distinguish between different elimination reaction mechanisms (E2, E1, E1cB) is often difficult and involves the study of chemical kinetics, kinetic isotope effects, substituent effects and computational chemistry. The reaction of 1-methyl-2-(2-fluoroethyl) pyridinium iodide with the hydroxide ion in water is an example of a 2-step E1cB reaction because the carbanion is stabilized by the resonance effect with the enamine.

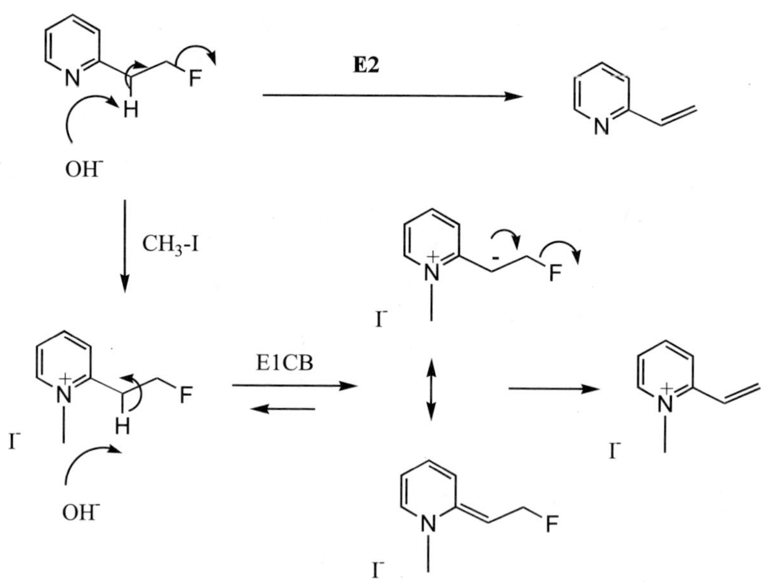

emphysema[emfisí:ma]: G. *emphusan*, *emphusēma,* puff up. 폐기종.

emulsion[imʌ́lʃən]: L. *emulgere*, *emulsion*, *e*, *ex*, out + *mulgere*, to milk, milk out. 유제, 유탁액. A mixture of two or more liquids that are normally immiscible (un-blendable). Emulsions are part of a more general class of two-phase systems of matter called colloids. Although the terms *colloid* and *emulsion* are sometimes used interchangeably, *emulsion* is used when both the dispersed and the continuous phase are liquid. In an emulsion, one liquid (the dispersed phase) is dispersed in the other (the continuous phase). Examples of emulsions include vinaigrettes(각성제 통), milk, and some cutting fluids for metal working. The photo-sensitive side of photographic film is an example of a *colloid*.

enantiomer[inæntíəmər]: G. *enantios*, opposite, + *meros*, part or portion. 거울상(이성질)체. In stereochemistry, an enantiomer is one of two stereoisomers that are mirror images of each other that are "non-superposable" (not identical), much as one's left and right hands are "the same" but opposite.

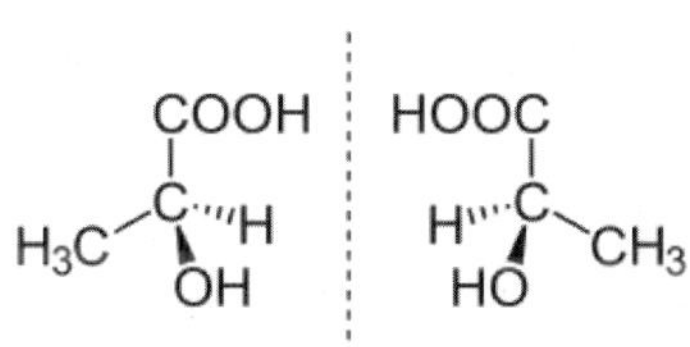

(*S*)-(+)-lactic acid (left) and (*R*)-(-)-lactic acid (right) are nonsuperposable mirror images of each other.

endergonic[endərgάnik / -gɔ́n-]: G. *endo-*, within + *ergon*, work + -ic. 자유에너지 증가의, (반응이) 에너지 흡수성의. Absorbing energy in the form of work.

endergonic reaction: 자유에너지 증가 반응. Not spontaneous. By thermodynamic standards, work, a form of energy, is defined as moving from the surroundings (the external region) to the system (the internal region). Thus, an endergonic process, as contrasted with an exergonic process, is one wherein the system absorbs energy from the surroundings. As a result, during an endergonic process, energy is put into the system. If the transformation occurs at constant pressure and temperature, $\Delta G > 0$. An endergonic reaction is a chemical reaction that absorbs energy in the form of work. A good example of an endergonic reaction is photosynthesis. Also, in metabolism, an endergonic reaction is anabolic, meaning, the break down of energy. In metabolism, catabolic and anabolic are coupled by ATP. Since endergonic reactions require energy, they form molecular bonds, which decreases entropy. These bonds are energy carriers, so the formation of these bonds requires some energy input.

endo-[éndou]: G > L. *end(o)- < endon,* within, inner. 내부, 속. A substituent is endo if it is closer to the longer or more unsaturated bridge.

endo-exo isomerism: a special type of isomerism found in organic compounds with a substituent on a bridged ring system. The prefix endo is reserved for the isomer with the substituent located closest, or "anti," to the longest bridge. The prefix exo is reserved for the isomer with the substituent located furthest, or "syn," to the longest bridge.

Exo Endo

endocyclic: situated within the ring.

endorphin[endɔ́:rfin]: 1970s: blend of **endo**genous mo**rphine**. 내부로부터 발생하는 모르핀. A morphine-like substance originating from within the body. 엔도르핀(척추 동물에 존재하는 팹티드 군으로 진통작용을 한다). Endogenous opioid(합성 진통 마취제) peptides that function as neurotransmitters. They are produced by the pituitary(뇌하수체의) gland and the hypothalamus[시상하부(視床下部)] in vertebrates during exercise, excitement, pain, consumption of spicy food, love and orgasm, and they resemble the opiates in their abilities to produce analgesia and a feeling of well-being. When a nerve impulse reaches the spinal cord, endorphins that prevent nerve cells from releasing more pain signals are released. Immediately after injury, endorphins allow animals to feel a sense of power and control over themselves that allows them to persist with activity for an extended time.

endothermic[endoueə́:rmik]: G. *endon-*, within, inner, inside + *thermē*, heat, hot + *-ic*. absorption of heat. 흡열(吸熱)(적). A process or reaction in which the system absorbs energy from the surroundings in the form of heat. Hence it refers to a reaction that needs heat. The opposite of an endothermic process is an exothermic process, one that releases energy in the form of heat. Thus in each term(endothermic & exothermic) the prefix refers to where heat goes as the reaction occurs. The concept is frequently applied in physical sciences to, for example, chemical reactions, where thermal energy (heat) is converted to chemical bond energy.

energy[énərdʒi]: F. < G. ė*nergie* < *energeia*, *en-*, in, within + *ergon*, work, activity, operation. 에너지. An indirectly observed quantity. It is often understood as the ability a physical system has to do work on other physical systems. Since work is defined as a force acting through a distance (a length of space), energy is always equivalent to the ability to exert pulls or pushes against the basic forces of nature, along a path of a certain length. The total energy contained in an object is identified with its mass, and energy (like mass), cannot be created or destroyed. When matter (ordinary material particles) is changed into energy (such as energy of motion, or into radiation), the mass of the system does not change through the transformation process. However, there may be mechanistic limits as to how much of the matter in an object may be changed into other types of energy and thus into work, on other systems. Energy, like mass, is a scalar physical quantity. In the International System of Units (SI), energy is measured in joules, but in many fields other units, such as kilowatt-hours and kilocalories, are customary. All of these units translate to units of work, which is always defined in terms of forces and the distances that the forces act through.

enkephalin[enkéfəlin]: G. *enkephalos*, *en-*, in + *kephalē*, head + *-in*, brain. 엔케펄린(모르핀과 비슷한 작용을 하는 내인성 펩티드). Pentapeptide involved in regulating nociception(조직손상) in the body. The enkephalins are termed endogenous ligands, or specifically endorphins, as they are internally derived and bind to the body's opioid receptors. Discovered in 1975, two forms of enkephalin were revealed, one containing leucine ("leu"), and the other containing methionine ("met"). Both are products of the proenkephalin gene.

- Met-enkephalin is Tyr-Gly-Gly-Phe-**Met**.
- Leu-enkephalin has Tyr-Gly-Gly-Phe-**Leu**.

Fig. Met-enkephalin 3D structure, alpha-carbons shown as balls and labeled by residue.

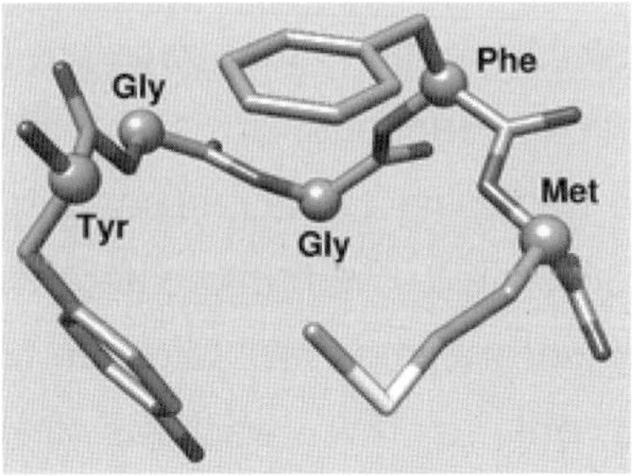

enthalpy[énθælpi]: G. *enthalpein*, *en*, in + *thalpein*, to warm, to heat, warm in. 엔탈피, 열함량(열 에너지를 나타내는 상태량의 하나). A thermodynamic property of a system: the sum of its internal energy and the product of its volume and the pressure exerted on it. (denoted as H, change in enthalpy ΔH is frequently a more useful value than H itself).

H = E + pV where (all units given in SI)
H is the enthalpy (in joules),
E is the internal energy (in joules),
p is the pressure of the system, (in pascals), and
V is the volume, (in cubic meters).

entropy[éntrəpi]: G. *entropia, en*, in, inside + *tropē*, turn, conversion, transformation, tuming toward. 엔트로피. A measure of the number of random ways in which a system may be arranged, often taken to be a measure of "disorder" (the higher the entropy, the higher the disorder). Thermodynamic entropy has the dimension of energy divided by temperature, and a unit of joules per kelvin (J/K) in the International System of Units.

enzyme[énzaim]: G. *enzumos*, *en-*, within + *zume*, leaven, leavened. 효소(酵素). Proteins that catalyze (*i.e.*, increase the rates of) chemical reactions. In enzymatic reactions, the molecules at the beginning of the process are called substrates, and the enzyme converts them into different molecules, called the products. Since enzymes are selective for their substrates and speed up only a few reactions from among many possibilities, the set of enzymes made in a cell determines which metabolic pathways occur in that cell. Like all catalysts, enzymes work by lowering the activation energy ($E_a^{\ddagger}$) for a reaction, thus dramatically increasing the rate of the reaction.

Most enzyme reaction rates are millions of times faster than those of comparable un-catalyzed reactions. As with all catalysts, enzymes are not consumed by the reactions they catalyze, nor do they alter the equilibrium of these reactions. However, enzymes do differ from most other catalysts by being much more specific. Enzymes are known to catalyze about 4,000 biochemical reactions. A few RNA molecules called ribozymes also catalyze reactions, with an important example being some parts of the ribosome.

ephedrine[ifédrin, éfidri:n]: late 19th cent. ephedra, L. *equisetum, equus,* horse + *saeta,* bristle + -ine. 에페드린(천식 치료용). A sympathomimetic (교감신경 자극성의) amine commonly used as a stimulant, appetite suppressant, concentration aid, decongestant(충혈 완화제), and to treat hypotension[저혈압(증)] associated with anaesthesia.

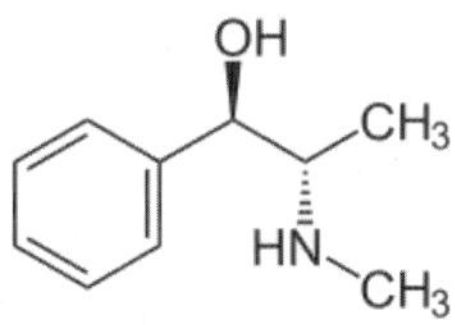

epi[épi, epə]: G. *ep(i),* upon, at, or close upon (a point of space or time), on the ground or occasion of, in addition. 바깥, 외(外).

epimer: 에피머. Diastereomers that differ in configuration of only one stereogenic center. Diastereomers are a class of stereoisomers that are non-superposable, non-mirror images of one another. The sugars α-glucose and β-glucose are epimers. In α-glucose, the -OH group on the first (anomeric) carbon is in the direction opposite the methylene group on carbon C-6 (in the axial position). In β-glucose, the -OH group is oriented in the same direction as the methylene group (in the equatorial position). These two molecules are both epimers and anomers.

Fig. α-Glucose and β-glucose

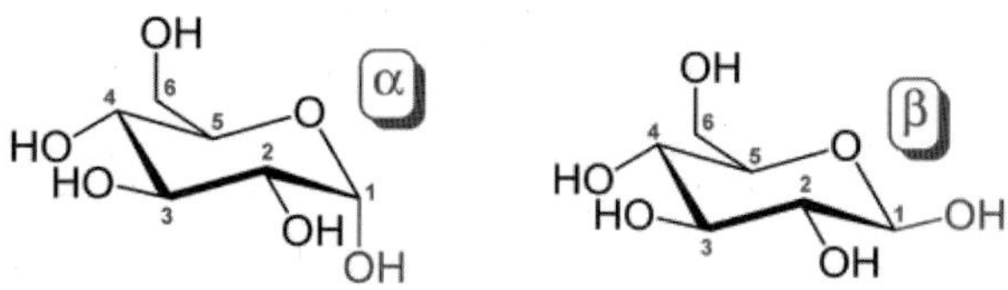

epidemic[epədémik]: F. < G. ėpidėmie, ėpidėmique < epidēmios, *epidēmia*, *epi-*, upon + *dēmos,* the people, prevalence of disease. 유행성의, 전염성의, 전염병. It occurs when new cases of a certain disease, in a given human population, and during a given period, substantially exceed what is expected based on recent experience. Epidemiologists often consider the term outbreak to be synonymous to epidemic, but the general public typically perceives outbreaks to be more local and less serious than epidemics. An epidemic may be restricted to one locale, however if it spreads to other countries or continents and affects a substantial number of people, it may be termed a pandemic (세계적인 유행병).

epidemiology[epədi:miáləd͡ʒi / -ɔ́l-]: G. *epidēmia*, prevalence of disease + -logy, *logos*, word, science. 전염병학, 역학(疫學). The study of the distribution and patterns of health-events, health-characteristics and their causes or influences in well-defined populations. It is the cornerstone method of public health research, and helps inform policy decisions and evidence-based medicine by identifying risk factors for disease and targets for preventive medicine. Epidemiologists are involved in the design of studies, collection and statistical analysis of data, and interpretation and dissemination(유포) of results (including peer review and occasional systematic review). Major areas of epidemiological study include outbreak investigation, disease surveillance and screening (medicine), biomonitoring, and comparisons of treatment effects such as in clinical trials. Epidemiologists rely on a number of other scientific disciplines such as biology (to better understand disease processes), biostatistics (to make efficient use of the data and draw appropriate conclusions), and exposure assessment and social science disciplines (to better understand proximate and distal risk factors, and their measurement).

epinephrine[epənéfri(:)n]: 19th cent. *epi-*, above + G. *nephros*, kidney + -ine. 에피네프린, 아드레날린(부신에서 분비되는 호르몬). A hormone and a neurotransmitter. It increases heart rate, constricts blood vessels, dilates air passages and participates in the fight-or-flight response of the sympathetic nervous system. Chemically, epinephrine is a catecholamine, a monoamine produced only by the adrenal glands from the amino acids phenylalanine and tyrosine.

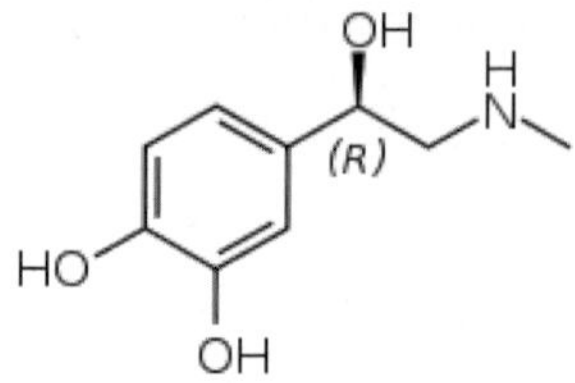

epoxy[ipάksi, epάk- / epɔ́k-]: early 20th cent. *epi-*, in addition + *oxy-,* oxygen. Oxygen addition. 에폭시의, 에폭시 수지의. Epoxy or polyepoxide is a thermosetting polymer formed from reaction of an epoxide "resin" with polyamine "hardener". Epoxy has a wide range of applications, including fiber-reinforced plastic materials and general purpose adhesives.

epoxide (oxirane): a cyclic ether with three ring atoms. This ring approximately defines an equilateral triangle, which makes it highly strained. The strained ring makes epoxides more reactive than other ethers. Simple epoxides are named from the parent compound ethylene oxide or oxirane, such as in chloromethyloxirane. As a functional group epoxides feature the epoxy prefix such as in the compound 1,2-epoxycycloheptane which can also be called cycloheptene epoxide, or simply cycloheptene oxide.

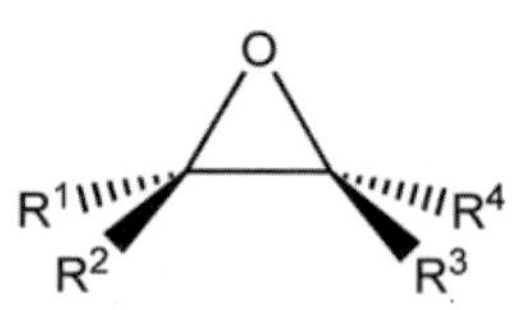

equation[ikwéiʒən, -ʃən]: L. *aequare*, *aequation*, make equal. 반응식, 방정식. A symbolic representation of the changes which occur in a chemical reaction, expressed in terms of the formulae of the molecules or other species involved.

equator[ikwéitər]: L. *aequare*, *aequator,* make equal, equalizing. 적도(선).

equillibrium[i:kwəlíbriəm, ek-]: L. *aequilibrium, aequi,* equal + *libra*, balance. 화학반응의 평형, (물리적 힘의) 균형, 평형. A state in which a process and its reverse are occurring at equal rates so that no overall change is taking place.

equivalent[ikwívələnt]: L. *aequivqlere*, *aequivalent, aequi-,* equally + *valere,* be worth, being of equal worth. (원자가) 당량(當量), 등가, 대등체.

symbol: eq or Eq. Molar equivalent: a unit of amount of substance used in chemistry and the biological sciences. The molar equivalent is formally defined as the amount of a substance which will either:

- react with or supply one mole of hydrogen ions (H^+) in an acid–base reaction; or
- react with or supply one mole of electrons in a redox reaction. The mass of one equivalent of a substance is called its equivalent weight.

A historical definition, used especially for the chemical elements, describes an equivalent as the amount of a substance that will react with one gram of hydrogen, or with eight grams of oxygen, or with 35.5 grams (1.25 oz) of chlorine, or displaces any of the three. In practice, the amount of a substance in equivalents often has a very small magnitude, so it is frequently described in terms of milliequivalents (mEq or meq), the prefix milli denoting that the measure is divided by 1000. Very often, the measure is used in terms of milliequivalents of solute per litre of solvent (or milliNormal, where mEq/L = mN).

eradicate[irǽdəkeit]: L. *eradicare*, *eradicat*, *e-*, *ex-*, out + *radix, radic*, root, torn up by the roots. - 을 근절하다, 박멸하다.

erectile[iréktl / -tail]: L. *erigere, erect*, *e-*, *ex-*, out + *regere*, to direct, set up. 직립성의, 발기(勃起)성의.

erythro-[iríθrou, -rə]: G. *eruthros*, red. 붉은.

erythro enantimers: the pair of enantiomers with similar groups on the same side as drawn in a Fischer projection.

erythrocyte: 적혈구

Fig. D-erythrose

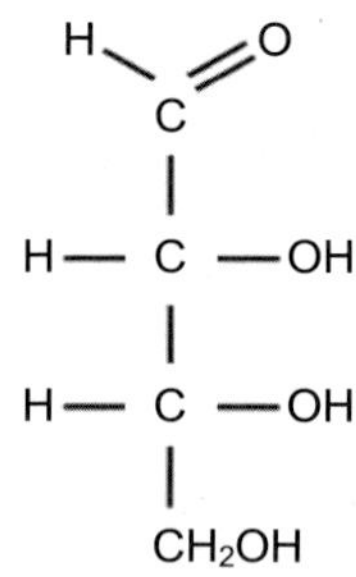

essential[isénʃəl, es-]: L. *essentia*, *essentialis,* absolutely necessary, extremely important. 필수의, 없어서는 안 되는.

essential amino acid: an amino acid that hunmans must obtain from their diet because they cannat synthesize it at all or cannot synthesize it in an adequate amounts.

Essentiality vs. conditional essentiality in humans

Essential	Nonessential
Isoleucine	Alanine
Leucine	Arginine*
Lysine	Aspartate
Methionine	Cysteine*
Phenylalanine	Glutamate
Threonine	Glutamine*
Tryptophan	Glycine*
Valine	Proline*
Histidine	Serine
Asparagine*	Tyrosine*
Selenocysteine**	Pyrrolysine**

(*) Essential only in certain cases.

(**) Truly unclassified. Added to sustain the 22 Numbers of Essential Amino Acids.

ester[éstər]: D. E*ssig*, vinegar + *Āther*, ether. 에스테르. Chemical compounds derived by reacting an oxoacid with a hydroxyl compound such as an alcohol or phenol. Esters are usually derived from an inorganic acid or organic acid in which at least one -OH (hydroxyl) group is replaced by an -O-alkyl (alkoxy) group, and most commonly from carboxylic acids and alcohols. That is, esters are formed by condensing an acid with an alcohol.

Esters are ubiquitous. Most naturally occurring fats and oils are the fatty acid esters of glycerol. Esters with low molecular weight are commonly used as fragrances and found in essential oils and pheromones. Phosphoesters form the backbone of DNA molecules. Nitrate esters, such as nitroglycerin, are known for their explosive properties, while polyesters are important plastics, with monomers linked by ester moieties.

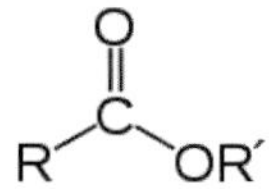

estrogen[éstrədʒən]: oestrogens(영). 1920s. oestrus, G. *oistros*, gadfly, frenzy, sexual desire + -Gen, *genēs*, born, to generate). 에스트로겐(여성 발정 호르몬 물질). A group of steroid compounds, named for their importance in the estrous(발정의) cycle, and functioning as the primary female sex hormones. Estrogens are synthesized in all vertebrates as well as some insects. Estrogens are used as part of some oral contraceptives (피임약), in estrogen replacement therapy for postmenopausal(폐경 후의) women, and in hormone replacement therapy for trans women. Like all steroid hormones, estrogens readily diffuse across the cell membrane. Once inside the cell, they bind to and activate estrogen receptors which in turn modulate the expression of many genes. The three major naturally occurring estrogens in women are estrone (E1), estradiol (E2), and estriol (E3). Estradiol (E2) is the predominant form in nonpregnant females, estrone is produced during menopause(폐경기), and estriol is the primary estrogen of pregnancy. In the body these are all produced from androgens through actions of enzymes.

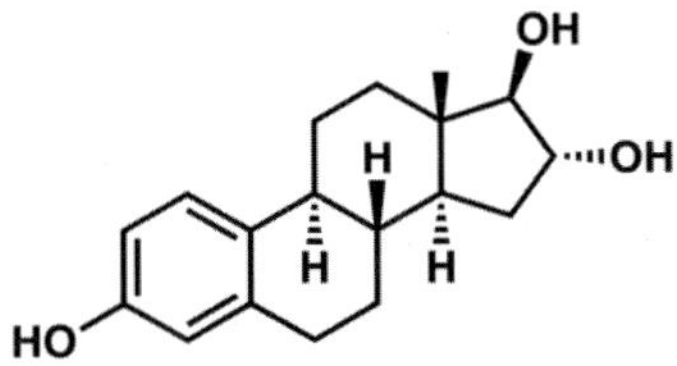

Estriol (E3)

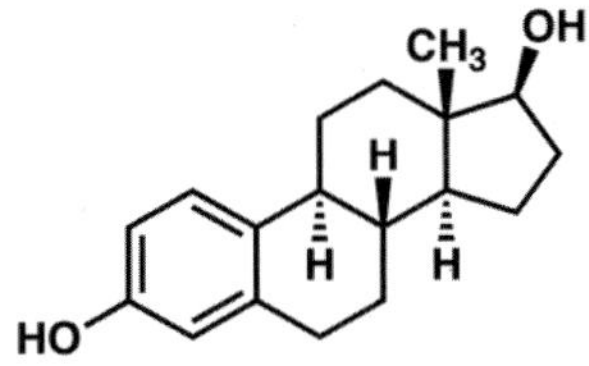

Estradiol (E2)

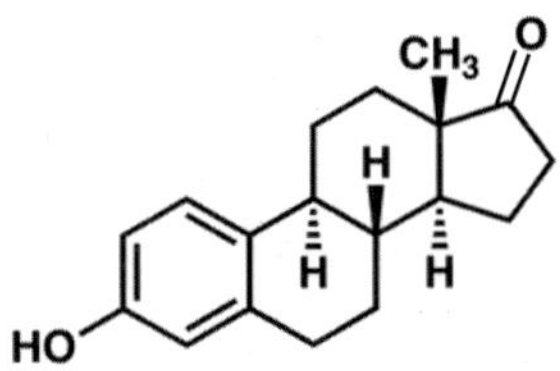

Estrone (E1)

ethanol[éθənɔ:l, -nɔl]: ethyl alcohol, pure alcohol, grain alcohol, or drinking alcohol, 20th cent. blend of ethane and alcohol. 에탄올. A volatile, flammable, colorless liquid. It is a powerful psychoactive drug and one of the oldest recreational drugs. Best known as the type of alcohol found in alcoholic beverages, it is also used in thermometers, as a solvent, and as an alcohol fuel. In common usage, it is often referred to simply as alcohol or spirits. Ethanol is a straight-chain alcohol, and its molecular formula is C_2H_5OH. Its empirical formula is C_2H_6O. It is a constitutional isomer of dimethyl ether. Ethanol is often abbreviated as EtOH, using the common organic chemistry notation of representing the ethyl group (C_2H_5) with Et. The fermentation of sugar into ethanol is one of the earliest organic reactions employed by humanity. In modern times, ethanol intended for industrial use is also produced from by-products of petroleum refining. Ethanol has widespread use as a solvent of substances intended for human contact or consumption, including scents, flavorings, colorings, and medicines. In chemistry, it is both an essential solvent and a feedstock for the synthesis of other products. It has a long history as a fuel for heat and light, and more recently as a fuel for internal combustion engines.

ether[í:θər]: G. *aithēr*, upper air. 에테르, 에틸 에테르(용매, 마취약용). A class of organic compounds that contain an ether group — an oxygen atom connected to two alkyl or aryl groups — of general formula R–O–R. A typical example is the solvent and anesthetic diethyl ether, commonly referred to simply as "ether" (CH_3-CH_2-O-CH_2-CH_3).

eukaryote (= eucaryote)[ju:kǽriout]: 1960s. *eu-*, easily(formed) + *karyo-*, kernel, nut + *-ote.* 진핵생물. A unicellular or multicellular body whose cell or cells contain a nucleus.

evolution[evəlú:ʃən, i:v-]: L. *evolvere*, *evolution*, opening out, development. 진화, 진화론.

excite[iksáit]: F. < L. *exciter* < *excitare, exciere*, call out or forth. (원자, 분자)를 여기(勵起)시키다.

exclusion[iksklú:zən]: L. *excludere*, *exclusion*, *ex-*, out + *claudere,* to shut, shut out. 제외, 배제, 추방.

excrete[ikskrí:t]: L. *excernere, excrete*, *ex-*, out + *cernere*, sift. -을 배설하다. 분비하다. (of a living organism or cell) separate and expel as waste (a substance, especially a product of metabolism).

exergonic[exsərgάnik/-gɔ́n-]: G. *exergonic*, *exō*, outside + *-ergonic*, *ergon*, work. (반응이) 에너지 방출성의. Releasing energy in the form of work. By thermodynamic standards, work, a form of energy, is defined as moving from the system (the internal region) to the surroundings (the external region). Thus, an exergonic process, as contrasted with an endergonic process, is one that releases energy from the system, of which it is a part, to the surroundings. As a result, during an exergonic process, energy is released out of the system. If the transformation occurs at constant pressure and temperature, $\Delta G < 0$. When Gibbs free energy is less than 0, the reaction is exergonic (releases energy).

exhale[ekshéil]: F. < L. *exhaler* < *exhalare*, *ex-*, out + *halare*, breathe, breathe out. give off (vapour or fumes). 발산하다, 숨을 토해내다.

exo-[eksou, sə]: G, *exo, ex-*, external, from outside. 외부, 바깥. endo-exo isomerism 참조.

exothermic[eksəθə́:rmik]: G. *ex-,* outside + *thermein* to heat, outside heating. 발열(성)의. In thermodynamics, the term describes a process or reaction that releases energy from the system, usually in the form of heat, but also in the form of light (e.g. a spark, flame, or explosion), electricity (e.g. a battery), or sound (e.g. burning hydrogen). The opposite of an exothermic process is an endothermic process, one that absorbs energy in the form of heat. The concept is frequently applied in the physical sciences to chemical reactions, where chemical bond energy is converted to thermal energy (heat).

expel[ikspél]: L. *expellere*, *ex-*, out + *pellere*, to drive, force (someone) to leave a place. - 을 내쫓다, 추방하다.

experiment[ikspérəmənt]: L. *experiri, experimentum, ex-*, out of, away from + *periri*, try, attemt, risk, away from risk. 실험.

exponent[ikspóunənt]: L. *exponere, exponent*, putting out. 지수(指數)(X^3에서 3).

fabric[fǽbrik]: F. < L. *fabrique* < *faber, fabrica,* something skilfully produced. 직물, 옷감, 짜는법, 구조, 건축물.

fat[fæt]: Old English. *faēt,* D. *feist*, Dutch. *vet*, well fed, plump, fatty, oily. 지방(脂肪)의, 기름이 많은, 살찐.

fatty acid: a carboxylic acid with a long unbranched aliphatic tail (chain), which is either saturated or unsaturated. Most naturally occurring fatty acids have a chain of four to 28 carbons. Fatty acids are usually derived from triglycerides or phospholipids. When they are not attached to other molecules, they are known as "free" fatty acids. Fatty acids are important sources of fuel because their metabolism yield large quantities of ATP. Many cell types can use either glucose or fatty acids for this purpose. In particular, heart and skeletal muscle prefer fatty acids. The brain cannot use fatty acids as a source of fuel; it relies on glucose or on ketone bodies.

Fig. Butyric acid: a short chain fatty acid

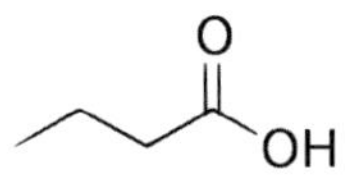

fetus[fi:təs]: L. *fetus*, pregnancy, childbirth, offspring. (임신 9주 이후의) 태아, 배(胚).

flavin(e)[fléivin]: L. *flavin*, *flavus*, yellow + *-in*. 플라빈(동, 식물 조직 중에 분포되어 있는 황색 색소). The common name for a group of organic compounds based on pteridine (a chemical compound composed of fused pyrimidine and pyrazine rings), formed by the tricyclic heteronuclear organic ring isoalloxazine. The biochemical source is the vitamin riboflavin. The flavin moiety is often attached with an adenosine diphosphate to form flavin adenine dinucleotide (FAD), and, in other circumstances, is found as flavin mononucleotide (or FMN), a phosphorylated form of riboflavin. It is in one or the other of these forms that flavin is present as a prosthetic group in flavoproteins. The flavin group is capable of undergoing oxidation-reduction reactions, and can accept either one electron in a two-step process or two electrons at once. Reduction is made with the addition of hydrogen atoms to specific nitrogen atoms on the isoalloxazine ring system.

Fig. Equilibrium between the oxidized (left) and totally reduced (right) forms of flavin.

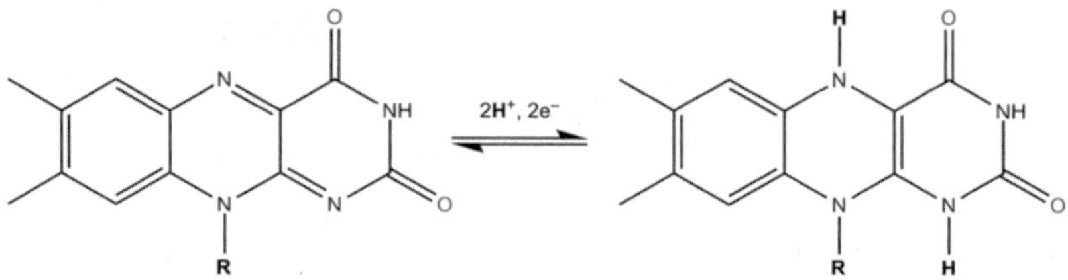

fluorescence[fluərésns, flɔ:r-]: F. *fluoresces,* on the pattern of opalescence. 오팔색으로 빛나는 모양, 형광, 발광, 형광성.

forensic[fərénsik]: L. *forum, fores*, *forensis*, in open court, public. 범죄 과학의, 법의학의, 범죄과학.

formal[fɔ́:rməl]: L. *forma, formalis,* shape, mould. 형식적인, 틀에 박힌, 알맹이가 없는.

formal charge: the charge assigned to an atom in a molecule, assuming that electrons in a chemical bond are shared equally between atoms, regardless of relative electronegativity. The formal charge of any atom in a molecule can be calculated by the following equation:

$$FC = V - N - \frac{B}{2}$$

Where V is the number of valence electrons of the atom in isolation (atom in ground state); N is the number of non-bonding valence electrons on this atom in the molecule; and B is the total number of electrons shared in covalent bonds with other atoms in the molecule.

When determining the correct Lewis structure (or predominant resonance structure) for a molecule, the structure is chosen such that the formal charge (without sign) on each of the atoms is minimized.

Formal charge is a test to determine the efficiency of electron distribution of a molecule. This is significant when drawing structures.

Fig. O_3 and NO_3^-

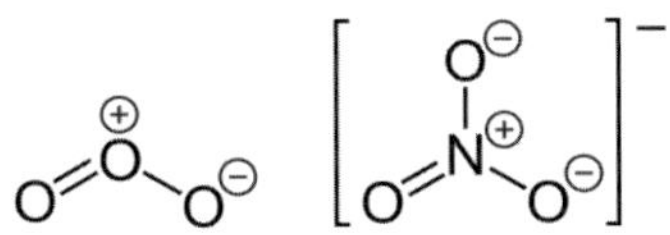

formation[fɔ:rmeiʃən]: L. *formare*, *formation,* to form. 생성, 형성, 조성(造性).

formic[fɔ:mik]: L. *formica*, ant. 개미의, 포름산의.

formic acid: 포름산. (systematically called methanoic acid), The simplest carboxylic acid. Its formula is HCOOH or HCO_2H. It is an important intermediate in chemical synthesis and occurs naturally, most notably in the venom(독액, 독) of bee and ant stings. In fact, its name comes from the Latin word for ant, formica, referring to its early isolation by the distillation of ant bodies. An ester or salt derived from formic acid is referred to as formate.

fossil[fάsəl / fɔ́s-]: F. < L. *fossile* < *fodere, fossilis,* dig, dug up. 화석.

fragment[frǽgmənt]: L. *frangere, fragmentum,* to break. 조각, 파편, 부분, 소량, 남은 것.

fragmentation: a type of chemical dissociation. Fragmentation of a molecule can take place by a process of heterolysis or homolysis. It is a phenomenon observed in mass spectrometry where it is used as a tool to find the structural formula of a molecule, process called structural elucidation. It can occur in the ion source (in-source fragmentation) where it is generally not a desired effect. Ion source conformation is an important criterium in the level of fragmentation observed. Desired fragmentation is made in the collision zone (post-source fragmentation) of a tandem mass spectrometer. It is a part of gas phase ion chemistry and there are different types of mass
fragmentation: collision-induced dissociation (CID),
electron-capture dissociation (ECD),
electron-transfer dissociation (ETD),
electron-detachment dissociation (EDD),
photodissociation, particularly infrared multiphoton dissociation (IRMPD) and blackbody infrared radiative dissociation (BIRD),
surface-induced dissociation (SID),
charge remote fragmentation,
Higher-energy C-trap dissociation (HCD). The certain structures favour fragmentation the α-cleavage and the McLafferty rearrangement are two examples for the often observed fragmentations.

frequency[frí:kwənsi]: L. *frequens*, *frequent*, *frequentia,* crowded, frequent. 진동수, 빈도(수), 주파수, 도수. The number of occurrences of a repeating event per unit time. It is also referred to as temporal frequency. The period is the duration of one cycle in a repeating event, so the period is the reciprocal of the frequency. For periodic waves, frequency has an inverse relationship to the concept of wavelength; simply, frequency is inversely proportional to wavelength λ (lambda). The frequency f is equal to the phase velocity v of the wave divided by the wavelength λ of the wave:

$$f = \frac{v}{\lambda}.$$

In the special case of electromagnetic waves moving through a vacuum, then $v = c$, where c is the speed of light in a vacuum, and this expression becomes:

$$f = \frac{c}{\lambda}.$$

When waves from a monochrome source travel from one medium to another, their frequency remains exactly the same — only their wavelength and speed change.

front[frʌnt]: F. < L. *fronter, front* < *frons, front*, forhead, front. 경계, 국경, 맨 앞부분, 전선(前線).

<u>frontier orbitals</u>: HOMO (highest occupied molecular orbital) and LUMO (lowest unoccupied molecular orbital). The energy difference between the HOMO and LUMO is termed the band gap. This may serve as a measure of the excitability of a molecule: all else being equal, the smaller the energy, the

more easily it will be excited. Roughly, the HOMO level is to organic semiconductors and quantum dots what the valence band is to inorganic semiconductors. The same analogy exists between the LUMO level and the conduction band. The energy difference between the HOMO and LUMO level is regarded as band gap energy.

fructose[frʌ́ktous]: L. *fructus*, fruit + *-ose*, on the pattern of (gluc)ose. 과당(果糖). Fruit sugar, is a simple monosaccharide found in many plants. It is one of the three dietary monosaccharides, along with glucose and galactose, that are absorbed directly into the bloodstream during digestion. Pure, dry fructose is a very sweet, white, odorless, crystalline solid and is the most water-soluble of all the sugars. From plant sources, fructose is found in honey, tree and vine fruits, flowers, berries and most root vegetables. In plants, fructose may be present as the monosaccharide and/or as a component of sucrose. Commercially, fructose is usually derived from sugar cane, sugar beets and corn and there are 3 commercially important forms. Crystalline fructose is the monosaccharide, dried and ground, and of high purity. Sucrose is the third form.

Fig. D-fructose

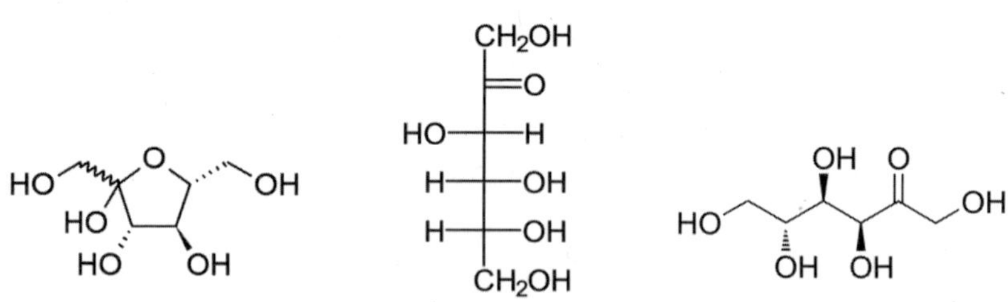

fume[fju:m]: F. < L. *fumer* < *fumare,* to smoke. 가스, 연기, 증기.

fuming sulfuric acid (=Oleum): 발연황산. Refers to a solution of various compositions of sulfur trioxide in sulfuric acid or sometimes more specifically to disulfuric acid (also known as pyrosulfuric acid). Oleums can be described by the formula $y SO_3.H_2O$ where y is the total molar sulfur trioxide content. The value of *y* can be varied, to include different oleums. They can also be described by the formula $H_2SO_4.x SO_3$ where x is now defined as the molar free sulfur trioxide content. Oleum is generally assayed according to the free SO_3 content by weight. It can also be expressed as a percentage of sulfuric acid strength; for oleum concentrations, that would be over 100%. For example, 10% Oleum can also be expressed as $H_2SO_4.0.1 SO_3$, $1.0225 SO_3.H_2O$ or 102.25% Sulfuric Acid. The conversion between % acid and % oleum is: % Acid = 100 + 18/80% oleum. A value for *x* of 1 gives us the empirical formula $H_2S_2O_7$ for disulfuric (pyrosulfuric) acid. Pure disulfuric acid itself is a solid at room temperature, melting at 36°C and rarely used either in the laboratory or industrial processes.

function[fʌ́ŋkʃən]: F. < L. *fonction* < *fungi*, *function,* perform. 기능, 작용.

functional group: 작용기. Specific groups of atoms within molecules that are responsible for the characteristic chemical reactions of those molecules. The same functional group will undergo the same or similar chemical reaction(s) regardless of the size of the molecule it is a part of. However, its relative reactivity can be modified by nearby functional groups.

Fig. Benzyl acetate has an ester functional group (bold), an acetyl moiety and a benzyl alcohol moiety. Other divisions can be made.

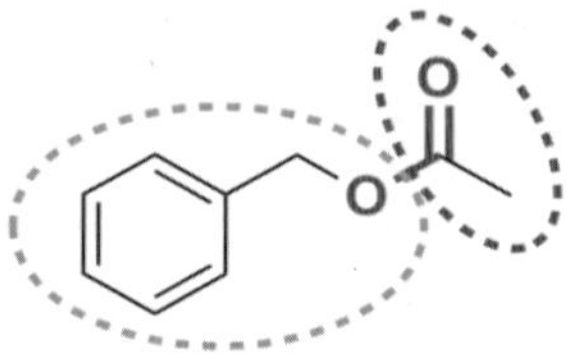

fuse[fju:z]: L. *fundere*, *fus,* poured, melted. - 을 녹이다, 융해(融解)시키다, 녹다, 융해하다. (전기) 퓨즈.

galactose[gəlǽktous]: G. *gala, galaktos*, milk + *-ose*, on the pattern of *(gluc)ose*. 갈락토오스. A type of sugar that is less sweet than glucose. It is considered a nutritive sweetener because it has food energy. It is a C-4 epimer of glucose.

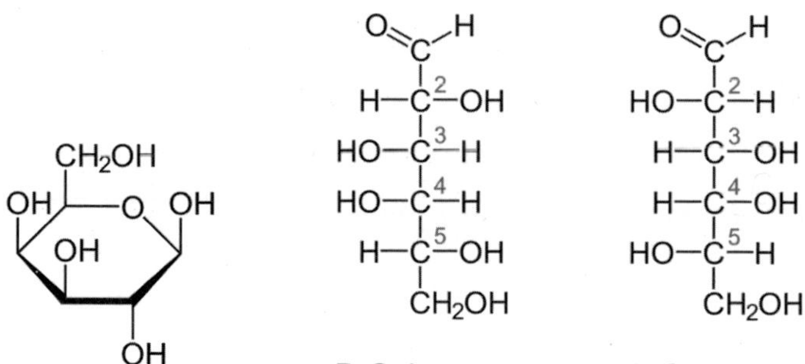

D-Galactose L-Galactose

gasoline[gǽsəli:n]: G. *khaos*, chaos + -ol, L. *oleum*, oil + -ine, L. feminine from *-ina.* Gas. 가솔린, 휘발유. A transparent petroleum-derived liquid that is primarily used as a fuel in internal combustion engines. It consists mostly of organic compounds obtained by the fractional distillation of petroleum, enhanced with a variety of additives. Some gasolines also contain ethanol as an alternative fuel. In North America, the term "gasoline" is often shortened in colloquial usage to "gas", whereas most current or former Commonwealth nations use the term "petrol". Under normal ambient conditions its material state is liquid, unlike liquefied petroleum gas or "natural gas".

gauche[gouʃ]: French. *gauche*, left. 고우시, 뒤틀린, 비대칭(非對稱)의.

gauch conformer: a staggered conformer in which the largest substituents are gauche to each other.

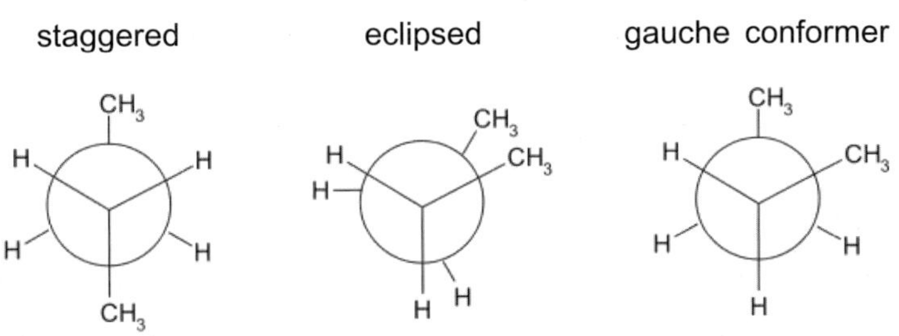

gauch interaction: the interaction between two atoms or groups that are gauch to each other.

gemini[dʒémənài, -ni]: L. pl. of *geminus*, twin. 쌍둥이. Refers to the relationship between two functional groups that are attached to the same atom. The prefix gem is applied to a chemical name to denote this relationship, as in a *gem*-dibromide. The following example shows the conversion of a cyclohexyl methyl ketone to a *gem*-dichloride through a reaction with phosphorus pentachloride. This *gem*-dichloride can then be used to synthesize an alkyne.

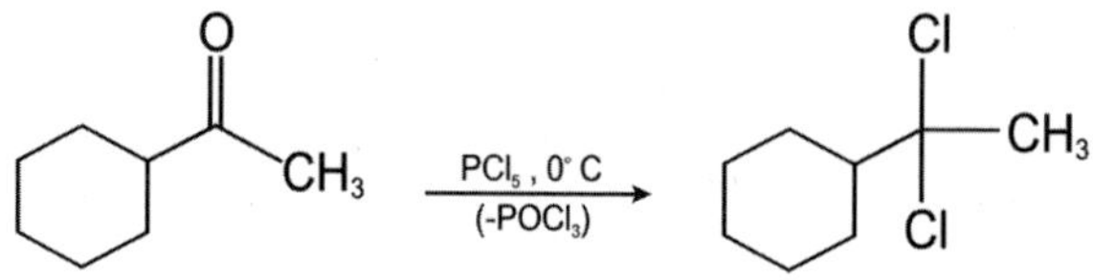

geminal coupling: the mutual splitting of two nonidentical protons bonded to the same carbon in nmr spectroscopy.

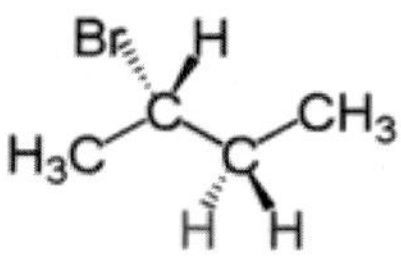

J_{ab} = 12 Hz

gene[dʒíːn]: D. < G. *gen* < *pangen*, *pan*, all + *genos*, race, kind, offspring. 유전자, 유전 인자. A unit of heredity in a living organism. It normally resides on a stretch of DNA that codes for a type of protein or for an RNA chain that has a function in the organism. All living things depend on genes, as they specify all proteins and functional RNA chains. Genes hold the information to build and maintain an organism's cells and pass genetic traits to offspring, although some organelles (e.g. mitochondria) are self-replicating and are not coded for by the organism's DNA.

gene theraphy: a technique that inserts a synthetic gene into the DNA of an organism that is defective in that gene.

genetic engineering: recombinant DNA technology.

generic[dʒinérik]: F. < L. *générique* < *genus, generic*, stock, race. 속(屬)의, (상표명이 아닌) 일반명(~name), 통칭적인.

genome[dʒí:noum]: 1930s. gene, D. < G. *gen* < *pangen*, *pan*, all + *genos*, race, kind, offspring + chromosome, G. *khrōma*, colour + *sōma*, body. 게놈. The entirety of an organism's hereditary information. It is encoded either in DNA or, for many types of virus, in RNA. The genome includes both the genes and the non-coding sequences of the DNA/RNA.

geodesic[dʒi:ədésik, -di:s-]: G. > L. *geodaesia* < *geōdaisia*, *gē*, earth + *daiein*, to divide, + -ic, G. > L. *-icus* < *-ikos*. geodesy. 측지(測地)(학)의, 측지선.

geometric[dʒi:əmétrik]: L. < G. *geometricus* < *geōmetrēs, geōmetrikos*, *gē*, earth + *metres*, measure. 기하학(상)의.

geometrical isomer: a form of stereoisomerism describing the orientation of functional groups within a molecule. In general, such isomers contain double bonds, which cannot rotate, but they can also arise from ring structures, wherein the rotation of bonds is greatly restricted. *Cis* and *trans* isomers occur both in organic molecules and in inorganic coordination complexes. The term "geometric isomerism" is considered an obsolete synonym of "*cis-trans (or E,Z)* isomerism" by IUPAC. It is sometimes used as a synonym for general stereoisomerism; the correct term for non-optical stereoisomerism is diastereomerism.

cis-2-pentene

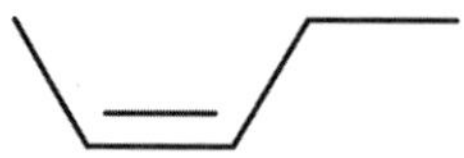

trans-2-pentene

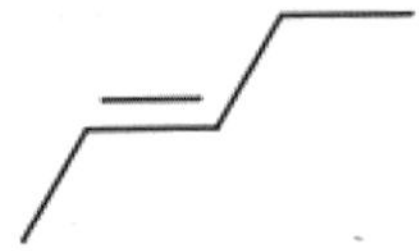

geometry[ʤiámətri, -ɔm-]: L. < G. *geometria, geometriae* < *geōmetria*, *gē,* earth + *metria,* measure. 기하학(幾何學), 평면(입체) 도형, 기하학적 배열, 구조, 조합.

glaucoma[glɔ:kóumə, glau-]: G. *glaukos*, *glaukōma,* bluish-green, bluish-grey (because of the grey-green haze in the pupil). 녹내장(綠內障). An eye disorder in which the optic nerve suffers damage, permanently damaging vision in the affected eye(s) and progressing to complete blindness if untreated. It is often, but not always, associated with increased pressure of the fluid in the eye (aqueous humour, 수액). The term 'ocular hypertension' is used for cases having constantly raised intraocular pressure (IOP, 안구압력) without any associated optic nerve damage. Conversely, the term 'normal' or 'low tension glaucoma' is suggested for the typical visual field defects when associated with a normal or low IOP.

glucose[glú:kous]: G. *glukus,* sweet + *-osus*, sugar. sweet sugar. 글루코오스, 포도당, 물엿. Also known as D-glucose, dextrose, or grape sugar. A monosaccharide (or simple sugar) also known as grape sugar, blood sugar, or corn sugar, is a very important carbohydrate in biology. The living cell uses it as a source of energy and metabolic intermediate. Glucose is one of the main products of photosynthesis and starts cellular respiration in both procaryotes (bacteria) and eucaryotes (animals, plants, fungi, and protists). Glucose exists in several different molecular structures, but all of these structures can be divided into two families of mirror-images (stereoisomers). Only one set of these isomers exists in nature, those derived from the "right-handed form" of glucose, denoted D-glucose. D-glucose is often referred to as dextrose. The term dextrose is derived from dextrorotatory glucose. Solutions of dextrose rotate polarized light to the right. Starch and cellulose are polymers derived from the dehydration of D-glucose. The other stereoisomer, called L-glucose, is hardly ever found in nature. The mirror-image of the molecule, L-glucose, cannot be metabolized by cells in the biochemical process known as glycolysis.

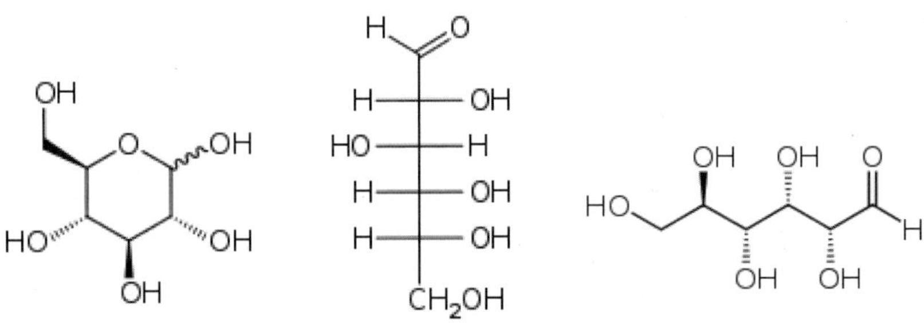

glutamine[glú:təmi:n, -min]: 19th cent. blend of glutamic acid + amine. 글루타민. One of the 20 amino acids encoded by the standard genetic code. It is not recognized as an essential amino acid but may become conditionally essential in certain situations, including intensive athletic training or certain gastrointestinal disorders. Its side-chain is an amide formed by replacing the side-chain hydroxyl of glutamic acid with an amine functional group. Therefore, it can be considered the amide of glutamic acid.

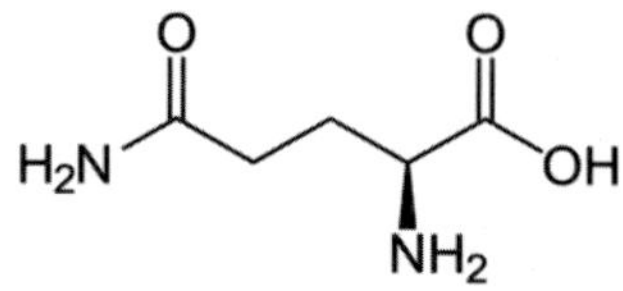

glycerin[glísərin]: glycerol or glycerine. F. < G. *glycerine* < *glukeros*, sweet. 글리세린. An organic compound. It is a colourless, odourless, viscous liquid that is widely used in pharmaceutical formulations. Glycerol has three hydrophilic hydroxyl groups that are responsible for its solubility in water and its hygroscopic nature. The glycerol substructure is a central component of many lipids. Glycerol is sweet-tasting and of low toxicity.

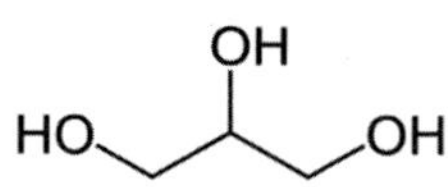

glycine[gláisi:n]: mid 19th cent. G. *glukeros*, sweet, + -ine. 글리신, 아미노아세트산(단맛이 있는 무색 결정). Abbreviated as **Gly** or **G,** glyconol or aminoacetic acid. An organic compound with the formula NH_2CH_2COOH. Glycine is the smallest of the 20 amino acids commonly found in proteins.

glycol[gláikɔ:l, -kɑl / -kɔl]: mid 19th cent. glycerine + -ol (originally intended to designate a substance intermediate between glycerine and alcohol). ethylene glycol. 글리콜(자동차의 부동액 · 용재). $HOCH_2CH_2OH$.

glycolysis: the series of reactions that converts D-glucose into two molecules of pyruvate. 해당(解糖)작용, 당분해.

glycoprotein: glucoprotein, glycopeptide. A protein that is covalently bonded to a polysaccharide. 당단백질.

glycoside[gláikəside]: the acetal of sugar. 배당체(配當體).

glycosidic bond: a type of covalent bond that joins a carbohydrate (sugar) molecule to another group, which may or may not be another carbohydrate. A glycosidic bond is formed between the hemiacetal group of a saccharide (or a molecule derived from a saccharide) and the hydroxyl group of some organic compound such as an alcohol.

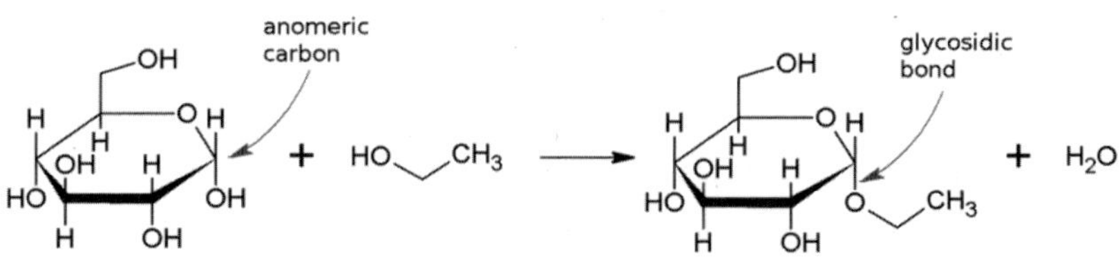

gout[gaut]: F. < L. *goute < gutta*, drop (because gout was believed to be caused by the dropping of diseased matter from the blood into the joints). 통풍(痛風).

graft[grǽft / grɑ́:ft]: F. < G. *grafe < graphion,* stylus, writing implement (with reference to the tapered tip of the scion). 융합, 합체(合體).

graft copolymers: a special type of branched copolymer in which the side chains are structurally distinct from the main chain. The illustration depicts a special case where the main chain and side chains are composed of distinct homopolymers. However, the individual chains of a graft copolymer may be homopolymers or copolymers. Note that different copolymer sequencing is sufficient to define a structural difference, thus an A-B diblock copolymer with A-B alternating copolymer side chains is properly called a graft copolymer. For example, suppose we perform a free-radical polymerization of styrene in the presence of polybutadiene, a synthetic rubber, which retains one reactive C=C double bond per residue. We get polystyrene chains growing out in either direction from some of the places where there were double bonds, with a one-carbon rearrangement. Or to look at it the other way around, the result is a polystyrene backbone with polybutadiene chains growing out of it in both directions. This is an interesting copolymer variant in that one of the ingredients was a polymer to begin with.

graphite[grǽfait]: D. < G. *graphit < graphein,* write (because of its use as pencil lead), to draw, to scratch *+ ite,* graphite. 흑연, 탄소의 결정형 동소체의 하나. Graphite has a layered, planar structure. In each layer, the carbon atoms are arranged in a hexagonal lattice with separation of 0.142 nm, and the distance between planes is 0.335 nm. The two known forms of graphite, *alpha* (hexagonal) and *beta* [rhombohedral, 사방(斜方) 육면체], have very similar physical properties (except that the graphene layers stack slightly differently). The hexagonal graphite may be either flat or buckled. The alpha form can be converted to the beta form through mechanical treatment and the beta form reverts to the alpha form when it is heated above 1300°C. The layering contributes to its lower density.

Fig. Ball-and-stick model of graphite (2 graphene layers)

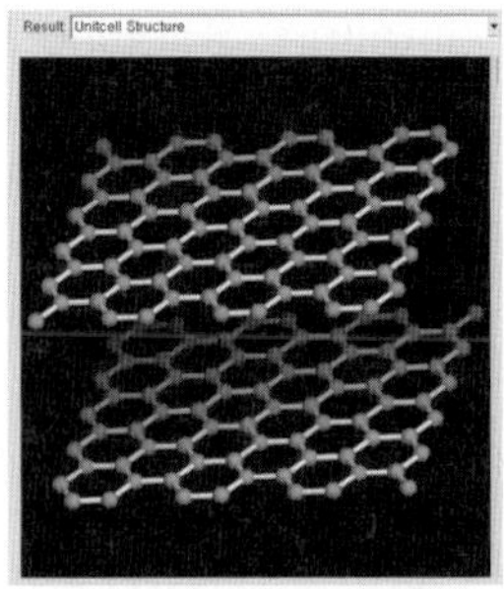

gypsum[dʒípsəm]: G. *gupsos, gypsos,* chalk, plaster. 석고, 깁스. A very soft sulfate mineral composed of calcium sulfate dihydrate, with the chemical formula $CaSO_4 \cdot 2H_2O$. It is found in alabaster(설화석고), a decorative stone used in Ancient Egypt. It is the second softest mineral on the Mohs Hardness Scale. It forms as an evaporite mineral and as a hydration product of anhydrite.

halogen[hǽlədʒən / -dʒen]: G. *hals, halo-*, salt + *-gen*. 할로겐. A series of nonmetal elements from Group 17 IUPAC Style (formerly: VII, VIIA) of the periodic table, comprising fluorine (F), chlorine (Cl), bromine (Br), iodine (I), and astatine (At). The artificially created element 117, provisionally referred to by the systematic name ununseptium, may also be a halogen. The group of halogens is the only periodic table group which contains elements in all three familiar states of matter at standard temperature and pressure.

<u>halogenation</u>: a chemical reaction that incorporates a halogen atom into a molecule. More specific descriptions exist that specify the type of halogen: fluorination, chlorination, bromination, and iodination. Specific halogenation methods are the Hunsdiecker reaction (from carboxylic acids) and the Sandmeyer reaction (arylhalides).

<u>Hunsdiecker reaction</u>:

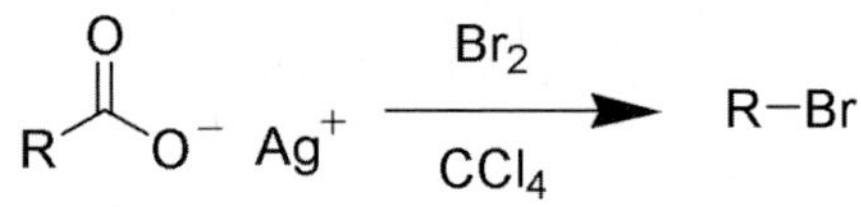

<u>Sandmeyer reaction:</u>

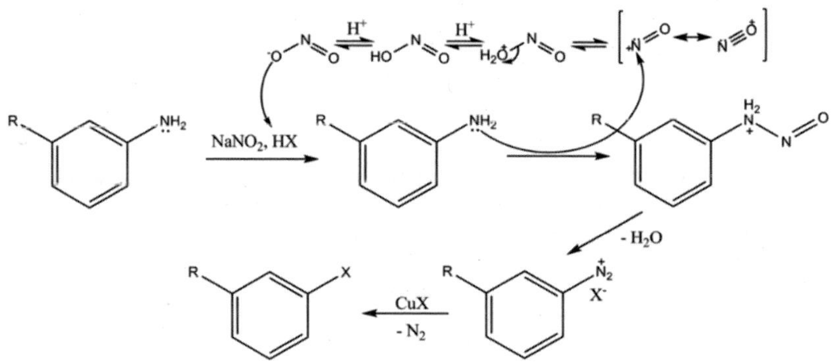

halothane[hæləθeìn]: 1950s; blend of halogen and ethane. 할로테인(흡입식 전신 마취제). Trademarked as Fluothane, is an inhalational general anesthetic. Its IUPAC name is 2-bromo-2-chloro-1,1,1-trifluoroethane. It is the only inhalational anesthetic agent containing a bromine atom; there are several other halogenated anesthesia agents which lack the bromine atom and do contain the fluorine and chlorine atoms present in halothane. It is colorless and pleasant-smelling, but unstable in light. It is packaged in dark-colored bottles and contains 0.01% thymol as a stabilizing agent. Halothane is a *core* medicine in the World Health Organization's "Essential Drugs List", which is a list of minimum medical needs for a basic health care system.

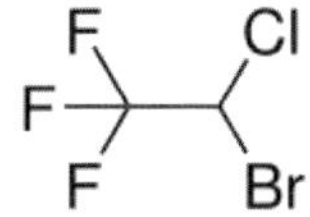

helium[hí:liəm]: G. *hēlios,* sun. **He**. 헬륨. Its existence was inferred from an emission line in the sun's spectrum. The chemical element with atomic number 2 and an atomic weight of 4.002602. It is a colorless, odorless, tasteless, non-toxic, inert, monatomic gas that heads the noble gas group in the periodic table. Its boiling and melting points are the lowest among the elements and it exists only as a gas except in extreme conditions. Next to hydrogen, it is the second most abundant element in the universe, and accounts for 24% of the elemental mass of our galaxy.

helix[hí:liks]: G. *helix*, twisted, curved, spiral ornament. 나(사)선. (pl: helixes or helices): A type of smoothspace curve, i.e. a curve in three-dimensional space. It has the property that the tangent line at any point makes a constant angle with a fixed line called the axis. Examples of helixes are coil springs and the handrails of spiral staircases. A "filled-in" helix – for example, a spiral ramp – is called a helicoid(나선상의). Helices are important in biology, as the DNA molecule is formed as two intertwined helices, and many proteins have helical substructures, known as alpha helices.

hemi-[hémi]: G. > L. *hēmi* > *semi,* half. 반, 2분의 1, 30분.

<u>hemiacetals and hemiketals</u>: compounds of the general formula $R_1R'_1C(OH)OR_2$, where R_2 is not hydrogen. These types of compounds are formed from carbonyl compounds and alcohols, i.e. hemiacetals from aldehydes, hemiketals from ketones. Thus, for hemiacetals, either R1 or R'1 must be a hydrogen, while for hemiketals, neither will be.

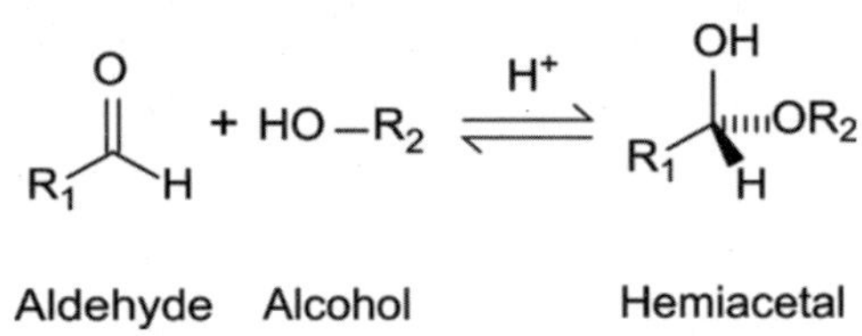

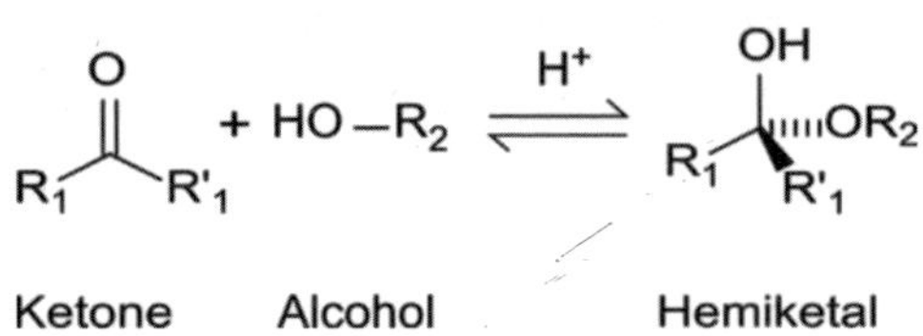

hemoglobin[hi:məglóubin, hem-]: G. *haima*, blood + *globin* a kind of protein. 헤모글로빈, 혈색소. A related family of proteins. Also rendered as haemoglobin and abbreviated **Hb** or **Hgb**, is the iron-containing oxygen-transport metalloprotein in the red blood cells of all vertebrates, with the exception of the fish family Channichthyidae (white-blooded fish), as well as the tissues of some invertebrates. Hemoglobin in the blood carries oxygen from the respiratory organs (lungs or gills) to the rest of the body (i.e., the tissues) where it releases the oxygen to burn nutrients to provide energy to power the functions of the organism, and collects the resultant carbon dioxide to bring it back to the respiratory organs to be dispensed from the organism.

heparin[hépərin]: G. *hēpar,* liver + *-in*. 헤파린(간장, 폐에서 혈액의 응고를 막는 물질). A highly-sulfated glycosaminoglycan, is widely used as an injectable anticoagulant, and has the highest negative charge density of any known biological molecule.

hepatitis[hepətáitis]: G. *hēpar*, liver + *-itis, ites,* inflammation, disease. 간염(肝炎). A medical condition defined by the inflammation of the liver and characterized by the presence of inflammatory cells in the tissue of the organ. The condition can be self-limiting (healing on its own) or can progress to fibrosis (섬유조직의 과다 증식, scarring) and cirrhosis(간경변). Hepatitis may occur with limited or no symptoms, but often leads to jaundice(황달), anorexia (poor appetite) and malaise. Hepatitis is acute when it lasts less than six months and chronic when it persists longer. A group of viruses known as the hepatitis viruses cause most cases of hepatitis worldwide, but it can also be due to toxins (notably alcohol, certain medications, some industrial organic solvents and plants), other infections and autoimmune diseases.

heroin[hérouin]: L. *heros,* hero (because of its effects on the user's self-esteem). 헤로인(진통제; 중독성 마약). Diacetylmorphine, also known as diamorphine, or, especially in older literature, as morphine diacetate, is a semi-synthetic opioid drug synthesized from morphine, a derivative of the opium poppy(양귀비). It is the 3,6-diacetyl ester of morphine (*di-acetyl-morphine*) and a morphine prodrug. The white crystalline form is commonly the hydrochloride salt diacetylmorphine hydrochloride. As with other opioids, diacetylmorphine is used as both an analgesic and a recreational drug. Frequent and regular administration is associated with tolerance and physical dependence, which may develop into addiction.

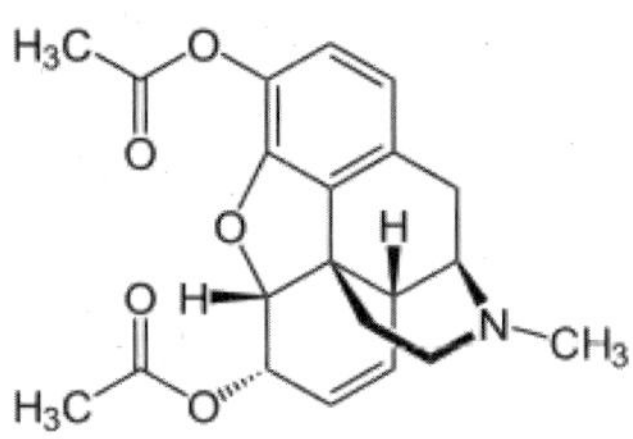

hetero-[hétərou]: G. *heteros, hetera, heteron,* another, different (of different kinds). 탄소 이외의 원자의.

<u>heteroatom</u>: in the nomenclature of organic chemistry, a heteroatom is any atom that is not carbon or hydrogen. It is typically, but not exclusively, nitrogen, oxygen, sulfur, phosphorus, boron, chlorine, bromine, or iodine.

<u>heterogeneous</u>: 불균일한. Describe an object or system consisting of multiple items having a large number of structural variations. It is the opposite of homogeneous.

heterolysis (=heterolytic fission): chemical bond cleavage of a neutral molecule generating a cation and an anion. In this process the two electrons that make up the bond are assigned to the same fragment. The more electronegative fragment receives both electrons.

$$A : B \rightarrow A^{+} + B^{-}$$

The energy involved in this process is called heterolytic bond dissociation energy. Bond cleavage is also possible by a process called homolysis. In heterolysis additional energy is required to separate the ion pair. An ionising solvent helps reduce this energy.

histamine[hístəmi:n, -min]: early 20th cent. Histidine + amine. 히스타민(자궁수축, 혈관 확장, 혈압 강하 작용을 한다). An organic nitrogen compound involved in local immune responses as well as regulating physiological function in the gut and acting as a neurotransmitter. Histamine triggers the inflammatory response.

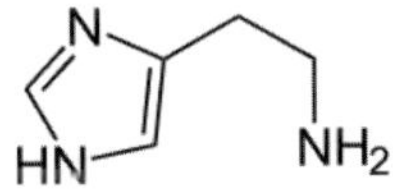

histidin(e)[hístədi:n, -din]: G. *histos*, web, tissue + -ide + -ine. 히스티딘. Abbreviated as **His or H**, an essential amino acid, has a positively charged imidazole functional group. It is the one of the 22 proteinogenic amino acids. Histidine is an essential amino acid in humans and other mammals. It was initially thought that it was only essential for infants, but longer-term studies established that it is also essential for adult humans.

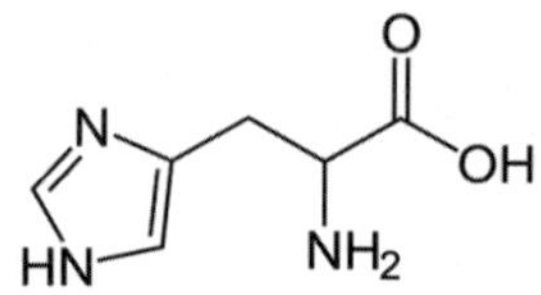

homo-[hóumou]: G. *homos, home, homon,* of the same, common. 동일 원자의.

<u>homogeneous</u>: 균일한. Perfectly uniform in their composition, as opposed to heterogeneous. A homogeneous mixture does not scatter light and its particles can not be separated by filtration. Its particle size is about 10^{-9}cm.

homolog[hóuməlɔ:g, -lɑg, hɑ́m-]: G. *homologos*, *homos*, same + *-logo*, ratio, portion, agreeing consistent. 동족체. A chemical compound from a series of compounds that differ only in the number of repeated structural units. A homolog is a special case of an analog. Examples are alkanes and compounds with alkyl functional groups that differ in the length of their side chain (the repeating unit being methylene group $-CH_2-$) or a peptide with one amino acid replaced by another.

<u>homolysis (homolytic fission)</u>: chemical bond dissociation of a neutral molecule generating two free radicals. That is, two electrons that are involved in the bond are distributed one by one to the two species.

$$A{-}B \rightarrow A\bullet + B\bullet$$

Such reactions can be induced by irradiation in the UV region, diffused sunlight or peroxide. The energy involved in this process is called bond dissociation energy.

homotopic groups: equivalent groups in a chemical compound. Two groups A and B are homotopic if the molecule remains the same (including stereochemically) when the groups are interchanged with the remaining parts of the molecule fixed. Homotopic atoms are always identical, in any environment. Homotopic NMR-active nuclei have the same chemical shift in an NMR spectrum. For example, the four hydrogen atoms of methane (CH_4) are homotopic with one another, as are the two hydrogens or the two chlorines in dichloromethane (CH_2Cl_2).

hormone[hɔ́:rmoun]: G. *horman*, *hormōn*, to stimulate, to excite, impel, set in motion. 호르몬. Chemical released by a cell in one part of the body, that sends out messages that affect cells in other parts of the organism. Only a small amount of hormone is required to alter cell metabolism. It is essentially a chemical messenger that transports a signal from one cell to another. All multicellular organisms produce hormones; plant hormones are also called phytohormones. Hormones in animals are often transported in the blood. Cells respond to a hormone when they express a specific receptor for that hormone. The hormone binds to the receptor protein, resulting in the activation of a signal transduction mechanism that ultimately leads to cell type-specific responses.

Fig. Epinephrine (adrenaline), a catecholamine-type hormone

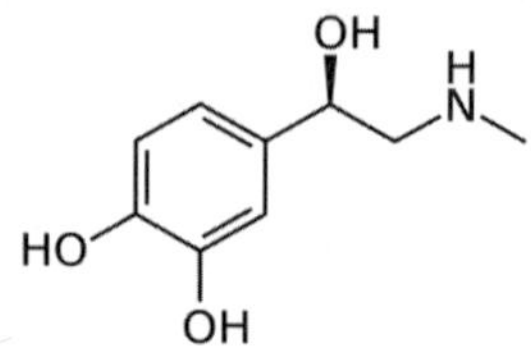

human[*h*jú:mən, jú:- /hjú:]: F. < L. *humain(e)* < *homo*, *humanus*, man, human being. 인간의, 인간다운, 인간성의, 인간.

human genome: the total DNA of a human cell, the genome of Homo sapiens, which is stored on 23 chromosome pairs. 22 of these are autosomal(성 염색체 이외의 상(常) 염색체) chromosome pairs, while the remaining pair is sex-determining. The haploid human genome occupies a total of just over 3 billion DNA base pairs.

hybrid[háibrid]: L. *hybrida*, offspring of a tame sow (암돼지) and wild boar (멧돼지), child of a freeman and slave, etc. 혼성.

hybridisation (or hybridization): the concept of mixing atomic orbitals to form new hybrid orbitals suitable for the qualitative description of atomic bonding properties. Hybridised orbitals are very useful in the explanation of the shape of molecular orbitals for molecules. It is an integral part of valence bond theory. Although sometimes taught together with the valence shell electron-pair repulsion (VSEPR) theory, valence bond and hybridization are in fact not related to the VSEPR model.

sp^3 hybridization

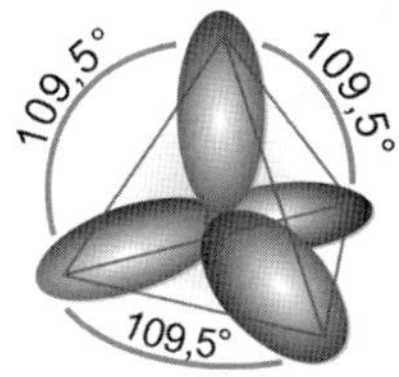

sp^2 hybridization

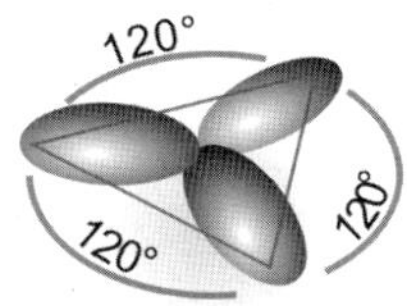

hydrate[háidreit]: G. *hudōr*, water. 수화물(水化物).

hydration: addition of water to a compound.

hydrazine: late 19th cent.: *hydrogen* + *azo-* + *-ine*. An inorganic compound with the formula N_2H_4. It is a colourless flammable liquid with an ammonia-like odor. Hydrazine is highly toxic and dangerously unstable unless handled in solution. Approximately 260,000 tons are manufactured annually. Hydrazine is mainly used as a foaming agent in preparing polymer foams, but significant applications also include its uses as a precursor to polymerization catalysts and pharmaceuticals. Additionally, hydrazine is used in various rocket fuels and to prepare the gas precursors used in air bags. Hydrazine is used within both nuclear and conventional electrical power plant steam cycles to control concentrations of dissolved oxygen in an effort to reduce corrosion.

hydride: a binary compound of hydrogen with a metal. 수소화물

1,2-hydride shift: a rearrangement involving a hydrogen atom, or the movement of a hydride ion from one carbon to an adjacent carbon.

hydride ion: a negatively charged hydrogen.

hydr(o)-: 모음 앞에서는 hyd-, G. *hydor, hydoratos*, water.

hydrocarbon: a compound that contains only carbon and hydrogen, such as any of those which are the chief components of petroleum and natural gas.

hydrophilic: 친수성(親水性)의.

hydrophobic: 소수성(疏水性)의.

hydrous: 물을 함유하는.

hydroxide: 수산화물(水酸化物).

hydroxide ion: OH^-, 수산화 이온

hydrogen[háidrədʒən]: G. *hudro-*, water + *-genes*, -gen. **H**. 수소. The chemical element with atomic number 1. With an average atomic weight of 1.00794 u (1.007825 u for hydrogen-1), hydrogen is the lightest and most abundant chemical element, constituting roughly 75% of the Universe's chemical elemental mass. Non-remnant stars are mainly composed of hydrogen in its plasma state. Naturally occurring elemental hydrogen is relatively rare on Earth. The most common isotope of hydrogen is protium (name rarely used, symbol ^1H) with a single proton and no neutrons. In ionic compounds it can take a negative charge (an anion known as a hydride and written as H^-), or as a positively charged species H^+. The latter cation is written as though composed of a bare proton, but in reality, hydrogen cations in ionic compounds always occur as more complex species. Hydrogen forms compounds with most elements and is present in water and most organic compounds. It plays a particularly important role in acid-base chemistry with many reactions exchanging protons between soluble molecules. As the simplest atom known, the hydrogen atom has been of theoretical use. For example, as the only neutral atom with an analytic solution to the Schrödinger equation, the study of the energetics and bonding of the hydrogen atom played a key role in the development of quantum mechanics. Hydrogen gas (now known to be H_2) was first artificially produced in the early 16th century, via the mixing of metals with strong acids. In 1766–81, Henry Cavendish was the first to recognize that hydrogen gas was a discrete substance, and that it produces water when burned, a property which later gave it its name, which in Greek means "water-former." At standard temperature and pressure, hydrogen is a colorless, odorless, nonmetallic, tasteless, non-toxic, highly combustible diatomic gas with the molecular formula H_2. Industrial production is mainly from the steam reforming of natural gas, and less often from more energy-intensive hydrogen production methods like the electrolysis of water. Most hydrogen is employed near its production site, with the two largest uses being fossil fuel processing (e.g., hydrocracking) and ammonia production, mostly for the fertilizer market. Hydrogen is a concern in metallurgy(야금) as it can embrittle(부서지기 쉽게, 무르게하다) many metals, complicating the design of pipelines and storage tanks.

hydrogenation: 수소화 (반응), 수소 첨가 (반응). To treat with hydrogen, also a form of chemical reduction, is a chemical reaction between molecular hydrogen (H_2) and another compound or element, usually in the presence of a catalyst. The process is commonly employed to reduce or saturate organic compounds. Hydrogenation typically constitutes the addition of pairs of hydrogen atoms to a molecule, generally an alkene. An illustrative example of a hydrogenation reaction is the addition of hydrogen to maleic acid to form succinic acid. Numerous important applications of this petrochemical are found in pharmaceutical and food industries. Hydrogenation of unsaturated fats produces saturated fats and, in some cases, trans fats.

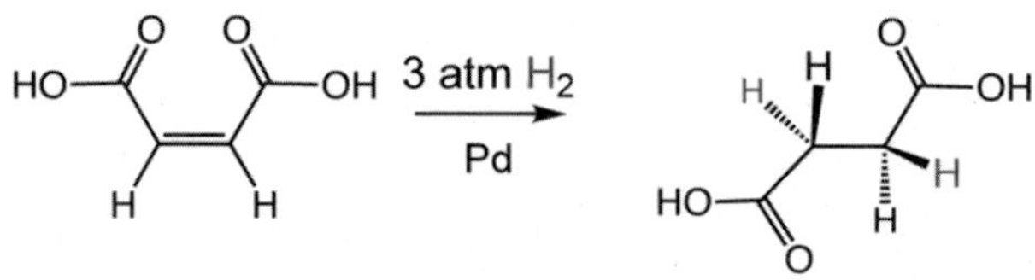

hyperconjugation[haipərkɑndʒugéiʃən / kɔn-]: 1940s. *hyper-*, abbreviation of *hyperactive*, extremly active + *conjugare*, *conjugation*, join together. The interaction of the electrons in a sigma bond (usually C–H or C–C) with an adjacent empty (or partially filled) non-bonding p-orbital or antibonding π orbital or filled π orbital, to give an extended molecular orbital that increases the stability of the system. Only electrons in bonds that are β to the positively charged carbon can stabilize a carbocation by hyperconjugation.

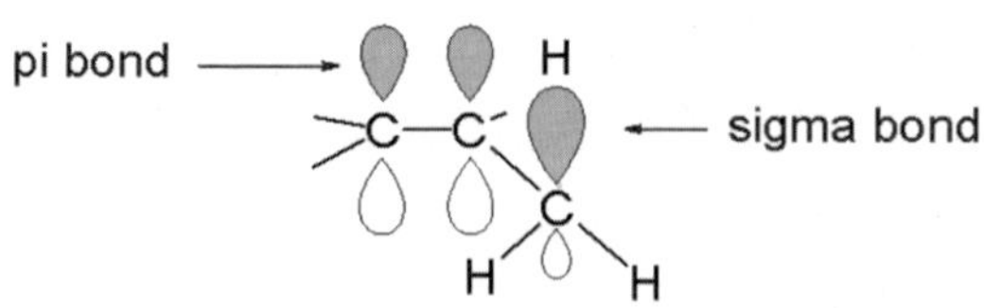

hypertension[haipərténʃən]: G. *hyper*, over, extremly active + L. *tendere*, *tension,* stretch. 과도한 긴장, 고혈압(증), 긴장 항진(증). High blood pressure, or arterial hypertension: a chronic medical condition in which the blood pressure in the arteries is elevated. This requires the heart to work harder than normal to circulate blood through the blood vessels. Blood pressure involves two measurements, systolic (심장 수축의) and diastolic (심장 이완의), which depend on whether the heart muscle is contracting (systole) or relaxed between beats (diastole). Normal blood pressure at rest is within the range of 100-140mmHg systolic (top reading) and 60-90mmHg diastolic (bottom reading). High blood pressure is said to be present if it is persistently at or above 140/90 mmHg. Hypertension is classified as either primary (essential) hypertension or secondary hypertension; about 90–95% of cases are categorized as "primary hypertension" which means high blood pressure with no obvious underlying medical cause. The remaining 5–10% of cases (secondary hypertension) are caused by other conditions that affect the kidneys, arteries, heart or endocrine system. Hypertension is a major risk factor for stroke, myocardial infarction (heart attacks), heart failure, aneurysms of the arteries (e.g. aortic aneurysm), peripheral arterial disease and is a cause of chronic kidney disease. Even moderate elevation of arterial blood pressure is associated with a shortened life expectancy. Dietary and lifestyle changes can improve blood pressure control and decrease the risk of associated health complications, although drug treatment is often necessary in people for whom lifestyle changes prove ineffective or insufficient.

imide[imaid, imid]: French. *arbitary alteration of amide*, *ammonia + -ide*. 이미드. A functional group consisting of two carbonyl groups bound to nitrogen. Being highly polar, imides exhibit good solubility in polar media. The N-H center for imides derived from ammonia is acidic and can participate in hydrogen bonding. Unlike the structurally related acid anhydrides, they resist hydrolysis and some can even be recrystallized from boiling water.

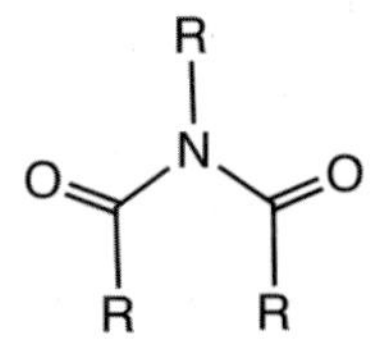

in vivo[invíːvou]: L. in a living things, within the living. 생체 조건 내에서의. Experimentation using a whole, living organism as opposed to a partial or dead organism, or an in vitro ("within the glass", i.e., in a test tube or petri dish) controlled environment. Animal testing and clinical trials are two forms of *in vivo* research. *In vivo* testing is often employed over *in vitro* because it is better suited for observing the overall effects of an experiment on a living subject.

inclusion[inklúːʒən]: L. *includere*, *inclusion,* shut in. 포함, 함유, 함유물.

inclusion compound: a complex in which one chemical compound (the "host") forms a cavity in which molecules of a second "guest" compound are located. The definition of inclusion compounds is very broad, extending to channels formed between molecules in a crystal lattice in which guest molecules can fit. If the spaces in the host lattice are enclosed on all sides so that the guest species is 'trapped' as in a cage, the compound is known as a clathrate. In molecular encapsulation a guest molecule is actually trapped inside another molecule.

Fig. Example of a inclusion complex consisting of a p-xylylenediammonium bound within a cucurbituril.

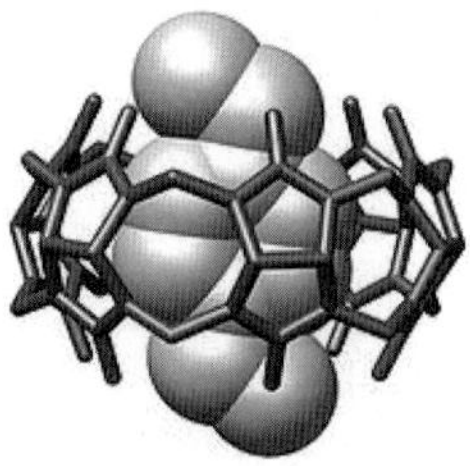

indigo[índigou]: Portuguese. < G. *indigo* < *indikos*, *indikon,* indian (dye). 인디고(암청색의 물감); (진한) 남색. A color named after the purple dye derived from the plant Indigofera tinctoria and related species. The color is placed on the electromagnetic spectrum between about 420 and 450 nm in wavelength, placing it between blue and violet. Although traditionally considered one of seven divisions of the rainbow or the optical spectrum, modern color scientists do not usually recognize indigo as a separate division and generally classify wavelengths shorter than about 450 nm as violet.

indole[índoul]: Portuguese. < G. *indigo < indikos*, *indikon,* indian + L. *oleum*, oil. indigo oil. 인돌(저온도에서 녹는 무색 결정; 향료·시약용). An aromatic heterocyclic organic compound. It has a bicyclic structure, consisting of a six-membered benzene ring fused to a five-membered nitrogen-containing pyrrole ring. Indole is a popular component of fragrances and the precursor to many pharmaceuticals. Compounds that contain an indole ring are called indoles. The indolic amino acid tryptophan is the precursor of the neurotransmitter serotonin.

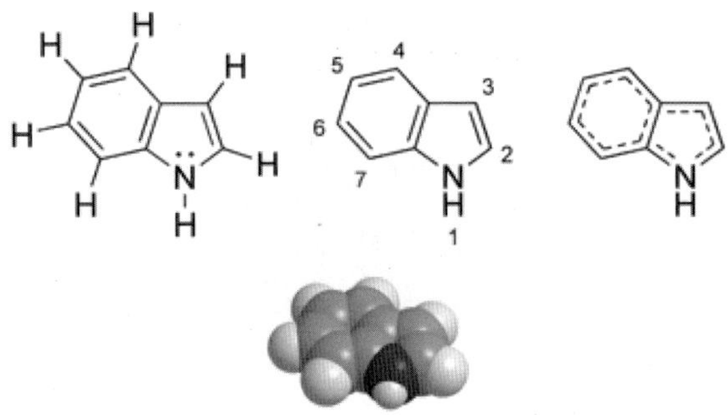

induce[ind*j*ú:s / -djú:s]: F. < L. *enduire* < *inducere*, *in-*, into + *ducere*, to lead, to lead. 유도하다.

induced-dipole: the London dispersion force otherwise known as quantum induced instantaneous polarization (one of the three types of van der Waals forces) is caused by instantaneous changes in the dipole of atoms, caused by the location of the electrons in the atoms' orbitals.

inflammation[infləméiʃən]: L. *inflammare*, *in-*, into, + *flamma*, flame. 점화(點火), 인화, 연소, 염증.

inflection[inflékʃən]: L. *inflectere*, *inflexion*, bend in, curve. 변곡(變曲), 굴곡, 각(角).

inflection (inflexion) point: 변곡점. On a curve at which the curvature changes sign. The curve changes from being concave upwards (positive curvature) to concave downwards (negative curvature), or vice versa.

Fig. Plot of $y = x^3$ with inflection point of (0,0), also a saddle point.

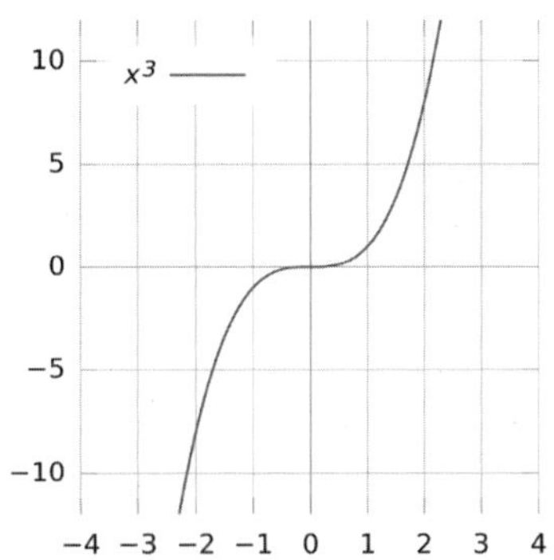

influenza[influénzə]: Italian. < L. *influence* < *influentia*, an outbreak of an epidemic. 인플루엔자, 유행성 감기(독감). Commonly referred to as the flu, is an infectious disease caused by RNA viruses of the family Orthomyxoviridae (the influenza viruses), that affects birds and mammals. The most common symptoms of the disease are chills, fever, sore throat, muscle pains, severe headache, coughing, weakness/fatigue and general discomfort.

infra-[ínfrə]: L. below, beneath.

infrared[ínfrəred]: L. *infra-*, below, beneath + *red*, *rufus, ruber*, red. 적외선(赤外線).

<u>infrared light</u>: electromagnetic radiation with a wavelength between 0.7 and 300 micrometres, which equates to a frequency range between approximately 1 and 430 THz (1 THz = 10^{12} Hz). IR wavelengths are longer than that of visible light, but shorter than that of terahertz radiation microwaves. Bright sunlight provides an irradiance of just over 1 kilowatt per square meter at sea level. Of this energy, 527 watts is infrared radiation, 445 watts is visible light, and 32 watts is ultraviolet radiation.

inhibit[inhíbit]: L. *inhibere*, *in-*, in + *habere*, hold, hinder. (화학반응을) 억제하다.

<u>inhibitor</u>: 억제제

corrosion inhibitor: a substance that decreases the rate of metal oxidation.

enzyme inhibitor: a substance that binds to an enzyme and decreases the enzyme's activity.

reaction inhibitor: a substance that prevents or decreases the rate of a chemical reaction.

initiation[iniʃiéiʃən]: L. *initium*, *initiare, initiation,* beginning. 개시(開始), 착수, 입문.

<u>initiation reactions</u>: free radicals are produced.

Step 1 of initiation process - sunlight (UV light) provides enough energy to break the diatomic halogen's bond - this is photodissociation.

Step 2 of initiation process - the bond splits equally and each atom gets to keep one electron - this is called homolytic fission. The atom becomes a highly reactive free radical because of its unpaired electron.

initiation step: chain initiation, usually by means of an initiator which starts the chemical process. Typical initiators include any organic compound with a labile group: e.g. azo (-N=N-), disulfide (-S-S-), or peroxide (-O-O-). Two examples are benzoyl peroxide and AIBN.

insomnia[insάmniə/-sɔ́m-]: L. *insomnis,* sleepless, *in*, not + *somnus*, somni,* sleep, not sleep. (특히 만성적인) 불면, 불면증.

*Somnus: (로마 신화) 잠의 신(밤의 여신 Nox의 아들, 죽음의 신 Mors의 형제, 그리스 신화의 Hypnos에 해당).

inter-[íntər]: L. *in-*, into + *terra*, earth. between, among, mutually, during, together, within.

intermolecular: 분자간의, 분자 사이에 있는 (에 일어나는).

intermolecular forces: forces that act between stable molecules or between functional groups of macromolecules.

1. London dispersion forces (Instantaneous dipole/ induced dipole): known as quantum induced instantaneous polarization (one of the three types of van der Waals forces) is caused by instantaneous changes in the dipole of atoms, caused by the location of the electrons in the atoms' orbitals.

2. Dipole-Dipole interactions (Keesom interactions after Willem Hendrik Keesom): caused by permanent dipoles in molecules. When one atom is covalently bonded to another with a significantly different electronegativity, the electronegative atom draws the electrons in the bond nearer to itself, becoming slightly negative. Conversely, the other atom becomes slightly positive. Electrostatic forces are generated between the opposing charges and the molecules align themselves to increase the attraction (reducing potential energy).
3. Hydrogen bonds: a stronger form of dipole-dipole interactions, caused by highly electronegative atoms. They only occur between hydrogen and oxygen, fluorine or nitrogen, and are the strongest intermolecular force. The high electronegativities of F, O and N create highly polar bonds with hydrogen, which leads to strong bonding between hydrogen atoms on one molecule and the lone pairs of F, O or N atoms on adjacent molecules. The high boiling point of water is an effect of the extensive hydrogen bonding between the molecules:

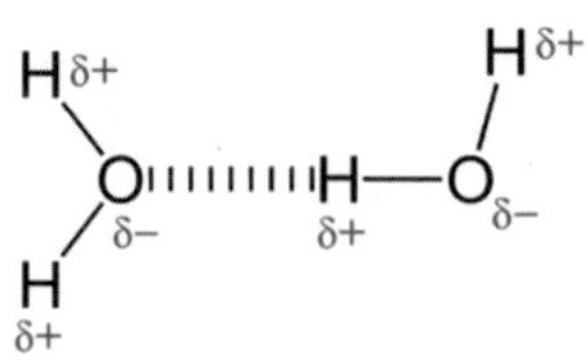

Relative strength of forces

Bond type	Dissociation energy (kcal)
Covalent	400
Hydrogen bonds	12-16
Dipole-dipole	2.0-0.5
London (van der Waals) Forces	<1

intermolecular reaction: a reaction that take place between two molecules.

intermediate[intərmí:diət]: L. *intermediatus*, *intermedius*, *inter-*, between + *medius*, middle, reaction intermediate. 중간 생성물(화합물). A molecular entity that is formed from the reactants (or preceding intermediates) and reacts further to give the directly observed products of a chemical reaction. Most chemical reactions are stepwise, that is they take more than one elementary step to complete. An intermediate is the reaction product of each of these steps, except for the last one, which forms the final product. Reactive intermediates are usually short lived and are very seldom isolated. Also, owing to the short lifetime, they do not remain in the product mixture.

For example, consider this hypothetical stepwise reaction:

$$A + B \rightarrow C + D$$

The reaction includes these elementary steps:

$$A + B \rightarrow X^*$$

$$X^* \rightarrow C + D$$

The chemical species X* is an intermediate.

internal[intə́:rnl]: L. *internus*, *internalis*, inward, internal. 내부의, 내복(内服) 용의.

internal alkyne: an alkyne with the triple bond not at the end of the carbon chain. Symmetrical examples include diphenyl acetylene and 3-hexyne.

intimate[íntəmət]: L. *intimare, intimatus,* impress, make familiar. 사이가 좋은, 가까운.

intimate ion pair: 인접 이온쌍. The concept introduced by Saul Winstein describes the interactions between a cation, anion and surrounding solvent molecules. In ordinary aqueous solutions of inorganic salts an ion is completely solvated and shielded from the counterion. In less polar solvents two ions can still be connected to some extent. In a *tight* or *intimate* or *contact* ion pair there are no solvent molecules between the two ions. When solvation increases, ionic bonding decreases and a *loose* or *solvent-shared* ion pair results. The ion pair concept explains stereochemistry in solvolysis.

Fig. Schematic representations of ion-pairs

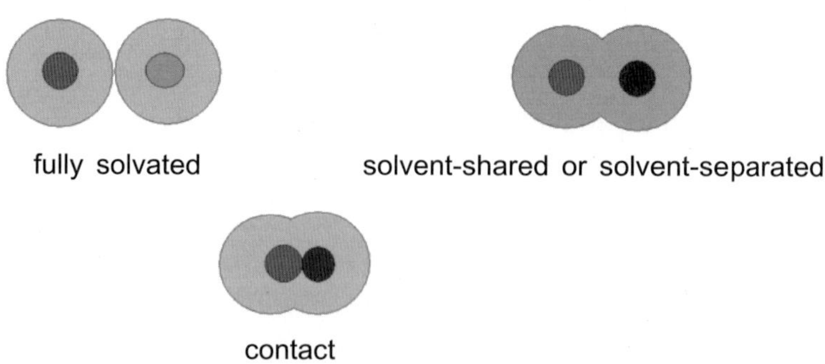

intra-[íntrə]: L. *intra-,* inside, within, intro.

intramolecular: 분자내의(에서 일어나는). Describes a process or characteristic limited within the structure of a single molecule; a property or phenomenon limited to the extent of a single molecule.

intramolecular reactions: a process or characteristic limited within the structure of a single molecule, a property or phenomenon limited to the extent of a single molecule.

Examples:

intramolecular hydride transfer: transfer of a hydride ion from one part to another within the same molecule.

intramolecular hydrogen bond: a hydrogen bond formed between two functional groups of the same molecule.

intron[íntrɑn / -trɔn]: 1970s: *intra-*, within + *-on*. 인트론. A stretch of base in DNA that contain no genetic information.

inversion[invə́:rʒən, -ʃən]: L. *invertere*, *inversion*, *in-*, into + *vertere*, to turn, turn out. 전화(轉化).

invertase: an enzyme that catalyzes the hydrolysis (breakdown) of sucrose (table sugar). The resulting mixture of fructose and glucose is called inverted sugar syrup. Related to invertases are sucrases. Invertases and sucrases hydrolyze sucrose to give the same mixture of glucose and fructose. Invertases cleave the O-C(fructose) bond, whereas the sucrases cleave the O-C(glucose) bond. 전화효소, 자당 분해효소.

invertebrate[invə́:rtəbrət, -breit]: L. *invertebrata*, *in-*, without + *vertebra*, jointed, 무척추 동물의.

iodine[áiədain / -di:n, aiədi:n]: F. < G. *iode* < *iōdēs*, *ion*, violet + *-eidēs,* like + -ine, violet-coloured. 요오드, 옥도(沃度). A chemical element with the symbol I and atomic number 53. Iodine and its compounds are primarily used in nutrition. Iodine's relatively high atomic number, low toxicity, and ease of attachment to organic compounds have made it a part of many X-ray contrast materials in modern medicine. Like the other halogens, iodine occurs mainly as a diatomic molecule I_2, not the atom. In nature, iodine is a relatively rare element, ranking 47th in abundance. It is the heaviest essential element utilized widely by life in biological functions. Its rarity in many soils has led to many deficiency problems in land animals and inland human populations, with iodine deficiency affecting about two billion people and being the leading preventable cause of mental retardation. As a component of thyroid hormones, iodine is required by higher animals. Radioisotopes of iodine are concentrated in the thyroid gland. This property of thyroid-concentration, along with its mode of beta decay, makes iodine-131 one of the most carcinogenic nuclear fission products.

ion[áiən, -ɑn]: G. *ienai*, go, going. 이온. An atom or molecule in which the total number of electrons is not equal to the total number of protons, giving it a net positive or negative electrical charge. The name was given by physicist Michael Faraday for the substances that allow a current to pass ("go") between electrodes in a solution, when an electric field is applied. An ion consisting of a single atom is an atomic or monatomic ion; if it consists of two or more atoms, it is a molecular or polyatomic ion.

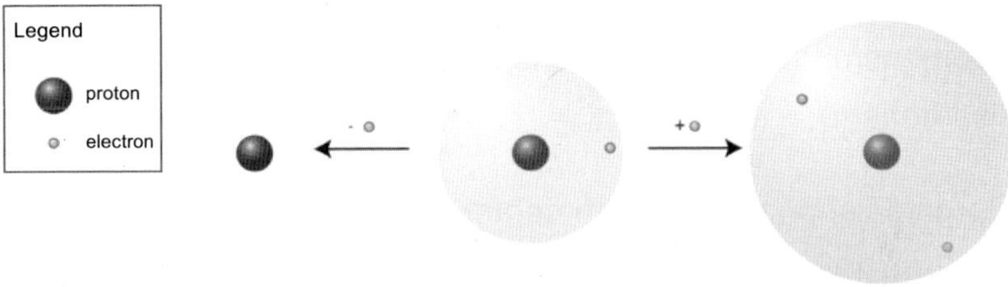

No. of protons	1	1	1
No. of electrons	0	1	2
Charge	+1	0	-1
Notation	H^+	H	H^-
Classification	**cation**	neutral (not an ion)	**anion**

is(o)-[áisou, -sə]: G. *isos, ise, ison*, equal. 동일한, 이성체.

<u>isoelectric point (pI)</u>: 등전점. Sometimes abbreviated to **IEP**, is the pH at which a particular molecule or surface carries no net electrical charge. Amphoteric molecules called zwitterions contain both positive and negative charges depending on the functional groups present in the molecule. The net charge on the molecule is affected by pH of their surrounding environment and can become more positively or negatively charged due to the loss or gain of protons (H^+). The pI is the pH value at which the molecule carries no electrical charge or the negative

and positive charges are equal.

isomer[áisəmər]: G. *isomerēs*, *isos*, equal + *méros,* a share, part. equal part. 이성질체. Compounds with the same molecular formula but different structural formulas. There are many different classes of isomers, like stereoisomers, enantiomers, geometrical isomers, et cetera. There are two main forms of isomerism: structural isomerism and stereoisomerism (spatial isomerism).

isoprene[áisəpri:n]: mid 19th cent.: *iso,* equal + *pr(opyl)ene,* fat, or 2-methyl-1,3-butadiene. 이소프렌(합성고무의 원료). A common organic compound with the formula $CH_2=C(CH_3)CH=CH_2$. It is present under standard conditions as a colorless liquid. It is the monomer of natural rubber and is a precursor to an immense variety of other naturally occurring compounds.

isotope[áisətoup]: G. *isos*, equal + *topos*, place (because the isotopes occupy the same place in the periodic table of elements), equal place. 동위원소. An element occupying the same place in the periodic table. different types of atoms of the same chemical element, each having a different number of neutrons. Correspondingly, isotopes differ in mass number (or number of nucleons) but not in atomic number.

Ex) helium-3, carbon-12, carbon-13, iodine-131 and uranium-238.

ketone[kí:toun]: D. *keton*, alteration of *aketon*, acetone. 케톤. A compound with the structure RC(=O)R', where R and R' can be a variety of atoms and groups of atoms. It features a carbonyl group (C=O) bonded to two other carbon atoms. Acetone is the simplest example of a ketone.

kinase[káineis, kín-]: G. *kinein*, to move + -ase. 키나아제[인산화(燐酸化)반응의 촉매가 되는 효소]. Alternatively known as a phosphotransferase, a type of enzyme that transfers phosphate groups from high-energy donor molecules, such as ATP, to specific substrates. The process is referred to as phosphorylation, not to be confused with phosphorolysis, which is carried out by phosphorylases. Phosphorylation is the transfer of a phosphate group to a molecule, not the reverse, i.e., phosphorolysis, the transfer of a molecular moiety to a phosphate group. An enzyme that removes phosphate groups is known as a phosphatase.

kinetic[kinétik, kai-]: G. *kinein, kinētikos,* to move. 운동의, 운동에 의한.

kinetic isotope effect(KIE): a dependence of the rate of a chemical reaction on the isotopic identity of an atom in a reactant. It is also called "isotope fractionation," although this term is somewhat broader in meaning. A KIE involving hydrogen and deuterium is represented as:

$$KIE = \frac{k_H}{k_D}$$

with k_H and k_D reaction rate constants. An isotopic substitution will greatly modify the reaction rate when the isotopic replacement is in a chemical bond that is broken or formed in the rate limiting step. In such a case, the change is termed a **primary** isotope effect. When the substitution is not involved in the bond that is breaking or forming, a smaller rate change, termed a **secondary** sotope effect is observed. Thus, the magnitude of the kinetic isotope effect can be used to elucidate the reaction mechanism. If other steps are partially rate-determining, the effect of isotopic substitution will be masked. Isotopic rate changes are most pronounced when the relative mass change is greatest since the effect is related to vibrational frequencies of the affected bonds. For instance, changing a hydrogen atom to deuterium represents a 100% increase in mass, whereas in replacing carbon-12 with carbon-13, the mass increases by only 8%. The rate of a reaction involving a C-H bond is typically 6 to 10 times faster than the corresponding C-D bond, whereas a ^{12}C reaction is only ~1.04 times faster than the corresponding ^{13}C reaction (even though, in both cases, the isotope is one atomic mass unit heavier).

<u>kinetic reaction control or thermodynamic reaction control</u>: in a chemical reaction can decide the composition in a reaction product mixture when competing pathways lead to different products and the reaction conditions influence the selectivity. The distinction is relevant when product **A** forms faster (which is called the kinetically controlled product) than product **B** because the activation energy for product **A** is lower than that for product **B**, yet product **B** is more stable (this is called the

thermodynamically controlled product). The conditions of the reaction, such as temperature, pressure, or solvent, affect which reaction pathway may be favored: either the kinetically controlled or the thermodynamically controlled one. Note this is only true if the activation energy of the two pathways differ, with one pathway having a lower E_a (energy of activation) than the other. Prevalence of thermodynamic or kinetic control determines the final composition of the product when these competing reaction pathways lead to different products. The reaction conditions as mentioned above influence the selectivity of the reaction - i.e., which pathway is taken.

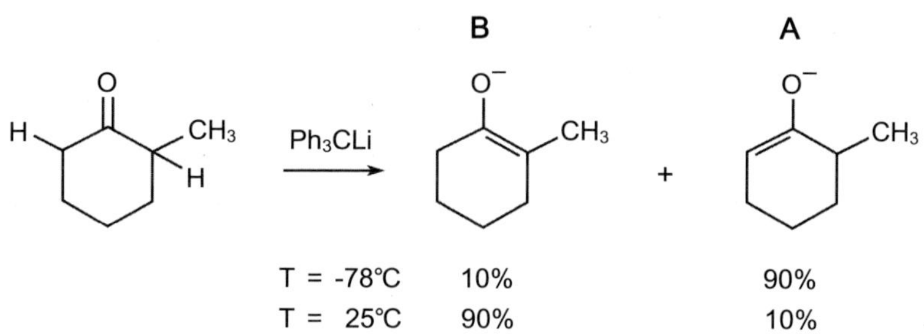

krypton[kríptɑn /-tɔn]: G. *kryptos*, *krypton*, hidden one. 크립톤 (불활성 기체상 원소의 하나). **Kr**. atomic number 36. It is a member of Group 18 and Period 4 elements. A colorless, odorless, tasteless noble gas, krypton occurs in trace amounts in the atmosphere, is isolated by fractionally distilling liquified air, and is often used with other rare gases in fluorescent lamps. Krypton is inert for most practical purposes. Krypton, like the other noble gases, can be used in lighting and photography.

lactic[lǽktik]: L. *lac, lact-,* milk + -ic. 젖산의, 유산의.

lactase: 락타아제(유당 분해 효소). Also known as lactase-phlorizin hydrolase, or LPH. A part of the β-galactosidase family of enzymes. A glycoside hydrolase involved in the hydrolysis of the disaccharide lactose into constituent galactose and glucose monomers. Lactase is present predominantly along the brush border(표피 조직 표면으로부터 조밀하게 줄지어 나 있는 미소 융털) membrane of the differentiated enterocytes(장내의 세포) lining the villi(소장 점막의 융모) of the small intestine. In humans, lactase is encoded by the Lactase gene. Lactase is essential for digestive hydrolysis of lactose in milk. Deficiency of the enzyme causes lactose intolerance. The optimum temperature for lactase is about 77 °F (25 °C) for its activity and has an optimum pH of 6.

lactic acid: 젖산. 유산. IUPAC systematic name: 2-hydroxy-propanoic acid, also known as milk acid. A chemical compound that plays a role in several biochemical processes. It has a hydroxyl group adjacent to the carboxyl group, making it an alpha hydroxy acid (AHA). In solution, it can lose a proton from the acidic group, producing the lactate ion $CH_3CH(OH)COO^-$. It is miscible with water or ethanol, and is hygroscopic. Lactic acid is chiral and has two optical isomers. One is known as L-(+)-lactic acid or (S)-lactic acid and the other, its mirror image, is D-(−)-lactic acid or (R)-lactic acid. L-(+)-Lactic acid is the biologically important isomer.

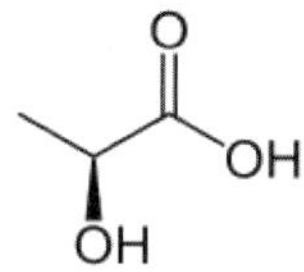

lactose: 락토오스, 유당(乳糖). A disaccharide sugar that is found most notably in milk and is formed from galactose and glucose. Lactose makes up around 2~8% of milk (by weight), although the amount varies among species and individuals. It is extracted from sweet or sour whey. It has a formula of $C_{12}H_{22}O_{11}$.

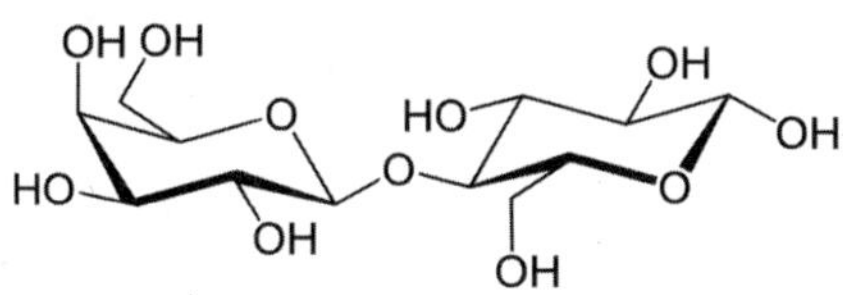

lambda[lǽmdə]: G. the eleventh letter of the Greek alphabet (Λ, λ) transliterated as L, l. 람다, 그리스어 알파벳의 열 한 번째 자(Λ, λ)의 명칭, 영어의 L, l에 해당. symbol for λ, wavelength.

latex[léiteks]: L. liquid, fluid. 유액, 라텍스(합성 고무, 플라스틱의 유탁액). The stable dispersion (emulsion) of polymer microparticles in an aqueous medium. Latexes may be natural or synthetic. Latex as found in nature is a milky fluid found in 10% of all flowering plants (angiosperms). It is a complex emulsion consisting of proteins, alkaloids, starches, sugars, oils, tannins, resins, and gums that coagulates on exposure to air.

lead[led]: D. *leiten,* plummet, solder. 납, 연필 심. A main-group element with the symbol **Pb** (from Latin: *plumbum*) and atomic number 82. Lead is a soft, malleable poor metal. It is also counted as one of the heavy metals. Metallic lead has a bluish-white color after being freshly cut, but it soon tarnishes to a dull grayish color when exposed to air. Lead has a shiny chrome-silver luster when it is melted into a liquid. Lead is used in building construction, lead-acid batteries, bullets and shots, weights, as part of solders, pewters, fusible alloys and as a radiation shield. Lead has the highest atomic number of all of the stable elements, although the next higher element, bismuth, has a half-life that is so long (much longer than the age of the universe) that it can be considered stable. Its four stable isotopes have 82 protons, a magic number in the nuclear shell model of atomic nuclei.

leave[lí:v]: D. *bleiben*, remain. 남은 것.

<u>leaving group</u>: 이탈기. A molecular fragment that departs with a pair of electrons in heterolytic bond cleavage. Leaving groups can be anions or neutral molecules. Common anionic leaving groups are halides such as Cl^-, Br^-, and I^-, and sulfonate esters, such as para-toluenesulfonate ("tosylate", TsO^-). Common neutral molecule leaving groups are water (H_2O), ammonia (NH_3), and alcohols (ROH)

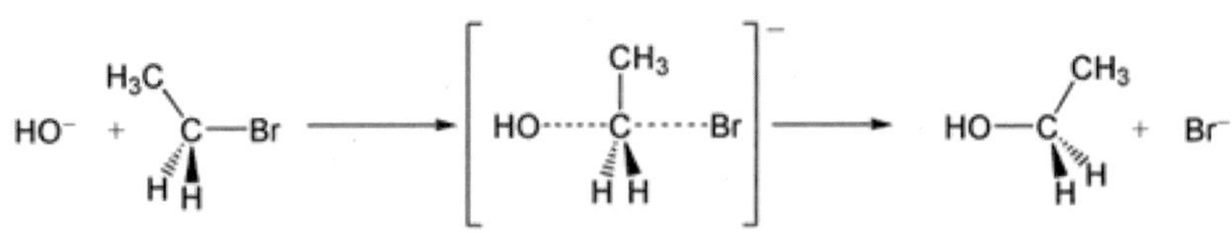

In this S_N2 reaction, bromide (Br^-) acts as the leaving group and hydroxide (OH^-) as the nucleophile. The ability of a leaving group to depart is correlated with the pKa of the conjugate acid, with lower pK_a being associated with better leaving group ability. The correlation is not perfect because leaving group ability is a kinetic phenomenon, relating to a reaction's rate, whereas pK_a is a thermodynamic phenomenon, describing the position of an equilibrium. Nevertheless, it is a general rule that more highly stabilized anions act as better leaving groups. Consistent with this rule, strong bases such as alkoxide (RO^-), hydroxide (HO^-), and amide (R_2N^-) are poor leaving groups.

Leaving groups ordered approximately in decreasing ability to leave

*$R\text{-}N_2^+$	diazonium salts
$R\text{-}OR'_2{}^+$	oxonium ions
$R\text{-}OSO_2C_4F_9$	nonaflates
$R\text{-}OSO_2CF_3$	triflates
$R\text{-}OSO_2F$	fluorosulfonates
R-OTs,R-OMs, etc.	tosylates, mesylates, and similar
R-I	iodides
R-Br	bromides
$R\text{-}OH_2^+$	(Conjugate acid of an alcohol)
R-Cl	chlorides, and acyl chloride when attached to carbonyl carbon

R-OHR'$^+$	Conjugate acid of an ether
R-ONO_2, R-$OPO(OH)_2$	nitrates, phosphates, and other inorganic esters
R-$SR'_2{}^+$	trialkylsulfornium Salts
R-$NR'_3{}^+$	tetraalkylammonium Salts
R-F	fluorides
R-OCOR	esters, and acid anhydrides when attached to carbonyl carbon
R-$NH_3{}^+$	ammonium salts
R-OAr	phenoxides
R-OH	alcohols, and carboxylic acids when attached to carbonyl carbon
R-OR	ethers, and esters when attached to carbonyl carbon

It is uncommon for groups such as H^- (hydrides), R_3C^- (alkyl anions, R=alkyl or H), or R_2N^- (amides, R=alkyl or H) to depart with a pair of electrons because of the instability of these bases.

leprosy[leprəsi]: L. *lepra*, *leprosus,* scaly. 나병, 한센병 (Hansen's* disease).

Hansen*: Gerhard Henrik Armaued Hansen (1841-1912), Norwegian physician.

leptin[léptin]: G. *leptos,* fine, thin + *-in.* 렙틴(체내 지방 용해 물질). A 16 kDa protein hormone that plays a key role in regulating energy intake and energy expenditure, including appetite and metabolism. It is one of the most important adipose derived hormones.

lethal[lí:θəl]: G. *lethum, lethalis,* death. 죽음을 초래하는, 치사(致死)의, 치명적인.

leucine[lú:si:n / ljú:-]: G. *leukos*, white + *-ine*. 류신(백색 결정성 아미노산). **Leu** or **L.** A branched-chain α-amino acid with the chemical formula $HO_2CCH(NH_2)CH_2CH(CH_3)_2$. It is an essential amino acid, which means that the human body cannot synthesize it. With a hydrocarbon side chain, leucine is classified as a hydrophobic amino acid. It has an isobutyl R group. Leucine is a major component of the sub units in ferritin, astacin and other 'buffer' proteins.

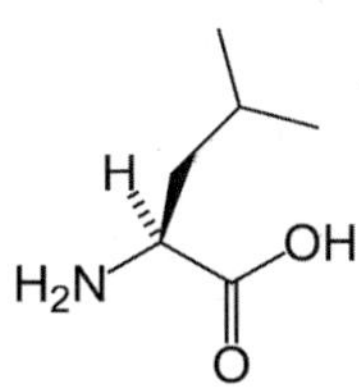

leukemia[lu:kí:mia]: G. *leukos*, white + *haima*, blood. 백혈병의. Type of cancer of the blood or bone marrow characterized by an abnormal increase of white blood cells. Leukemia is a broad term covering a spectrum of diseases. In turn, it is part of the even broader group of diseases affecting the blood, bone marrow and lymphoid system, which are all known as hematological neoplasms(혈액학적 종양).

levo-[lí:vou, -və]: L. *laev(o)-, lev(o)-,* left, turning counter clockwise. 왼쪽의, 좌(左)의, 시계반대방향 회전의.

levorotation and dextrorotation (also spelled laevorotation): refer, respectively, to the properties of rotating plane polarized light clockwise (for dextrorotation) or counterclockwise (for levorotation), seen by an observer whom the light is approaching. A compound with dextrorotation is called dextrorotatory or dextrorotary, while a compound with levorotation is called levorotatory or levorotary. Compounds with these properties are said to have optical activity and consist of chiral molecules. If a chiral molecule is dextrorotary, its enantiomer will be levorotary, and vice-versa. In fact, the enantiomers will rotate polarized light the same number of degrees, but in opposite directions. It is not possible to determine whether a given chiral molecule will be levorotatory or dextrorotatory directly from its configuration, except via detailed computer modeling. In particular, both "R" and "S" stereocenters have the ability to be dextrorotatory or levorotatory(좌선성, 左旋性).

levulose: fructose, a simple monosaccharide found in many foods. It is a white solid that dissolves readily in water. Honey, tree fruits, berries, melons, and some root vegetables contain significant amounts of the fructose derivative sucrose (table sugar). Sucrose is a disaccharide derived from the condensation of glucose and fructose.

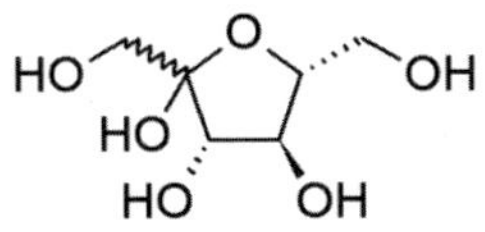

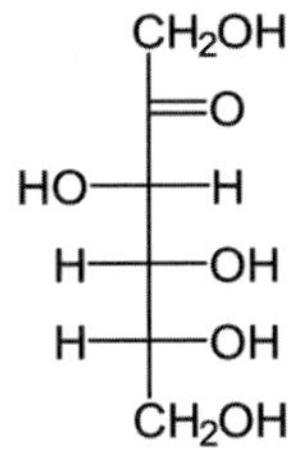

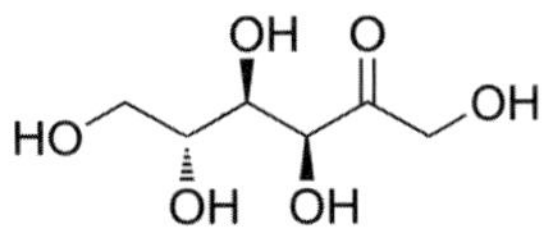

lidocaine[láidəkein]: 1940s. *(acetani)lid(e)* + *-caine(cocaine).* 리도카인(국소마취제로 쓰이는 결정화합물). Xylocaine, or lignocaine, A common local anesthetic and antiarrhythmic(항관절염의) drug. Lidocaine is used topically to relieve itching, burning and pain from skin inflammations, injected as a dental anesthetic or as a local anesthetic for minor surgery.

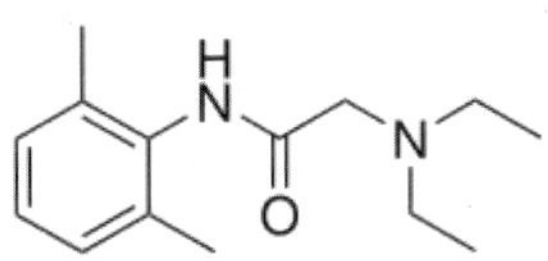

ligate[láigeit]: L. *ligare, ligation,* tied. - 을 잡아매다, [출혈하는 동맥따위를] 동여매다.

<u>ligation</u>: sharing of nonbonding electrons with a metal ions. In coordination chemistry, making a bond between a ligand and a Lewis acid.

light[lait]: D. < G. > L. *licht* < *leukos* > *lux*, light. 빛, 광선, (물리)가시광선. Or visible light, electromagnetic radiation that is visible to the human eye, and is responsible for the sense of sight. Visible light has wavelength in a range from about 380 nanometres to about 740 nm, with a frequency range of about 405 THz to 790 THz. In physics, the term *light* sometimes refers to electromagnetic radiation of any wavelength, whether visible or not. Primary properties of light are intensity, propagation direction, frequency or wavelength spectrum, and polarisation, while its speed in a vacuum, 299,792,458 meters per second (about 300,000 kilometers per second), is one of the fundamental constants of nature. Light, which is emitted and absorbed in tiny "packets" called photons, exhibits properties of both waves and particles. This property is referred to as the wave–particle duality. The study of light, known as optics, is an important research area in modern physics.

linear[líniər]: L. *linea, linearis,* a line. 직선의, 선형(線形)의.

LCAO (linear combination of atomic orbitals): the combination of atomic orbitals to produce a molecular orbitals. A quantum superposition of atomic orbitals and a technique for calculating molecular orbitals in quantum chemistry. In quantum mechanics, electron configurations of atoms are described as wave functions. In mathematical sense, these wave functions are the basis set of functions, the basis functions, which describe the electrons of a given atom. In chemical reactions, orbital wave functions are modified, i.e. the electron cloud shape is changed, according to the type of atoms participating in the chemical bond. It was introduced in 1929 by Sir John Lennard-Jones with the description of bonding in the diatomic molecules of the first main row of the periodic table, but had been used earlier by Linus Pauling for H_2^+.

lipid[lípid, laip-]: F. < G. *lipide < lipos*, fat. 지방질. A broad group of naturally occurring molecules which includes fats, waxes, sterols, fat-soluble vitamins (such as vitamins A, D, E and K), monoglycerides, diglycerides, phospholipids, and others. The main biological functions of lipids include energy storage, as structural components of cell membranes, and as important signaling molecules. Lipids may be broadly defined as hydrophobic or amphiphilic small molecules; the amphiphilic nature of some lipids allows them to form structures such as vesicles, liposomes, or membranes in an aqueous environment. Biological lipids originate entirely or in part from two distinct types of biochemical subunits or "building blocks": ketoacyl and isoprene groups. Using this approach, lipids may be divided into eight categories: fatty acyls, glycerolipids, glycerophospholipids, sphingolipids, saccharolipids and polyketides (derived from condensation of ketoacyl subunits); and sterol lipids and prenol lipids (derived from condensation of isoprene subunits).

Fig. Structures of some common lipids. At the top are oleic acid and the second structure is cholesterol. The third structure is a triglyceride composed of oleoyl, stearoyl, and palmitoyl chains attached to a glycerol backbone. At the bottom is the common phospholipid, phosphatidylcholine.

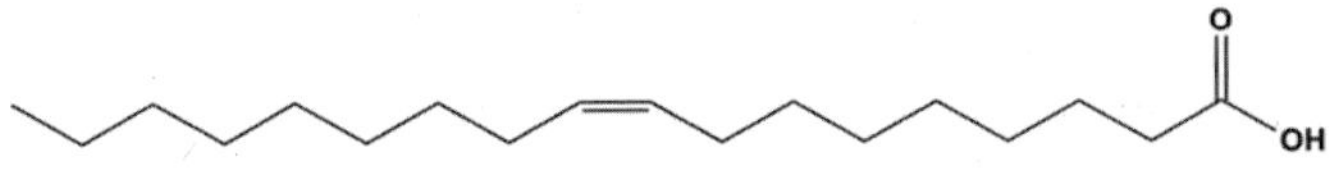

A free fatty acid

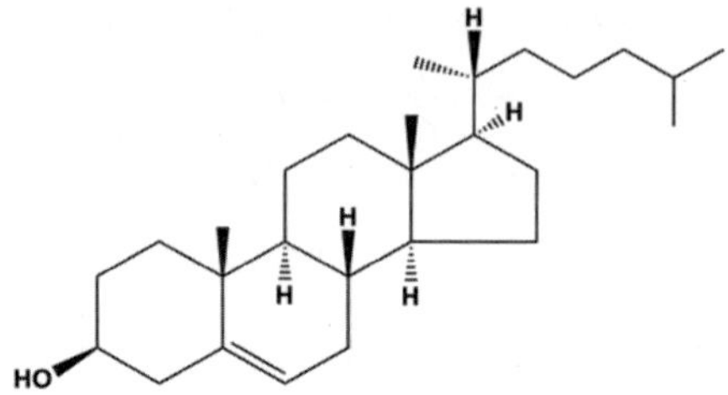

Cholesterol

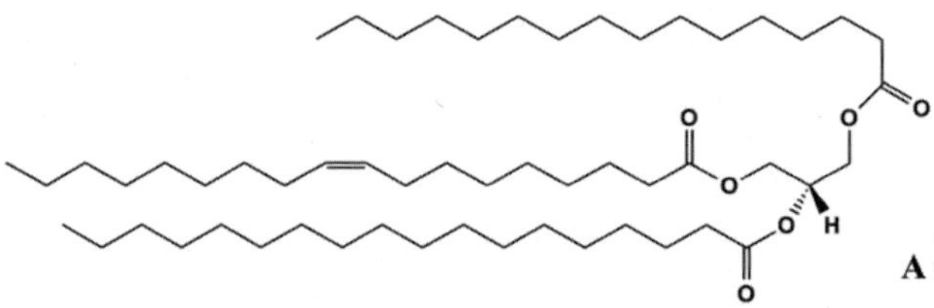

A triglyceride

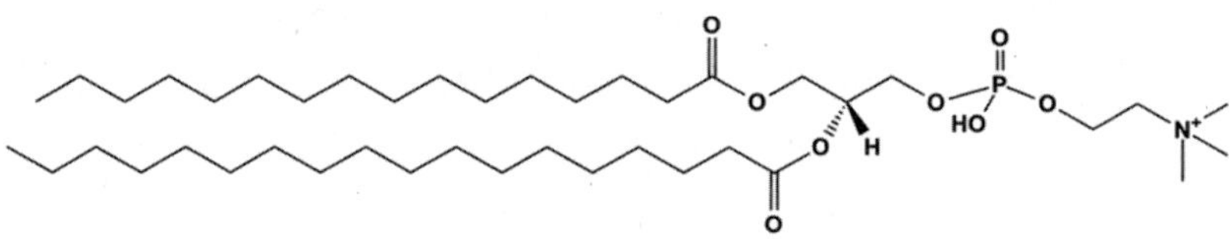

A phospholipid

lipo-[lípou, -pə, láip-]: *lipos*, fat. 지방.

lipoprotein: 지(방)질 단백질. A biochemical assembly that contains both proteins and lipids water-bound to the proteins. Many enzymes, transporters, structural proteins, antigens, adhesins and toxins are lipoproteins. Examples include the high density (HDL) and low density (LDL) lipoproteins which enable fats to be carried in the blood stream, the transmembrane proteins of the mitochondrion and the chloroplast, and bacterial lipoproteins.

lithium[líθiəm]: L. *lithia, altered from lithion*, stone, **Li**. 리튬(가장 가벼운 금속 원소). A soft, silver-white metal that belongs to the alkali metal group of chemical elements. It has the atomic number 3. Under standard conditions it is the lightest metal and the least dense solid element. Like all alkali metals, lithium is highly reactive and flammable. For this reason, it is typically stored in mineral oil. When cut open, lithium exhibits a metallic luster, but contact with moist air corrodes the surface quickly to a dull silvery gray, then black tarnish. Because of its high reactivity, lithium never occurs freely in nature, and instead, only appears in compounds, which are usually ionic. Lithium occurs in a number of pegmatitic(페그마타이트의, 굵은 결정을 가진 화성암의 총칭) minerals, but due to its solubility as an ion is present in ocean water and is commonly obtained from brines (소금물) and clays. On a commercial scale, lithium is isolated electrolytically from a mixture of lithium chloride and potassium chloride. The nuclei of lithium verge on instability, since the two stable lithium isotopes found in nature have among the lowest binding energies per nucleon of all stable nuclides. Because of its relative nuclear instability, lithium is less common in the solar system than 25 of the first 32 chemical elements even though the nuclei are very light in atomic weight. For related reasons, lithium has important links to nuclear physics. The transmutation of lithium atoms to helium in 1932 was the first fully man-made nuclear reaction, and lithium deuteride serves as a fusion fuel in staged thermonuclear weapons. Lithium and its compounds have several industrial applications, including heat-resistant glass and ceramics, high strength-to-weight alloys used in aircraft, lithium batteries and lithium-ion batteries. These uses consume more than half of lithium production. Trace amounts of lithium are present in all organisms. The element serves no apparent vital biological function, since animals and plants survive in good health without it. Nonvital functions have not been ruled out. The lithium ion Li^+ administered as any of several lithium salts has proved to be useful as a mood-stabilizing drug due to neurological effects of the ion in the human body.

lithium aluminum hydride: commonly abbreviated to **LAH**, is an inorganic compound with the chemical formula $LiAlH_4$. It was discovered by Finholt, Bond and Schlesinger in 1947. This compound is used as a reducing agent in organic synthesis, especially for the reduction of esters, carboxylic acids, and amides. The solid is dangerously reactive toward water, releasing gaseous hydrogen (H_2). Some related derivatives have been discussed for the hydrogen storage.

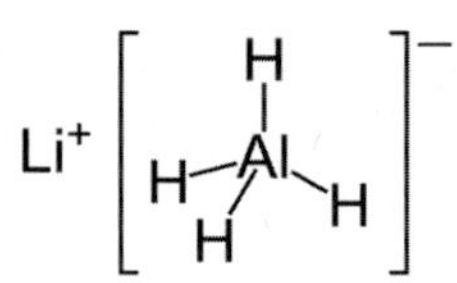

lithium diisopropylamide: the chemical compound with the formula $[(CH_3)_2CH]_2NLi$. Generally abbreviated LDA, it is a strong base used in organic chemistry for the deprotonation of weakly acidic compounds. The reagent has been widely accepted because it is soluble in non-polar organic solvents and it is non-pyrophoric. LDA is a non-nucleophilic base.

Potassium diisopropylamide (KDA) is a similar compound, but it has a potassium cation instead of a lithium cation. LDA is cheaper than KDA and is more widely used.

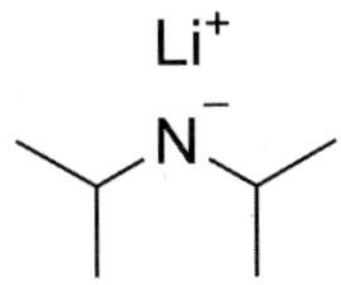

litmus[lítməs]: ON. *lit-mosi*, *litr*, dye + *mosi*, moss, moss used in dyeing. 리트머스(청색 색소). A water-soluble mixture of different dyes extracted from lichens(이끼), especially Roccella tinctoria. It is often absorbed onto filter paper. Blue litmus paper turns red under acidic conditions and red litmus paper turns blue under basic (i.e. alkaline) conditions, the color change occurring over the pH range 4.5-8.3 (at 25°C). Neutral litmus paper is purple in color. The mixture contains 10 to 15 different dyes. Acid-base indicators on litmus owe their properties to 7-hydroxy-phenoxazone chromophore.

Fig. 7-Hydroxyphenoxazone chromophore of litmus components

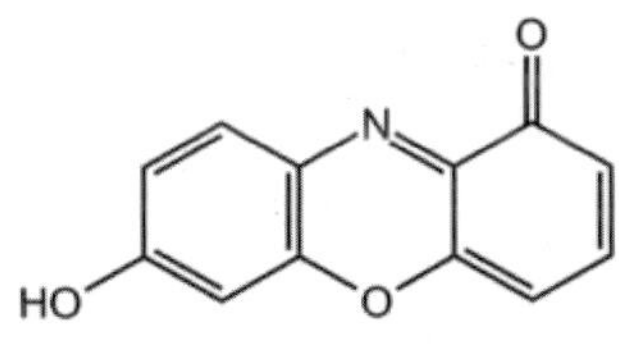

local[lóukəl]: L. *locus*, *localis*, pertaining to or concerned with 'place' or position in space. 지역적인, 국소(局所)의.

localized Electrons: Electrons that are restricted to a particular loyalty.

lock[lɑk / lɔk]: D. *loch*, hole. 구멍, 자물쇠.

lock-and-key model: enzymes are very specific, and it was suggested by the Nobel laureate organic chemist Emil Fischer in 1894 that this was because both the enzyme and the substrate possess specific complementary geometric shapes that fit exactly into one another. This is often referred to as "the lock and key" model. However, while this model explains enzyme specificity, it fails to explain the stabilization of the transition state that enzymes achieve.

Fig. Diagrams to show the induced fit hypothesis of enzyme action

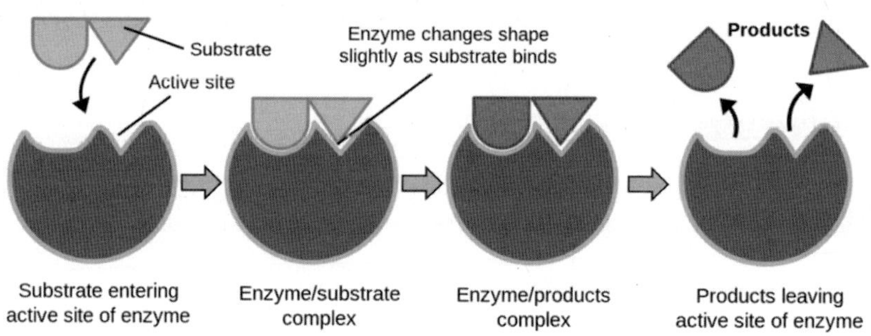

lone[loun]: late middle English: shortening of *alone*. 홀 전자의.

lone-pair electrons: unpaired electron. 비공유 전자쌍, 비결합 전자쌍. A valence electron pair without bonding or sharing with other atoms. They are found in the outermost electron shell of an atom, so lone pairs are a subset of a molecule's valence electrons. They can be identified by examining the outermost energy level of an atom — lone electron pairs consist of paired electrons as opposed to single electrons, which may appear if the atomic orbital is not full. Electron pairs are therefore considered lone pairs if two electrons are paired but are not used in chemical bonding. Thus, the number of lone electrons plus the number of bonding electrons equal the total number of valence electrons in a compound.

Fig. Lone pairs in hydroxide (shown as dots)

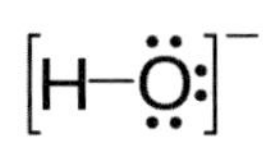

luciferin[lu:sífərin]: L. *lux, luc-*, light + *-fer,* bearing + -in. light-bringer. 루시페린(개똥벌레 따위의 체내에 있는 발광(發光)물질). A class of light-emitting biological pigments found in organisms that cause bioluminescence. The term is used generically to refer to any light-emitting molecule utilized by a luciferase or photoprotein.

Fig. The structure of firefly luciferin

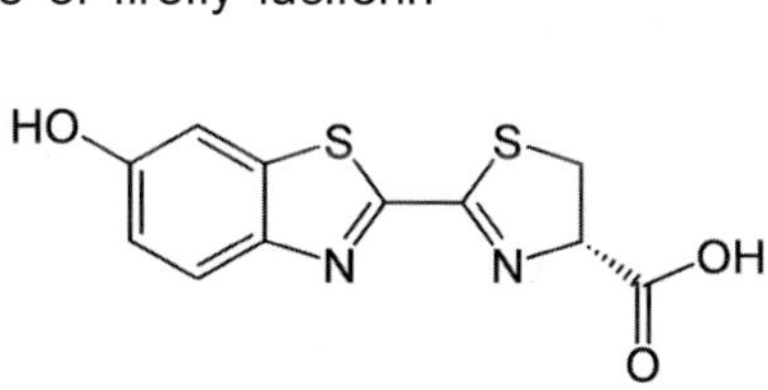

luminescence[lu:mənésns]: L. *lumen, lumin,* light + *-escence* (denoting a state). (열은 나지 않는) 발광(현상), (인광, 형광 따위). Emission of light by a substance not resulting from heat; it is thus a form of cold body radiation. It can be caused by chemical reactions, electrical energy, subatomic motions, or stress on a crystal. This distinguishes luminescence from incandescence(고온 발광, 백열광), which is light emitted by a substance as a result of heating. Historically, radioactivity was thought of as a form of "radio-luminescence", although it is today considered to be separate since it involves more than electromagnetic radiation. The term 'luminescence' was introduced in 1888 by Eilhard Wiedemann. The following are types of luminescence:

- Chemiluminescence, a result of a chemical reaction.
    - Bioluminescence, emission by a living organism.
    - Electrochemiluminescence, a result of an electrochemical reaction.
- Crystalloluminescence, produced during crystallization.
- Electroluminescence, a result of an electric current passed through a substance.
    - Cathodoluminescence, a result of being struck by an electron.
- Mechanoluminescence, a result of a mechanical action on a solid.
    - Triboluminescence, generated when bonds in a material are broken when that material is scratched, crushed, or rubbed.

  - Fractoluminescence, generated when bonds in certain crystals are broken by fractures.
  - Piezoluminescence, produced by the action of pressure on certain solid.
- Photoluminescence, a result of absorption of photons.
  - Fluorescence, photoluminescence in which the emitted photons are of lower energy than those absorbed.
  - Phosphorescence, fluorescence slightly delayed after initial absorption of radiation, (on a scale of seconds to hours).
- Radioluminescence, a result of bombardment by ionizing radiation. Sonoluminescence, a result of imploding bubbles in a liquid when excited by sound.
- Thermoluminescence, the re-emission of absorbed light when a substance if heated.

lysine[láisi(:)n]: D. *lysin*, based on lysis. 리신(필수 아미노산의 일종). **Lys** or **K,** An α-amino acid with the chemical formula $HO_2CCH(NH_2)(CH_2)_4NH_2$. It is an essential amino acid, which means that the human body cannot synthesize it. Its codons are AAA and AAG. Lysine is a base, as are arginine and histidine. The ε-amino group often participates in hydrogen bonding and as a general base in catalysis. (The ε-amino group, which is attached to the NH_3^+ group, is the fifth carbon down from the α-carbon, which is attached to the carboxyl (C=OOH) group.

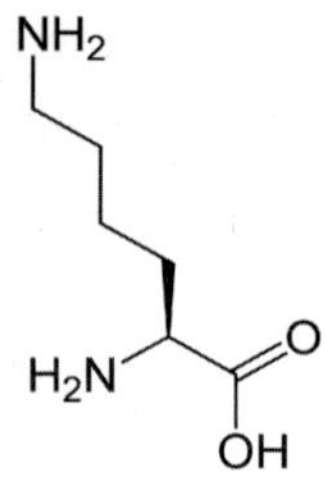

-lysis[ləsis]: G. *lusis,* loosening, breaking down. 분해, 용해.

<u>aminolysis</u>: any chemical reaction in which a molecule is split into two parts by reacting with a molecule of ammonia or an amine. An example of an aminolysis reaction is the replacement of a halogen in an alkyl group (R-X) by an amine (R'-NH_2) and the elimination of hydrogen halide (HX).

$$R\text{-}X + R'\text{-}NH_2 \rightarrow R\text{-}NH\text{-}R' + HX$$

Another common example is the reaction of a primary amine or secondary amine with a carboxylic acid or with a carboxylic acid derivative to form an amide. This reaction is widely used, especially in the synthesis of peptides. On the simple addition of an amine to a carboxylic acid, a salt of the organic acid and base is obtained. To overcome this, the carboxylic acid first needs to be "activated". This is usually done by converting the acid into a more reactive derivative (i.e. anhydride, acid halide) or by using a coupling agent. In some cases, high temperatures (>200 °C) can overcome salt formation by driving off water, without the need for "activation" of the carboxyl group. The downside to this simple reation is that the compounds may decompose at these elevated temperatures. The carboxylic acid derivatives can be esters, anhydrides, acid halides or any other activated species. The choice of activated carboxyl group or coupling agent can be very important in peptide synthesis, as using the wrong one can lead to racemization.

hydrolysis: 가수분해(반응). A chemical reaction during which molecules of water (H_2O) are split into hydrogen cations (H^+, conventionally referred to as protons) and hydroxide anions (OH^-) in the process of a chemical mechanism. It is the type of reaction that is used to break down certain polymers, especially those made by condensation polymerization. Such polymer degradation is usually catalysed by either acid, e.g., concentrated sulfuric acid (H_2SO_4), or alkali, e.g., sodium hydroxide (NaOH).

solvolysis: 가용매(加溶媒) 분해, 용매화 분해. A special type of nucleophilic substitution or elimination where the nucleophile is a solvent molecule. For certain nucleophiles, there are specific terms for the type of solvolysis reaction. For water, the term is hydrolysis; for alcohols, it is alcoholysis; for ammonia, it is ammonolysis; for glycols, glycolysis; for amines, aminolysis.

alcoholysis: an example of a solvolysis reaction is the reaction of a triglyceride with a simple alcohol such as methanol or ethanol to give the methyl or ethyl esters of the fatty acid, as well as glycerol. This reaction is more commonly known as a transesterification reaction due to the exchange of the alcohol fragments. Solvolysis of a chiral reactant results in the racemate (as expected in S_N1) but is often accompanied by Walden inversion. This is explained by postulating an intimate ion pair whereby the leaving anion is still in proximity to the carbocation and effectively shielding it from an attack by the nucleophile.

lysozyme[láisəzaim]: early 20th cent. *lysis + enzyme*, 라이소자임(계란 흰자, 콧물, 눈물 따위에 존재하는 박테리아 용해효소). Muramidase or N-acetylmuramide glycanhydrolase. Glycoside hydrolases, enzymes that damage bacterial cell walls by catalyzing hydrolysis of 1,4-beta-linkages between N-acetylmuramic acid and N-acetyl-D-glucosamine residues in a peptidoglycan and between N-acetyl-D-glucosamine residues in chitodextrins. Lysozyme is abundant in a number of secretions(분비액), such as tears, saliva(침), human milk, and mucus(점액).

macr(o)-[mǽkrou]: G. *markros, makra, makron,* large, long, 큰, 긴, 대규모의, 거시적인.

macromolecule: a very large molecule commonly created by polymerization of smaller subunits. In biochemistry, the term is applied to the four conventional biopolymers (nucleic acids, proteins, carbohydrates, and lipids), as well as non-polymeric molecules with large molecular mass such as macrocycles. The individual constituent molecules of macromolecules are called monomers.

magnesium[mæɡníːziəm] **Mg**: L. *magnesium, magnesii,* 마그네

슘. Atomic number 12, and common oxidation number +2. It is an alkaline earth metal and the eighth most abundant element in the Earth's crust, where it constitutes about 2% by mass, and ninth in the known universe as a whole.

magnet[mǽgnit]: G > L. *magnes, magnet*, < *magnēs lithos*, lodestone(천연자석), the stone of magnesia. 마그네시아의 돌, 자석. Magnesia, the southeastern area of Thessaley in central Greece.

magnetic anisotropy: 자기적 이방성(磁氣的 異方性). The direction dependence of a material's magnetic properties. In the absence of an applied magnetic field, a magnetically isotropic material has no preferential direction for its magnetic moment while a magnetically anisotropic material will align its moment with one of the easy axes. An easy axis is an energetically favorable direction of spontaneous magnetization that is determined by the sources of magnetic anisotropy listed below. The two opposite directions along an easy axis are usually equivalent, and the actual direction of magnetization can be either of them. Magnetic anisotropy is a prerequisite for hysteresis in ferromagnets[강자성체(强磁性體)]: without it, a ferromagnet is superparamagnetic [(초상자싱체(超常磁性體)].

magnetic field: 자기장(磁氣場). A mathematical description of the magnetic influence of electric currents and magnetic materials. The magnetic field at any given point is specified by both a direction and a magnitude (or strength); as such it is a vector field. The magnetic field is most commonly defined in terms of the Lorentz force it exerts on moving electric charges. Magnetic field can refer to two separate but closely related fields which are denoted by the symbols **B** and **H**. Magnetic fields are produced by moving electric charges and the intrinsic magnetic moments of elementary particles associated with a fundamental quantum property, their spin. In special relativity, electric and magnetic fields are two interrelated aspects of a single object, called the electromagnetic tensor; the split of this tensor into electric and magnetic fields depends on the relative velocity of the observer and charge. In quantum physics, the electromagnetic field is quantized and electromagnetic interactions result from the exchange of photons. In everyday life, magnetic fields are most often encountered as an invisible force created by permanent magnets which pulls on iron objects and attracts or repels other magnets. Magnetic fields are very widely used throughout modern technology, particularly in electrical

Fig. Magnetic field of an ideal cylindrical magnet with its axis of symmetry inside the image plane. The magnetic field is represented by magnetic field lines, which show the direction of the field at different points.

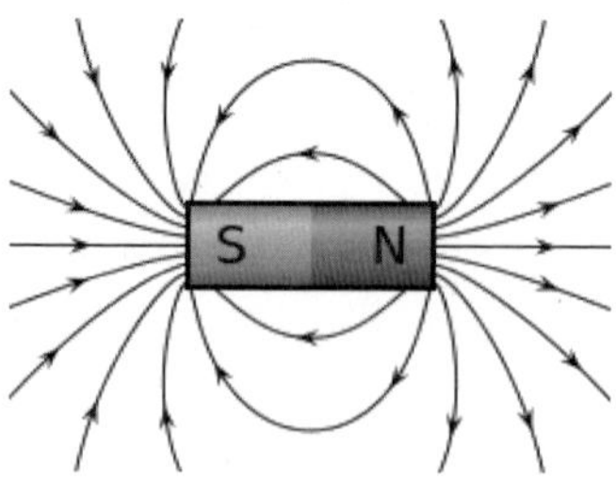

engineering and electromechanics. The Earth produces its own magnetic field, which is important in navigation. Rotating magnetic fields are utilized in both electric motors and generators. Magnetic forces give information about the charge carriers in a material through the Hall effect. The interaction of magnetic fields in electric devices such as transformers is studied in the discipline of magnetic circuits.

malathion[mæləəáiɑn]: 1950s: *(diethyl)mal(eate)* + *thio*, G. *theion*, sulphur + *-on*, suffix. 말라티온(액체 살충제). An organophosphate parasympathomimetic which binds irreversibly to cholinesterase. Malathion is an insecticide of relatively low human toxicity, however one recent study has shown that children with higher levels of organophosphate pesticide metabolites in their urine are more likely to have attention deficit hyperactivity disorder.

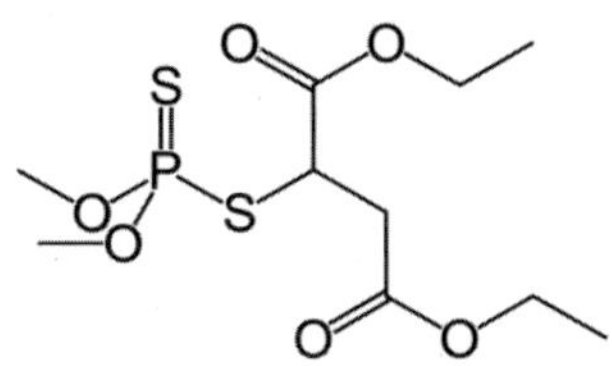

malic[mǽlik, méil-]: L. *malum,* apple. 사과.

maleic acid: an organic compound that is a dicarboxylic acid, a molecule with two carboxyl groups. Maleic acid is the cis-isomer of butenedioic acid, whereas fumaric acid is the *trans*-isomer. It is mainly used as a precursor to fumaric acid, and relative to its parent maleic anhydride, maleic acid has few applications.

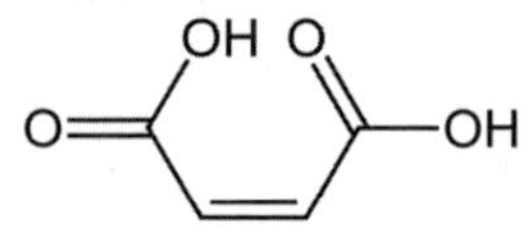

malic acid: an organic compound with the formula $HO_2CCH_2CHOHCO_2H$. This dicarboxylic acid is the active ingredient in many sour or tart foods. Malic acid is found mostly in unripe fruits. The salts and esters of malic acid are known as malates. The malate anion is an intermediate in the citric acid cycle.

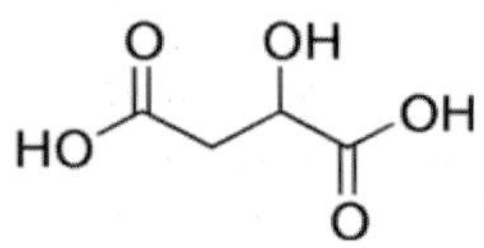

malonic acid: propanedioic acid, is a dicarboxylic acid with structure $CH_2(COOH)_2$. The ionised form of malonic acid, as well as its esters and salts, are known as malonates. For example, diethyl malonate is malonic acid's ethyl ester. The name originates from the Greek word *malum* meaning apple.

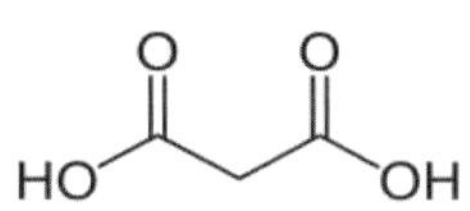

malignant[məlígnənt]: L. *malignant*, from the verb *malignare,* contriving maliciously. 악성의, 불치의.

maltose[mɔ́:ltous]: mid 19th cent. *malt*, barly + *-ose*, (gluc)ose. 말토오스, 맥아당. Malt sugar, a disaccharide formed from two units of glucose joined with an α(1→4)bond, formed from a condensation reaction. The isomer "isomaltose" has two glucose molecules linked through an α(1→6)bond. Maltose is the second member of an important biochemical series of glucose chains. Maltose is the disaccharide produced when amylase breaks down starch. It is found in germinating seeds such as barley as they break down their starch stores to use for food. It is also produced when glucose is caramelized.

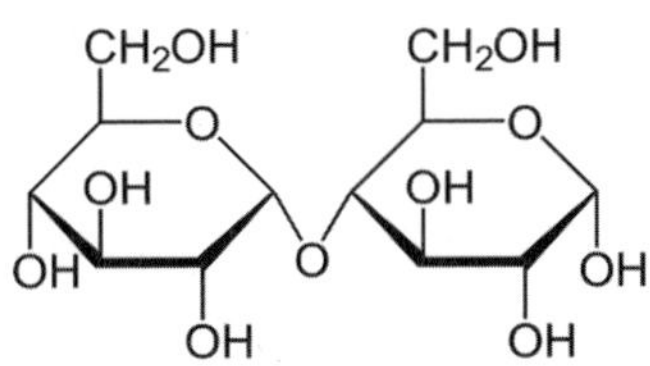

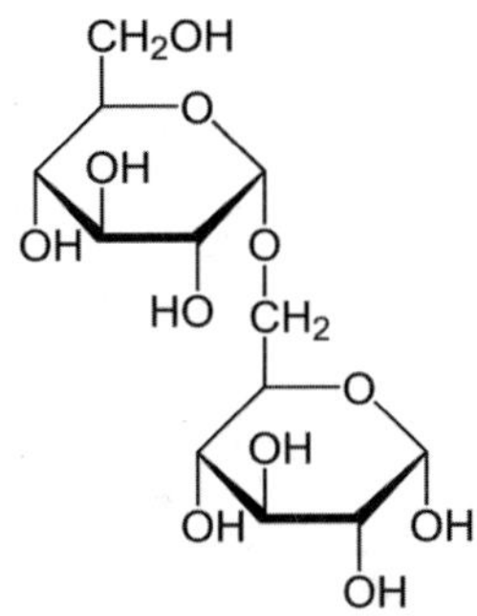

mammal[mǽməl]: L. *mammalia,* neuter plural of *mammalis* from *mamma*, breast. 포유동물.

mammalian[məméiliən, -ljən]: 포유류의.

manganese[mǽŋgəni:s, -ni:z] **Mn**: Italian. *manganese*. 망간. It has the atomic number 25. It is found as a free element in nature (often in combination with iron), and in many minerals. As a free element, manganese is a metal with important industrial metal alloy uses, particularly in stainless steels.

mannitol[mǽnətɔ:l / -tɔl]: 마니톨, 만나당 (manna ash에서 채취하는 흰 결정, 하제 따위로 사용). A white, crystalline organic compound with the formula [$C_6H_8(OH)_6$]. This polyol is used as an osmotic diuretic agent(삼투성 이뇨제) and a weak renal vasodilator(신장 혈관 확장제). It was originally isolated from the secretions of the flowering ash, called manna after their resemblance to the Biblical food, and is also referred to as mannite and manna sugar. In plants, it is used to induce osmotic stress.

material[mətíəriəl]: L. *materialis, materia,* matter. 구성 물질, 소재, 재료.

mechanism[mékənizm]: L. *mechanismus, mechanismi,* a natural or established process by which something takes place or is brought about. 메카니즘.

melanin[mélənin]: G. *melas*, *melan-*, black + *-in*, suffix. 멜라닌, 검은 색소. An ubiquitous natural pigment found in most organisms. In animals melanin pigments are derivatives of the amino acid tyrosine. The most common biological melanin is **eumelanin**. This is a brown-black polymer of dihydroxyindole carboxylic acids and their reduced forms. Another common form of melanin is pheomelanin, a cysteine-containing red-brown polymer of benzothiazine units largely responsible for red hair and freckles(주근깨, 작은 반점, 기미). The production of melanin in human skin is called melanogenesis. Production of melanin is stimulated by DNA damage from UVB-radiation, which acts as a photo-protectant. This occurs by means of a process called "ultrafast internal conversion". This property enables melanin to dissipate more than 99.9% of the absorbed UV radiation as heat (see photoprotection). This prevents the UVB radiation damage that is responsible for the formation of malignant melanoma and other skin cancers.

melanoma[melənóumə]: G. *melas*, *melan-*, black + *-oma*, suffix, tumors. 흑색소 세포종.

melatonin[melətóunin]: G. *melas*, black + *serotonin*, *serum,* whey(*h*wei), 유장(乳漿) + *tonic,* of or for stretching + *-in,* suffix). 송과선(松科腺)에서 분비되는 호르몬, 바이오 리듬 조절. N-acetyl-5-methoxytryptamine. A naturally occurring compound found in animals, plants, and microbes. In animals, circulating levels of melatonin vary in a daily cycle, thereby regulating the circadian(24시간 주기의) rhythms of several biological functions. Many biological effects of melatonin are produced through activation of melatonin receptors, while others are due to its role as a pervasive and powerful antioxidant, with a particular role in the protection of nuclear and mitochondrial DNA.

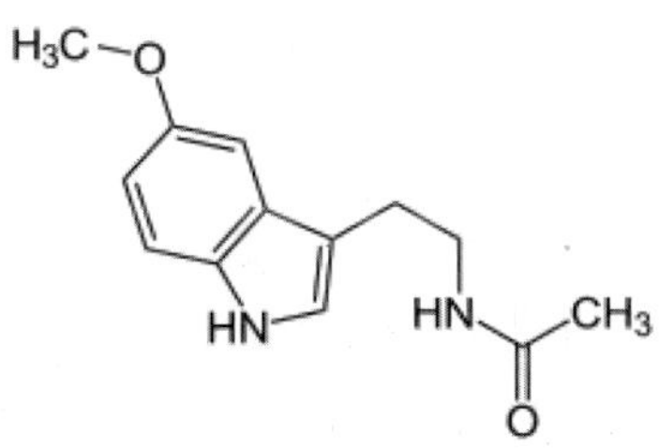

melt[melt]: G. meldein, to melt. 녹다.

melting point: 녹는점. The temperature at which it changes state from solid to liquid. At the melting point the solid and liquid phase exist in equilibrium. The melting point of a substance depends (usually slightly) on pressure, but is intended to be defined at standard atmospheric pressure if not otherwise specified. When considered as the temperature of the reverse change from liquid to solid, it is referred to as the freezing point(어는점) or crystallization point(응결점). Because of the ability of some substances to supercool, the freezing point is not considered to be a characteristic property of a substance. When the "characteristic freezing point" of a substance is determined, in fact the actual methodology is almost always "the principle of observing the disappearance rather than the formation of ice", that is, the melting point.

membrane[mémbrein]: L. *membrana, membrum*, limb. (동, 식물 조직의) 막, 얇은 막, 피막(皮膜).

menthol[ménθɔ:l / -θɔl]: L. *mentha*, mint + *-ol*, (*alcoh*)ol. 멘톨, 박하뇌(腦). An organic compound made synthetically or obtained from peppermint or other mint oils. It is a waxy, crystalline substance, clear or white in color, which is solid at room temperature and melts slightly above. The main form of menthol occurring in nature is (−)-menthol, which is assigned the (1R,2S,5R) configuration. Menthol has local anesthetic and counterirritant qualities, and it is widely used to relieve minor throat irritation.

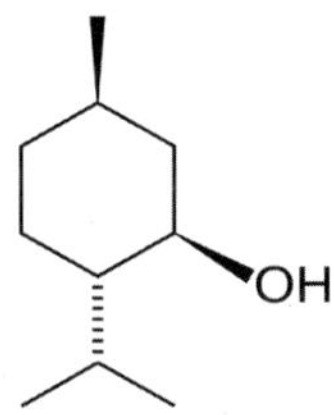

mercaptan[mərkǽptæn]: L. *mercurium captans*, capturing mercury, because the thiolate group bonds so strongly with mercury compounds. 메르캅탄 thiol. A organosulfur compound that contains a carbon-bonded sulfhydryl (-C-SH or R-SH) group (where R represents an alkane, alkene, or other carbon-containing moiety). Thiols are the sulfur analogue of alcohols, and the word is a portmanteau(혼성어) of "thio" + "alcohol," with the first word deriving from Greek, *thion* "sulfur". The -SH functional group itself is referred to as either a thiol group or a sulfhydryl group. Many thiols have strong odours resembling that of garlic. They are used as odourants to assist in the detection of natural gas (which in pure form is odourless), and the "smell of natural gas" is in fact due to the smell of the thiol used as the odourant[착취제(着臭劑)].

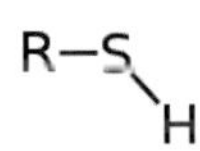

meso[mézou, mi:- / mésou]: G. *mesos*, middle. 모음 앞에서는 mes-, 메소.

meso compound (= meso isomer): a non-optically active member of a set of stereoisomers, at least two of which are optically active. This means that despite containing two or more stereocenters (chiral centers) it is not chiral. A meso compound is superimposable on its mirror image, and it does not produce a "(+)" or "(-)" reading when analyzed with a polarimeter. For example, there are 3 isomers of tartaric acid (depicted below), there is a meso compound (the 2R,3S and 2S,3R isomers are equivalent) and the optically active pair of levotartaric acid (D-(*S,S*)-(−)-tartaric acid) and dextrotartaric acid (L-(*R,R*)-(+)-tartaric acid). In the meso compound an internal plane of symmetry exists, bisecting the molecule which is not present in the non-meso compounds. That is, on rotating the meso compound by 180° on a plane perpendicular to the screen, the same stereochemistry is obtained, again this is not seen in the non-meso tartaric acid. (see Fischer projection).

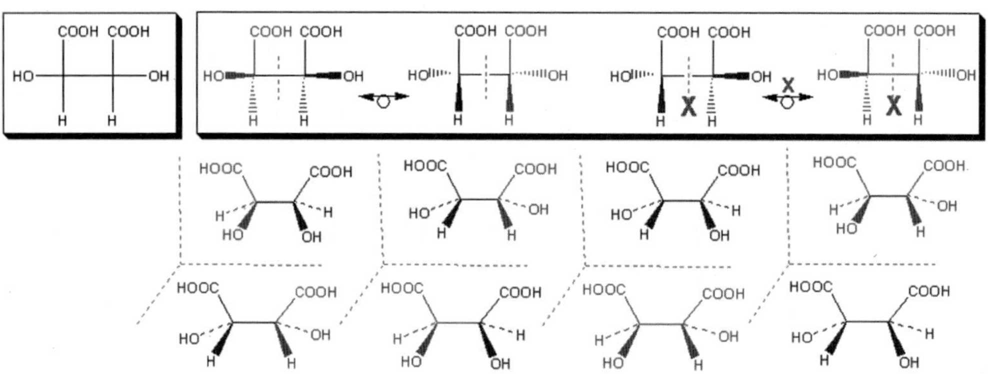

meta-[métə]: G. *met(a)*, 모음 앞에서는 *met-*, with, after, between. 사이.

meta-substitution: the substituents occupy positions 1 and 3.

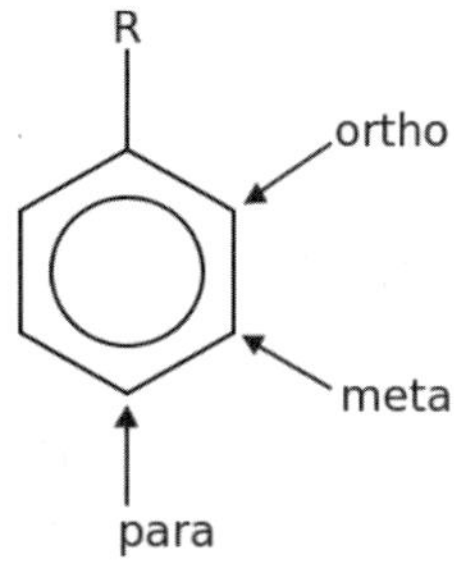

meta-directing substituents: a substituent that directs an incoming substituent meta to an existing substituents. Non-halogen groups with atoms that are more electronegative than carbon, such as a carboxylic acid group (CO_2H) draw substantial electron density from the *pi* system. These groups are strongly deactivating groups. Additionally, since the substituted carbon is already electron-poor, the resonance contributor with a positive charge on this carbon (produced by *ortho/para* attack) is less stable than the others. Therefore, these electron-withdrawing groups are *meta* directing.

metabolism[mətǽbəlizm]: G. *metabole, metaballein*, change, to change. 신진대사, 대사작용. The set of chemical reactions that happen in living organisms to maintain life. These processes allow organisms to grow and reproduce, maintain their structures, and respond to their environments. Metabolism is usually divided into two categories. Catabolism(분해대사, 이화작용) breaks down organic matter, for example to harvest energy in cellular respiration. Anabolism(합성대사, 동화작용) uses energy to construct components of cells such as proteins and nucleic acids.

metal[métl]: G > L. *metallum* < *metallon*, mine, quarry, or metal. 금속, 금속 원소, 금속 화합물.

metathesis[mətǽθəsis]: G. *metatithenai,* transpose, transposition, change the position of. 상호교환(반응), 치환, 복분해. Molecular process involving the exchange of bonds between the two reacting chemical species, which results in the creation of products with similar or identical bonding affiliations. This is represented by the general reaction scheme:

$$AX + BY \rightarrow BX + AY$$

These chemical species can either be ionic or covalent. When referring to precipitation reactions between solutions of ions in inorganic chemistry, these were formerly referred to as double displacement or double replacement reactions, and these terms are still encouraged. In other literature, the term double decomposition is frequently encountered. It seems that the term double decomposition is more specifically used when at least one of the substances does not dissolve in the liquor, as the ligand or ion exchange takes place in the solid state of the reactant,

i.e. $AX(aq) + BY(s) \rightarrow AY(aq) + BX(s)$.

<u>olefin metathesis (= transalkylidenation)</u>: 올래핀 상호교환 반응. An organic reaction that entails redistribution of alkylene fragments by the scission of carbon - carbon double bonds in olefins (alkenes). Its advantages include the creation of fewer sideproducts and hazardous wastes. Yves Chauvin, Robert H. Grubbs, and Richard R. Schrock shared the 2005

Nobel Prize in Chemistry for "the development of the metathesis method in organic synthesis". The reaction is catalyzed by metals such as nickel, tungsten, rhenium, ruthenium and molybdenum. The reaction consists of an alkene double bond cleavage, followed by a statistical redistribution of alkylidene fragments. The general scope is outlined by the following scheme:

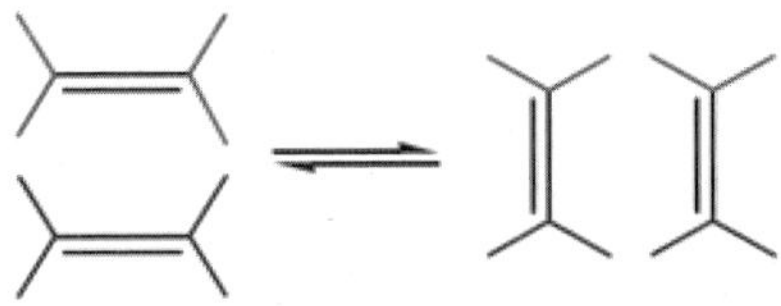

methane[méθein / mi:θ-]: G. *methy,* wine + L. *-anus*, origin. 메탄, 메테인, 소기(沼氣). A colourless, odourless, highly flammable gas, CH_4, the simplest of the alkanes, which is produced by decaying organic matter and in the guts of ruminants, occurs in natural gas and coal gas, and is used as a fuel.

methionine[meθáiəni:n, -nən]: 1920s, methyl + G. *theion*, sulphur, 메티오닌 (필수 아미노산의 하나). Abbreviated as **Met** or **M,** is an α-amino acid with the chemical formula $HO_2CCH(NH_2)CH_2CH_2SCH_3$. This essential amino acid is classified as nonpolar. This amino-acid is coded by the codon AUG, also known as the initiation codon, since it indicates mRNA's coding region where translation into protein begins.

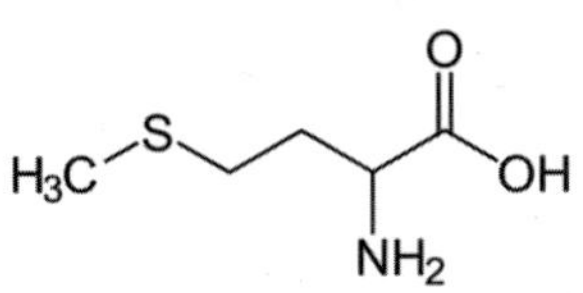

methylene[méθəli:n]: G. *methylene*, *methu*, wine + *hule*, wood + *-ene*, 메틸렌(기). $-CH_2$ group.

micelle[misél]: L. *micella, micellae, mica*, small + *-ella*, *diminutive suffix*, 지소 접미사. 미포(微胞), 미셀. An aggregate of surfactant molecules dispersed in a liquid colloid. A typical micelle in aqueous solution forms an aggregate with the hydrophilic "head" regions in contact with surrounding solvent, sequestering the hydrophobic single tail regions in the micelle center.

micro[máikrou]: G. *mikros, mikra, mikron,* small, little, short. 작은, 적은, 짧은.

microscope[máikrəskoup]: L. *microscopium*, *micro-*, small-scale + *scope*, look at. 현미경.

migraine[máigrein / mi:g-]: G. *hēmikrania*, *hēmi-,* half + *kranion*, skull. Pain on one side of the head. 편두통(偏頭痛).

moisture[mɔ́istʃər]: Anglo-Norman. *moisture,* liquid, vapor. 습기, 수분, (공기 중의) 수증기.

molality[moulǽləti]: German. *molal, mol < molekűl,* molecule + *-al,* suffix + L. *-(i)tas, -(i)tatum,* state, or condition. Molal concentration: the number of moles of solute per kilogram of solvent (not solution). A small letter **m** is used to abbreviate units of mol/kg. (용량) 몰랄 농도.

molar[móulər]: L. *molaris*, *moles*, mass + *-aris*, suffix, 몰(mole)의, 그램분자의.

molarity[moulǽrəti]: German. *molar, mol < molekűl,* molecule + *-ar,* suffix + L. *-(i)tas, -(i)tatum,* state, or condition. Molar concentration: the number of moles of a given substance per liter of solution. A capital letter **M** is used to abbreviate units of mol/L, (용적) 몰 농도.

mold (= mould)[mould]: L. *modulus,* a frame or template for producing mouldings. 형, 주형(鑄型).

molecule[mɑ́ləkju:l /mɔ́l-]: L. *molecula,* diminutive of Latin *moles*, mass. 분자.

<u>molecular orbital(**MO**)</u>: 분자궤도함수. A mathematical function describing the wave-like behavior of an electron in a molecule. This function can be used to calculate chemical and physical properties such as the probability of finding an electron in any specific region. The term "orbital" was first used in English by Robert S. Mulliken as the English translation of Schrödinger's 'Eigenfunktion'. It has since been equated with the "region" generated with the function. Molecular orbitals are usually constructed by combining atomic orbitals or hybrid orbitals from each atom of the molecule, or other molecular orbitals from groups of atoms. They can be quantitatively calculated using the Hartree-Fock or Self-Consistent Field (SCF) methods.

<u>HOMO and LUMO</u>: 최고 점유 분자 궤도함수와 최저 비점유 분자 궤도함수. Acronyms for highest occupied molecular orbital and lowest unoccupied molecular orbital, respectively. The energy difference between the HOMO and LUMO is termed

the HOMO-LUMO gap. HOMO and LUMO are sometimes referred to as frontier orbitals. Roughly, the **HOMO** level is to organic semiconductors what the valence band is to inorganic semiconductors and quantum dots. The same analogy exists between the LUMO level and the conduction band. A **SOMO** is a *singly occupied molecular orbital* such as half-filled HOMO of a radical.

Fig. HOMO/LUMO Diagram

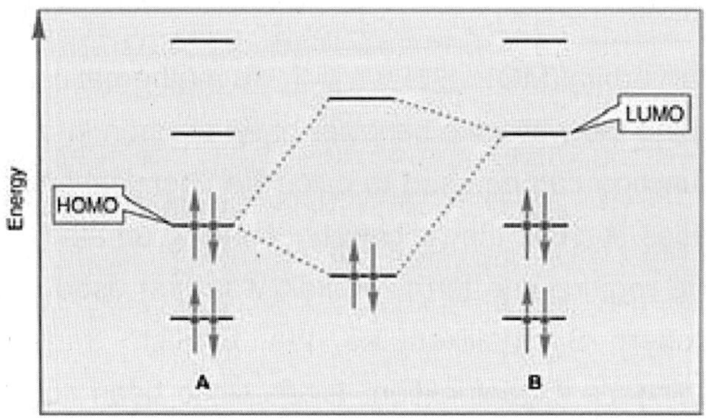

moment[móumənt]: L. *momentum*, *movimentum*, from *movere*, to move. 모멘트, 능률, 역률(力率), 벡터적(積).

mono-[mɑ́nou, -nə/ mɔ́n-]: G. *manos, mone, monon,* 모음 앞에서는 *mon-*, alone, only, left alone. 홀로, 고독한, 유일하게.

monomer: (*mono-*, alone + *merous*, having parts). 단량체(單量體). An atom or a small molecule that may bind chemically to other monomers to form a polymer. The term "monomeric protein" may also be used to describe one of the proteins making up a multiprotein complex. The most common natural monomer is glucose, which is linked by glycosidic bonds into polymers such as cellulose and starch, and is over 76% of the mass of all plant matter. Most often the term monomer refers to the organic molecules which form synthetic polymers, such as, for example, vinyl chloride, which is used to produce the polymer polyvinyl chloride (PVC).

monosaccharide: *mono-*, alone + *saccharide*, *saccharum*, sugar + *-ide*, suffix). 단당류(單糖類).

morphine[mɔ:rfin]: G. *morpheus, morphei,* the son of Somnus (god of sleep), the god of dreams + *ine*, suffix, The drug's sleep-inducing properties. 모르핀(마취약). Morphine is the most abundant alkaloid found in opium, the dried sap (latex) derived from shallowly slicing the unripe seedpods of the opium, or common and/or edible, poppy, Papaver somniferum. Morphine was the first active principle purified from a plant source and is one of at least 50 alkaloids of several different types present in opium, Poppy Straw Concentrate, and other poppy derivatives.

systematic (IUPAC) name:

(5α,6α)-7,8-didehydro-morphinan-3,6-diol.

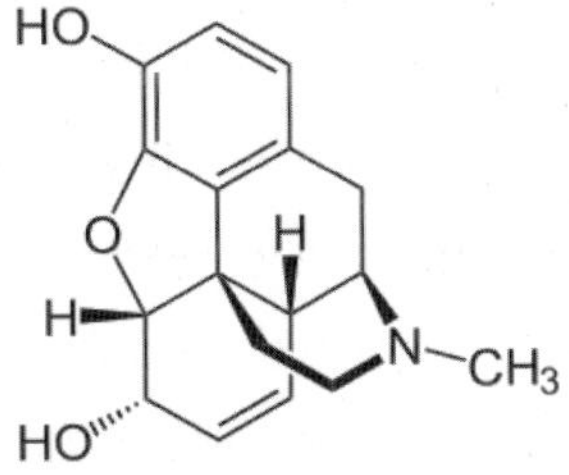

multiplex[mʌ́ltəpleks]: L. *multiplex*, *multus*, much, many + *plex* part, mutiple. 복합적인, 다중의.

multiplicity: 다중도(多衆度), 중복도(重複度). Multiplicity in quantum chemistry is used to distinguish between several degenerate wave functions that differ only in the orientation of their angular spin momenta. It is defined as $2S+1$, where S is the angular spin momentum. Multiplicity is the quantification of the amount of unpaired electron spin. Multiplicity is a result of Hund's rule which favors the single filling of degenerate (same energy) orbitals. The result is the filling of multiple orbitals with electrons or multiplicity. Multiplicity is calculated with the equation: $2S + 1$ (where S = ½ (# of unpaired electrons). When all electrons are paired $S = 0$, and the multiplicity = 2(0) + 1 = 1. This case is called a singlet. If a molecule has 1 unpaired electron $S = +½$ and $2S + 1 = 2$, which is called a doublet. Two unpaired electrons would result in a triplet, etc.

musk[mʌsk]: Sanskrit. *muska,* scrotum. 사향, 음낭. From the similarity in shape of the sac on the abdomen of a male musk deer in which musk is produced.

mutation[mju:téiʃən]: L. *mutare, mutation*, to change. 돌연변이. In molecular biology and genetics, mutations are changes in a genomic sequence: the DNA sequence of a cell's genome or the DNA or RNA sequence of a virus. They can be defined as sudden and spontaneous changes in the cell. Mutations are caused by radiation, viruses, transposons (전이인자의 하나) and mutagenic chemicals, as well as errors that occur during meiosis [(세포액의) 감수분열] or DNA replication. They can also be induced by the organism itself, by cellular processes such as hypermutation.

nan(o)-[nǽnə, néinə]: L. *nanus, nani,* dwarf (난쟁이, 왜소체). 10억분의 1. Prefixed to the name of a unit of measurement to denote a factor of 10^{-9} (one thousand-millionth).

naphtha[nǽfθə, nǽp-]: L. *naphtha, naphthae.* 나프타. Normally refers to a number of different flammable liquid mixtures of hydrocarbons, i.e. a distillation product from petroleum or coal tar boiling in a certain range and containing certain hydrocarbons, a broad term encompassing any volatile, flammable liquid hydrocarbon mixture.

naphthalene: (*naphtha* + -ine, suffix). 나프탈랜. Obtained as a distillation product of coal tar. A crystalline, aromatic, white, solid hydrocarbon with formula $C_{10}H_8$ and the structure of two fused benzene rings. It is best known as the traditional, primary ingredient of mothballs. It is volatile, forming an inflammable vapor, and readily sublimes at room temperature, producing a characteristic odor that is detectable at concentrations as low as 0.08 ppm by mass.

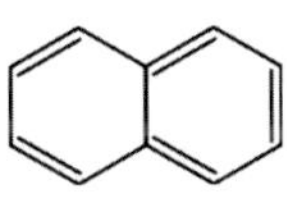

Naproxen[nə'prɒksən): L. *na-,* in naphthyl + *pr-* in propionic + *ox-* in oxy- + *-en* ene). A non-steroidal anti-inflammatory drug (NSAID) commonly used for the reduction of mild to moderate pain, fever, inflammation and stiffness caused by conditions such as osteoarthritis [(병리)골관절염], rheumatoid arthritis, psoriatic arthritis (건선 관절염), gout, ankylosing spondylitis (교착 척추염), menstrual cramps (생리통), tendinitis (건염), bursitis (혈액낭염), and the treatment of primary dysmenorrhea(생리불순). It works by inhibiting both the COX-1 and COX-2 enzymes. Systematic (IUPAC) name: (+)-(S)-2-(6-methoxynaphthalen-2-yl)propanoic acid

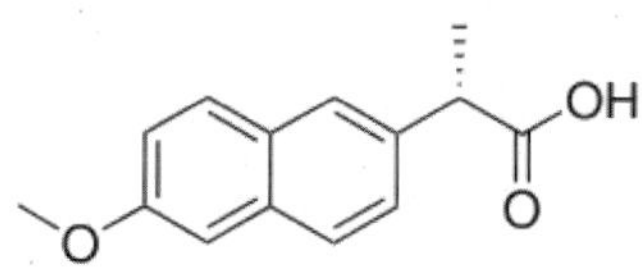

narcotic[nɑ:rkάtik / -kɔt-]: G. *narkōtikos, narkoun,* make numb. 마취성의, 마약의, 마약, 마취제.

nausea[nɔ́:ziə, -ʒə/ -siə]: L. *nausia*, from *naus*, ship. A feeling of sickness with an inclination to vomit. 구역질, 뱃멀미.

neon[ni:ɑn / -ən, ɔn]: G. *neos*, somthing new. 네온. The chemical element that has the symbol **Ne** and an atomic numer of 10. Although a very common element in the universe, it is rare on Earth. A colorless, inert noble gas under standard conditions, neon gives a distinct reddish-orange glow when used in either low-voltage neon glow lamps or in high-voltage discharge tubes or neon advertising signs. It is commercially extracted from air, in which it is found in trace amounts.

neoprene[ní:əpri:n]: G. *neo-*, new + *prene*, propyl + *ene,* suffix. A synthetic polymer resembling rubber, resistant to oil, heat, and weathering. 네오프렌(chloroprene을 중합하여 만드는 방유성(防油性) 고무. A family of synthetic rubbers that are produced by polymerization of chloroprene. Neoprene in general has good chemical stability, and maintains flexibility over a wide temperature range. It is used in a wide variety of applications, such as laptop sleeves, orthopedic braces (wrist, knee, etc.), electrical insulation, liquid and sheet applied elastomeric membranes or flashings, and car fan belts. A foamed neoprene containing gas cells is used as an insulation material, most notably in wetsuits. Foamed neoprene is also used in other insulation and shock-protection (packing) applications.

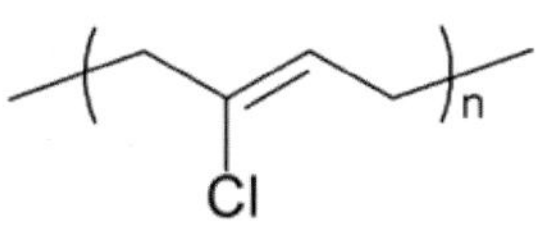

nerve[nə:rv]: G. *neuron*, nerve. 신경, 신경 섬유.

neuron[n*j*úəran/njúərɔn]: L. *neuron, neuri,* nerve. 신경단위, 신경세포.

neutral[n*j*ú:trəl / njú:-]: L. *neutralis*, *neuter*. An electrically neutral point, neither acid nor alkaline; having a pH of about 7, having neither a positive nor negative electrical charge. 중성의, 대전(帶電)하지않는.

neutron[n*j*ú:tran /njú:trɔn]: 20th cent. neutral, L. *neutralis*, *neuter*, + -on, suffix. 중성자. A subatomic hadron particle which has the symbol n or n^0, with no net electric charge and a mass slightly larger than that of a proton. With the exception of hydrogen, nuclei of atoms consist of protons and neutrons, which are therefore collectively referred to as nucleons (핵입자, 양성자와 중성자의 총칭). Neutrons are necessary within an atomic nucleus as they bind with protons via the strong force; protons are unable to bind with each other due to their mutual electromagnetic repulsion being stronger than the attraction of the strong force. The number of neutrons is the neutron number and determines the isotope of an element. For example, the abundant carbon-12 isotope has 6 protons and 6 neutrons, while the very rare radioactive carbon-14 isotope has 6 protons and 8 neutrons. While bound neutrons in stable nuclei are stable, free neutrons are unstable; they undergo beta decay with a mean lifetime of just under 15 minutes (881.5±1.5 s). Free neutrons are produced in nuclear fission and fusion.

nicotin[níkəti:n, -tin]: L. *nicotiana* (herba), tobacco (plant), named after Jean Nicot, a 16th-cent. French diplomat who introduced tobacco to France in 1560. 니코틴(담배에 함유된 유독성 알카로이드).

nitrogen[náitrədən]: 18th cent.: from French *nitrogène, nitre,* nitro + -gen, suffix. 질소(비금속 원소의 하나). A chemical element that has the symbol **N**, atomic number of 7 and atomic mass 14.00674 u. Elemental nitrogen is a colorless, odorless, tasteless, and mostly inertdiatomic gas at standard conditions, constituting 78.08% by volume of Earth's atmosphere. The element nitrogen was discovered as a separable component of air, by Scottish physician Daniel Rutherford, in 1772. Many industrially important compounds, such as ammonia, nitric acid, organic nitrates (propellants and explosives), and cyanides, contain nitrogen. The extremely strong bond in elemental nitrogen dominates nitrogen chemistry, causing difficulty for both organisms and industry in breaking the bond to convert the N_2 into useful compounds, but at the same time causing release of large amounts of often useful energy when the compounds burn, explode, or decay back into nitrogen gas.

node[noud]: L. *nodus*, knot. (진동체의) 정지점, 절(節), 교점(交點), 매듭. A point along a standing wave where the wave has minimal amplitude. For instance, in a vibrating guitar string, the ends of the string are nodes. that part of an orbital in which there is zero probability of finding an electron. The opposite of a node is an anti-node, a point where the amplitude of the standing wave is a maximum. These occur midway between the nodes.

Fig. A standing wave. The red dots are the wave nodes

nomenclature[nóumənkleitʃər, nouménklə-]: L. *nomenclatura*, *nomen*, name + *clatura*, calling, summoning. The devising or choosing of names for things, especially in a science or other discipline. (학술상의) 조직[체계]적 명명법.

nominal[nάmənl / nɔ́m-]: L. *nominalis, nomen, nomin,* name. 이름뿐인, 명목상의.

<u>nominal mass</u>: 명목 이온 질량. Mass rounded to the nearest whole number.

nona-[nɑ́nə, nóunə/ nɔ́nə]: L. nonus, ninth. 9, 9번째, 모음 앞에서는 non-.

nonpolar[nɑnpóulər / nɔn-]: *nonpolus*, *non-*, not + *polus*, end of an axis. 무극성.

normal[nɔ́:rməl]: L. *normalis, norma*, carpenter's square. 노말, 정상, 기준, 정규. (유기 화합물이) 직쇄상(直鎖狀)의.

<u>normal alkane (= straight-chain alkane)</u>: an alkane in which the carbons from a contiguous chain with no branches.

novocaine[nóuvəkein]: L. *novous*, new + -caine (from cocaine). 프로카인 (procaine, 국부마취제의 일종): 염산 프로카인. A local anesthetic drug of the amino ester group. It is used primarily to reduce the pain of intramuscular injection of penicillin, and it was also used in dentistry. Owing to the ubiquity of the trade name Novocain, in some regions procaine is referred to generically as novocaine. It acts mainly by being a sodium channel blocker. Procaine was first synthesized in 1905.

nucleophile[njú:kliəfail /njú:-]: L. *nucleus, nuclei,* kernel, inner part + *-phile*, loving, friendly, nucleus lover. 친핵체. A reagent that forms a chemical bond to its reaction partner (the electrophile) by donating both bonding electrons. Because nucleophiles donate electrons, they are by definition Lewis bases (see acid-base reaction theories). All molecules or ions with a free pair of electrons can act as nucleophiles. Neutral nucleophilic reactions with solvents such as alcohols and water are named solvolysis. A nucleophile is an electron-rich chemical reactant that is attracted by electron deficient compounds. Examples of nucleophiles are anions such as Cl^-, or a compound with a lone pair of electrons such as NH_3 (ammonia). Nucleophiles may take part in nucleophilic substitution, whereby a nucleophile becomes attracted to a full or partial positive charge. In general, in a row across the periodic table, the more basic the ion (the higher the pK_a of the conjugate acid), the more reactive it is as a nucleophile. In a given group, polarizability is more important in the determination of the nucleophilicity: the easier it is to distort the electron cloud around an atom or molecule, the more readily it will react. *e.g.,* the iodide ion (I^-) is more nucleophilic than the fluoride ion (F^-).

nucleophilic: 친핵(성). An adjective that describes the affinity of a nucleophile to the nuclei.

nucleophilicity (= nucleophile strength): 친핵성(도). Refers to the nucleophilic character. Nucleophilicity is often used to compare an atom's relative affinity to another's.

ambident nucleophile: 여러자리 친핵체, 양쪽자리 친핵체. One that can attack from two or more places, resulting in two or more products. For example, the thiocyanate ion (SCN^-) may attack from either the S or the N. For this reason, the S_N2 reaction of an alkyl halide with SCN^- often leads to a mixture of RSCN (an alkyl thiocyanate) and RNCS (an alkyl isothiocyanate).

nucleus[n*j*ú:kliəs/njú:-]: L. *nucleus, nuclei,* kernel, inner part. 핵.

nucleic acid: 핵산. Macromolecule composed of chains of monomeric nucleotides(핵산의 구성 성분). In biochemistry these molecules carry genetic information or form structures within cells. The most common nucleic acids are deoxyribonucleic acid (DNA) and ribonucleic acid (RNA).

nucleoside(s): 핵염기와 오탄당이 N-글리코사이드결합을 이룬 화합물. Glycosylamines consisting of a nucleobase (often referred to as simply base) bound to a ribose or deoxyribose sugar via a beta-glycosidic linkage. Examples of these include cytidine, uridine, adenosine, guanosine, thymidine and inosine.

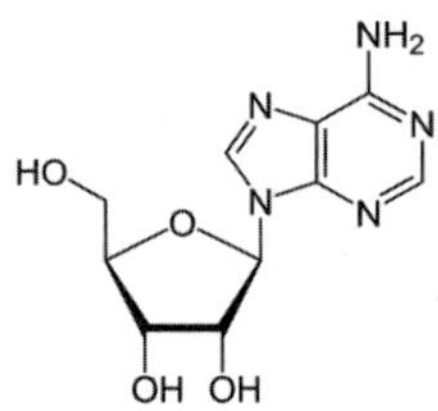
adenosine

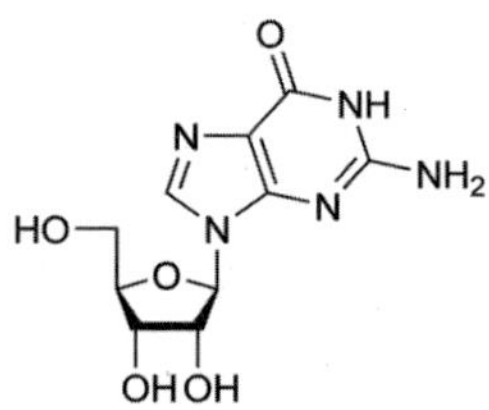
guanosine

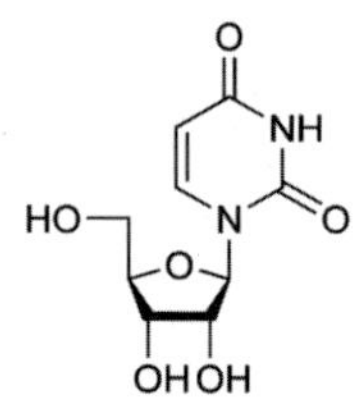
uridine

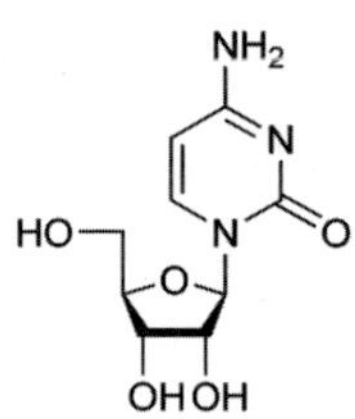
cytidine

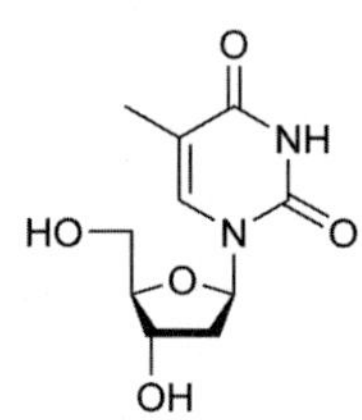
thymidine

nucleotide(s): 핵산의 구성 성분. Molecules that, when joined together, make up the structural units of RNA and DNA. In addition, nucleotides play central roles in metabolism. In that capacity, they serve as sources of chemical energy (adenosine triphosphate and guanosine triphosphate), participate in cellular signaling (cyclic guanosine monophosphate and cyclic adenosine monophosphate).

Fig. Structural elements of the most common nucleotides

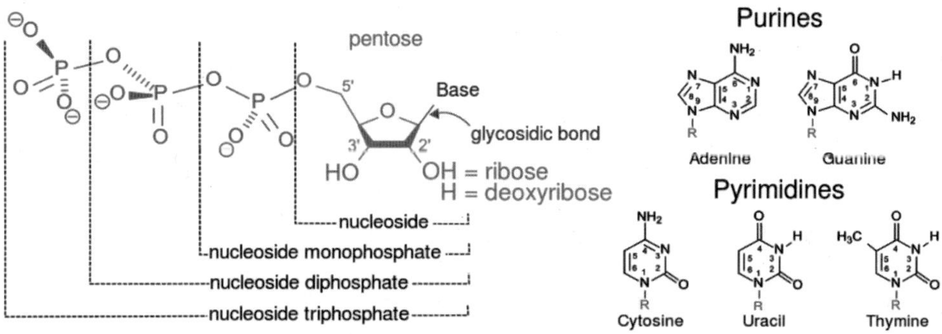

oct(o)-[ɑkt/ɔkt]: L. *octo-*, eight. 여덟.

octane[ɑ́ktein / ɔ́k-]: late 19th cent.: *octo-*, eight + *-ane*, on the pattern of words such as *-ene, ine*. 옥탄. A hydrocarbon and an alkane with the chemical formula C_8H_{18}, and the condensed structural formula $CH_3(CH_2)_6CH_3$. Octane has many structural isomers that differ by the amount and location of branching in the carbon chain. One of these isomers, 2,2,4-trimethylpentane (isooctane) is used as one of the standard values in the octane rating scale. As with all low-molecular weight hydrocarbons, octane and its isomers are very flammable. It and its isomers are components of gasoline (petrol).

octet[ɑktet / ɔk-]: *octo,* eight + *et,* diminutives suffix. 팔중항 (상태),

팔전자계 (핵외 전자의 배치). A stable group of eight electrons filling an electron shell of an atom.

octet rule: 팔전자계 규칙. A simple chemical rule of thumb that states that atoms tend to combine in such a way that they each have eight electrons in their valence shells, giving them the same electronic configuration as a noble gas. The rule is applicable to the main-group elements, especially carbon, nitrogen, oxygen, and the halogens, but also to metals such as sodium or magnesium. In simple terms, molecules or ions tend to be most stable when the outermost electron shells of their constituent atoms contain eight electrons.

Fig. The bonding in carbon dioxide (CO_2): all atoms are surrounded by 8 electrons, fulfilling the octet rule.

oil[ɔil]: L. *oleum*, (olive) oil. 기름, 유성물(油性物), 석유, 등유.

olefin[óuləfin]: French. *olefiant*, oil-forming (with reference to oily ethylene dichloride), 올레핀(에틸렌계 탄화수소 화합물). An unsaturated chemical compound containing at least one carbon-to-carbon double bond. The simplest acyclic alkenes, with only one double bond and no other functional groups, form an homologous series of hydrocarbons with the general formula C_nH_{2n}. The simplest alkene is ethylene (C_2H_4), which has the International Union of Pure and Applied Chemistry (IUPAC) name *ethene*. Alkenes are also called *olefins* (an archaic synonym, widely used in the petrochemical industry). For bridged alkenes, the Bredt's rule states that a double bond cannot be placed at the bridgehead of a bridged ring system, unless the rings are large enough. Aromatic compounds are often drawn as cyclic alkenes, but their structure and properties are different and they are not considered to be alkenes.

oleic[ouli:ik, ouli-]: L. oleum, oil. 기름의.

oleic acid: a fatty acid that occurs naturally in various animal and vegetable fats. It is an odorless, colourless oil, although commercial samples may be yellowish. In chemical terms, oleic acid is classified as a monounsaturated omega-9 fatty acid. It has the formula $CH_3(CH_2)_7CH=CH(CH_2)_7COOH$.

oligo-[áligou, -gə, əlíg- / ɔ́l-]: G. *oligos, oligoi*, small, few. little. (모음 앞에서는 olig-).

oligomer: 소중합체. A molecule that consists of a few monomer unit in contrast to a polymer that, at least in principle, consists of an unlimited number of monomers. Dimers, trimers, and tetramers are oligomers. Many oils are oligomeric, such as liquid paraffin. Plasticizers [가소제(可塑劑)] are oligomeric esters widely used to soften thermoplastics such as PVC. They may be made from monomers by linking them together, or by separation from the higher fractions of crude oil. Polybutene is an oligomeric oil used to make putty. Greek prefixes are often used to designate the number of monomer units in the oligomer, for example a tetramer being composed of four units and a hexamer of six. In biochemistry, the term oligomer is used for short, single-stranded nucleic acid fragments, such as DNA or RNA, or similar fragments of analogs of nucleic acids such as peptide nucleic acid or Morpholinos.

oligomerization: a chemical process that converts monomers to a finite degree of polymerization. The actual figure is a matter of debate, often a value between 10 and 100. When an oligomer forms as a result of chain transfer the oligomer is called a telomer and the process telomerization. A telomer is a region of highly repetitive DNA at the end of a linear chromosome.

opium[óupiəm]: G. *opion*, poppy(양귀비) juice, from *opos*, juice, from Indo-European root, water. 아편. The dried latex obtained from the opium poppy (Papaver somniferum). Opium contains up to 12% morphine, an alkaloid, which is frequently processed chemically to produce heroin for the illegal drug trade. The latex also includes codeine and non-narcotic alkaloids such as papaverine, thebaine and noscapine.

opsin[ɑ́psin/ɔp-]: G. *opsis,* seeing, viewing, appearance. 옵신. A group of light-sensitive 35-55 kDa membrane-bound G protein-coupled receptors of the retinylidene protein family found in photoreceptor cells of the retina.

rhodopsin: also known as visual purple, a pigment of the retina that is responsible for both the formation of the photoreceptor cells and the first events in the perception of light. Rhodopsins belong to the G-protein coupled receptor family and are extremely sensitive to light, enabling vision in low-light conditions. Exposed to light, the pigment immediately photobleaches, and it takes about 30 minutes to regenerate fully in humans.

visual phototransduction: a process by which light is converted into electrical signals in the rod cells, cone cells and photosensitive ganglion cells of the retina of the eye. The visual cycle is the biological conversion of a photon into an electrical signal in the retina. This process occurs via G-protein coupled receptors called opsins which contain the chromophore 11-*cis* retinal. 11-*cis* retinal is covalently linked to the opsin receptor via a Schiff base forming a retinylidene protein. When struck by a photon, 11-*cis* retinal undergoes photoisomerization to all-*trans* retinal which changes the conformation of the opsin GPCR leading to signal transduction cascades which causes closure of a cyclic GMP-gated cation channel, and hyperpolarization of the photoreceptor cell. Following isomerization and release from the opsin protein, all-*trans* retinal is reduced to all-*trans* retinol and travels back to the retinal pigment epithelium to be "recharged". It is first esterified by lecithin-retinol acyltransferase (LRAT) and then converted to 11-cis retinol by the isomerohydrolase RPE65. The isomerase activity of RPE65 has been shown, but it is still uncertain whether it also acts as a hydrolase. Finally, it is oxidized to 11-*cis* retinal before traveling back to the rod outer segment where it can again be conjugated to an opsin to form a new, functional visual pigment (rhodopsin).

Fig. visual cycle

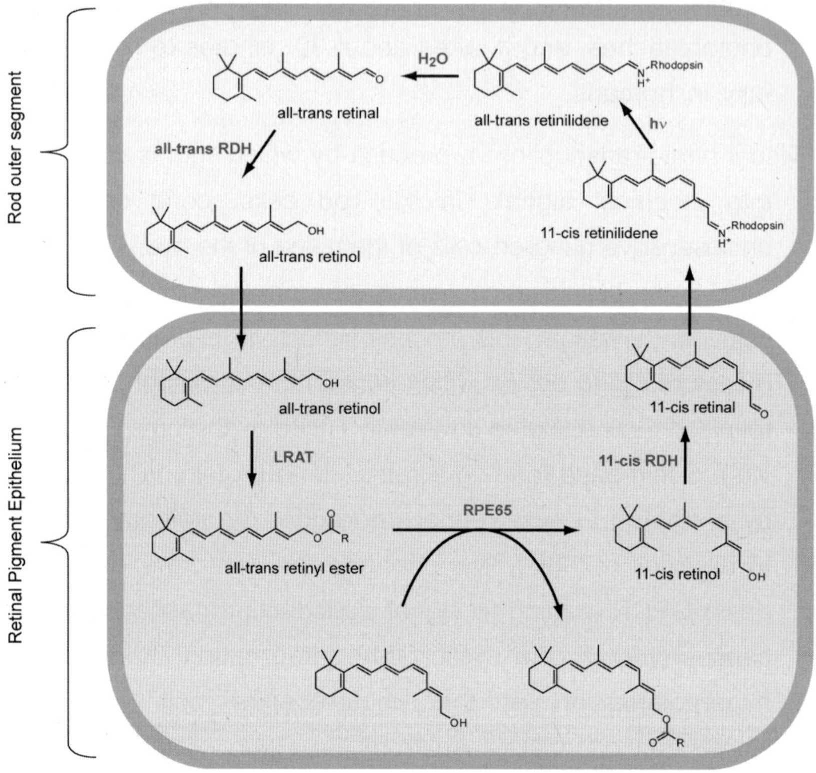

optic[ɑ́ptik/ɔ́p-]: L. *opticus, optica, opticum,* of or pertaining to sight. 시각(視覺)의, 시력의, 광학(光學)의.

<u>optics</u>: (단수취급)광학(光學)

<u>ophthalmic</u>: 눈의(ocular), 안과의, 안염의.

<u>optical activity: (+) and (−)</u>: an enantiomer can be named by the direction in which it rotates the plane of polarized light. If it rotates the light clockwise (as seen by a viewer towards

whom the light is traveling), that enantiomer is labeled (+). Its mirror-image is labeled (−). The (+) and (−) isomers have also been termed *d*- and *l*-, respectively (for dextrorotatory and levorotatory). Naming with *d*- and *l*- is easy to confuse with d- and l- labeling.

configuration: d- and l: an optical isomer can be named by the spatial configuration of its atoms. The d/l system does this by relating the molecule to glyceraldehyde. Glyceraldehyde is chiral itself, and its two isomers are labeled d and l. Certain chemical manipulations can be performed on glyceraldehyde without affecting its configuration, and its historical use for this purpose (possibly combined with its convenience as one of the smallest commonly used chiral molecules) has resulted in its use for nomenclature. In this system, compounds are named by analogy to glyceraldehyde, which, in general, produces unambiguous designations, but is easiest to see in the small biomolecules similar to glyceraldehyde. One example is the amino acid alanine, which has two optical isomers, and they are labeled according to which isomer of glyceraldehyde they come from. On the other hand, glycine, the amino acid derived from glyceraldehyde, has no optical activity, as it is not chiral (achiral). Alanine, however, is chiral. The d/l labeling is unrelated to (+)/(−); it does not indicate which enantiomer is dextrorotatory and which is levorotatory. Rather, it says that the compound's stereochemistry is related to that of the dextrorotatory or levorotatory enantiomer of glyceraldehyde — the dextrorotatory isomer of glyceraldehyde is, in fact, the d- isomer. Nine of the nineteen l-amino acids commonly found in proteins are dextrorotatory (at a wavelength of 589 nm), and d-fructose is also referred to as levulose because it is levorotatory. A rule of thumb for determining the d/l isomeric form of an amino acid is the "CORN" rule. The groups: COOH, R, NH_2 and H (where R is a variant carbon chain) are arranged around the chiral center carbon atom. Sighting with the hydrogen atom away from the viewer, if these groups are arranged clockwise around the carbon atom, then it is the d-form. If counter-clockwise, it is the l-form.

optical isomers: 광학이성질체. Two mirror images of a chiral molecule are called enantiomers or optical isomers. Pairs of enantiomers are often designated as "right-" and "left-handed."

optical purity (= enantiomeric excess): 광학 순도. A substance is a measure of how pure it is. In this case, the impurity is the undesired enantiomer (the "opposite-handed" mirror image of a chiral compound). The optical purity can be determined.

$$\text{optical purity (\%)} = \frac{[\alpha]_{\text{observed}}}{[\alpha]_{\text{maximal}}} \cdot 100$$

orbital[ɔ́:rbitl]: L. *orbitalis from orbita*, course, track, *orbitus,* circular, from *obis*, ring. 궤도함수.

atomic orbital: 원자 궤도함수. A mathematical function that describes the wave-like behavior of either one electron or a pair of electrons in an atom. This function can be used to calculate the probability of finding any electron of an atom in any specific region around the atom's nucleus.

organ[ɔ́:rgən]: L. *organon, organou,* tool, instrument, implement. (생물의) 기관(器官) (정부 등의) 기관(機關), 오르간, 악기.

organic[ɔ:rgǽnik]: L. *organicus, organica,* relating to or denoting compounds containing carbon. 유기적(有機的)인.

inorganic: (*in*, not + organic), relating to or denoting compounds which are not organic. 무기적(無機的)인.

ortho-[ɔ́:rθou, -θə]: G. *orthos*, straight, right. 옳은, 똑바른, 직립의, 수직의 (모음 앞에서는 orth-).

ortho/para directors: groups with unshared pairs of electrons, such as the amino group of aniline, are strongly *activating* and *ortho/para*-directing. Such activating groups donate those unshared electrons to the *pi* system.

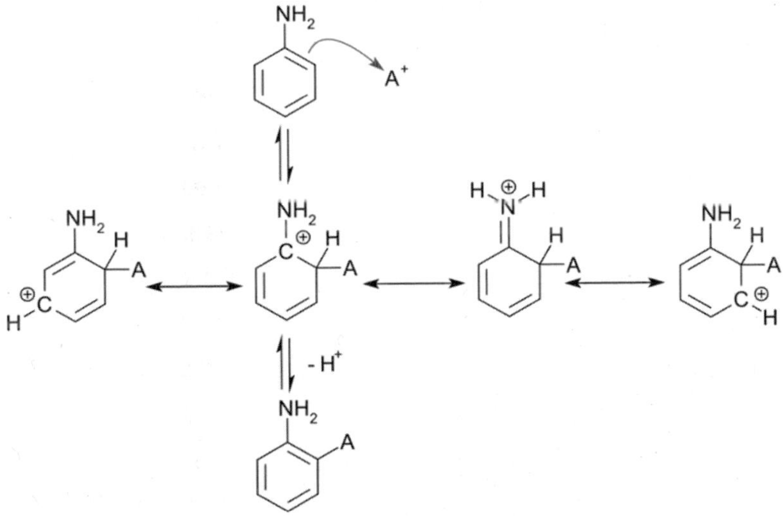

When the electrophile attacks the ortho and para positions of aniline, the nitrogen atom can donate electron density to the *pi* system (forming an iminium ion), giving four resonance structures (as opposed to three in the basic reaction). This substantially enhances the stability of the cationic intermediate.

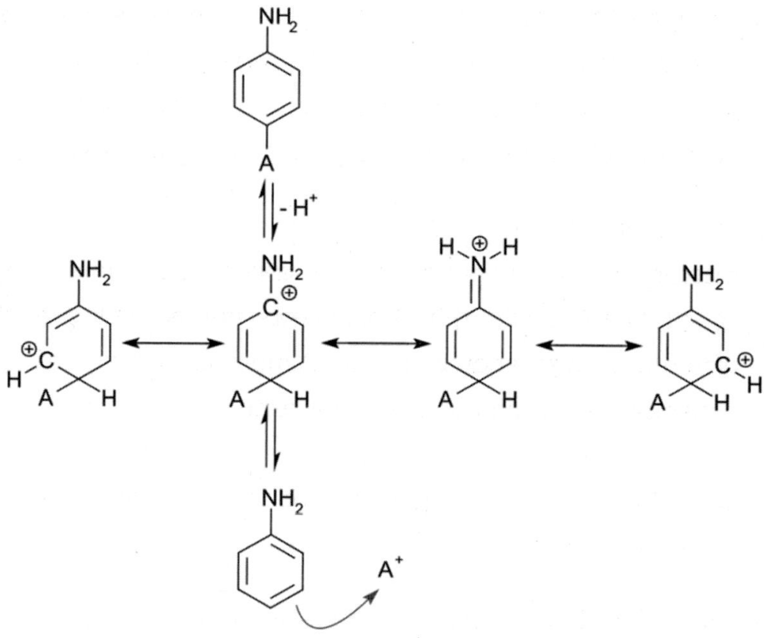

Compare this with the case when the electrophile attacks the *meta* position. In that case, the nitrogen atom cannot donate electron density to the *pi* system, giving only three resonance contributors. For this reason, the *meta*-substituted product is produced in much smaller proportion to the *ortho* and *para* products.

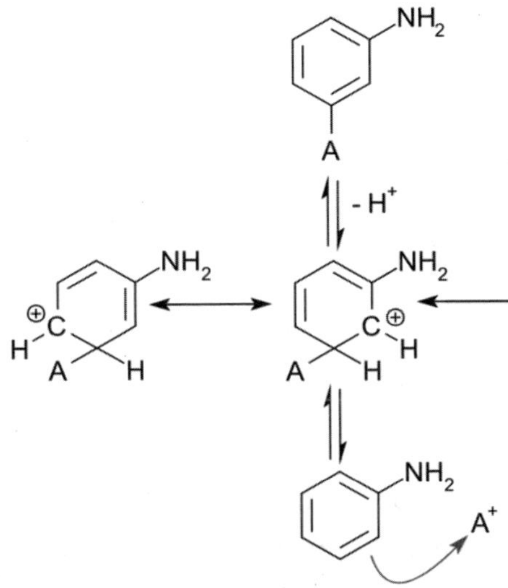

Other substituents, such as the alkyl and aryl substituents, may also donate electron density to the *pi* system; however, since they lack an available unshared pair of electrons, their ability to do this is rather limited. Thus they only weakly activate the ring and do not strongly disfavor the *meta* position. Halogens are *ortho/para* directors, since they possess an unshared pair of electrons just as nitrogen does. However, the stability this provides is offset by the fact that halogens are substantially more electronegative than carbon, and thus draw electron density away from the *pi* system. This destabilizes the cationic intermediate, and electrophilic aromatic substitution (EAS) occurs less readily. Halogens are therefore deactivating groups.

overtone[óuvərtoun]: L. < G. *overtone*, *over-*, above, over + *tone*, *tonus*, < *tonos*, tension, tone. 배음, 배진동(수).

<u>overtone band</u>: 배진동 띠. An absorption that occurs at a multiple of the fundamental absorption frequency in IR spectroscopy (2v, 3v).

oxalic[ɑksǽlik /ɔk-]: G. *oxus*, sour, because of its sharp-tasting leaves. 괭이밥(oxalis)의, 옥살산의.

<u>oxalic acid (= ethanedioic acid)</u>: the chemical compound with the formula that can be written in a number of equivalent ways, $C_2O_4H_2$, $C_2O_2(OH)_2$, and as HOOCCOOH. This colourless solid is a dicarboxylic acid. In terms of acid strength, it is about 3,000 times stronger than acetic acid. Its conjugate base, known as oxalate ($C_2O_4^{2-}$), is a reducing agent as well as a chelating agent for metal cations. Typically oxalic acid occurs as the dihydrate with the formula $C_2O_4H_2 \cdot 2H_2O$.

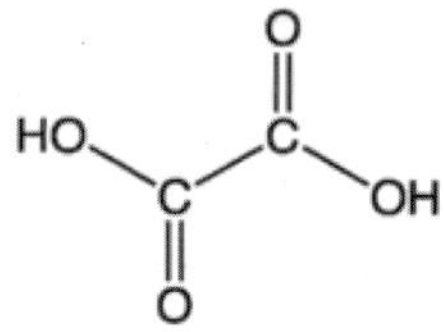

oxidation[ɑksədéiʃən / ɔk-]: French. *oxydation, oxider*, oxidize. 산화. The *loss* of electrons or an *increase* in oxidation state by a molecule, atom, or ion.

<u>oxidation reaction</u>: 산화반응. A reaction in which the number of C-H bonds decreases or the number of C-O, C-N, or C-X (x = a halogen) increases.

oxygen[ɑ́ksidʒən /ɔks-]: French. *oxygene*. 산소. Acidifying constituent. At first, it was held to be the essential component in the formation acids. The element with atomic number 8 and represented by the symbol **O**. Its name derives from the Greek roots ὀξύς (oxys) ("acid", literally "sharp", referring to the sour taste of acids) and -γενής (-genēs) ("producer", literally "begetter"), because at the time of naming, it was mistakenly thought that all acids required oxygen in their composition. At standard temperature and pressure, two atoms of the element bind to form dioxygen, a very pale blue, odorless, tasteless diatomic gas with the formula O_2. Oxygen is a member of the chalcogen(산소족) group on the periodic table and is a highly reactive nonmetallic element that readily forms compounds (notably oxides) with almost all other elements. Oxygen is a strong oxidizing agent and has the second highest electronegativity of all the elements (only fluorine has a higher electronegativity). By mass, oxygen is the third most abundant element in the universe after hydrogen and helium and the most abundant element by mass in the Earth's crust, making up almost half of the crust's mass. Free oxygen is too chemically reactive to appear on Earth without the photosynthetic action of living organisms, which use the energy of sunlight to produce elemental oxygen from water. Elemental O_2 only began to accumulate in the atmosphere after the evolutionary appearance of these organisms, roughly 2.5 billion years ago. Diatomic oxygen gas constitutes 20.8% of the volume of air.

ozone[óuzoun]: G. *ozein*, to smell. 오존. **O_3**, or trioxygen. A triatomic molecule, consisting of three oxygen atoms. It is an allotrope of oxygen that is much less stable than the diatomic allotrope (O_2). Ozone in the lower atmosphere is an air pollutant with harmful effects on the respiratory systems of animals and will burn sensitive plants; however, the ozone layer in the upper atmosphere is beneficial, preventing potentially damaging ultraviolet light from reaching the Earth's surface. Ozone is present in low concentrations throughout the Earth's atmosphere. It has many industrial and consumer applications.

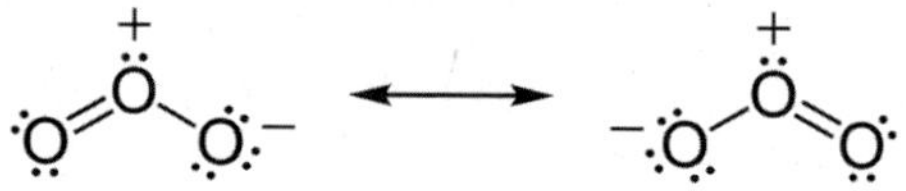

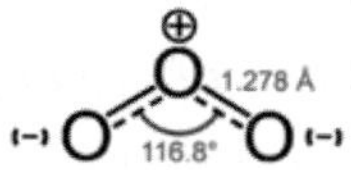

paclitaxel (Taxol): L. *taxus, taxi,* a yew(tree). 주목(朱木). Originally isolated from the bark of the pacific yew, Taxus brevifolia, which interferes with microtuble function and inhibits cell division.

Fig. Taxol

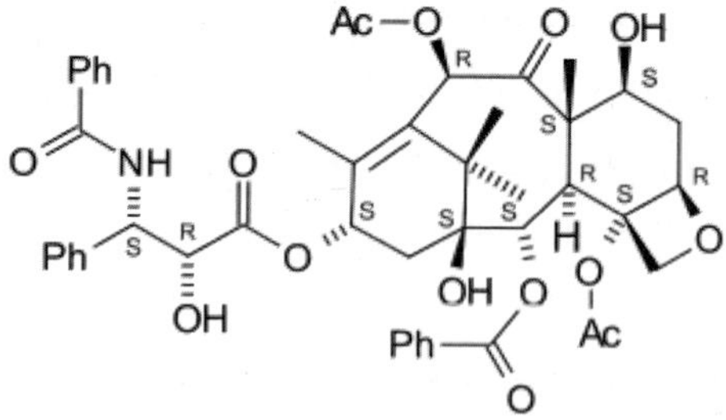

paediatric[pi:diǽtrik, ped-]: L. *paediatricus*, *paedo*, boy, child + -*iatros,* healer, physician. 소아과의, 소아과 의사의.

paediatrician (= pediatrician): 소아과 의사.

palladium[pəléidiəm]: L. *Pallas, Palladis*, (Greek mythology) the goddess Pallas, one of the names of Athena, (그리스 신화) 팔라스 여신, 아테나 여신의 다른 이름, 팔라듐(백금족의 희금속 원소). A chemical element with the chemical symbol **Pd** and an atomic number of 46. It is a rare and lustrous silvery-white metal discovered in 1803 by William Hyde Wollaston. He named it after the asteroid Pallas, which was itself named after the epithet of the Greek goddess Athena, acquired by her when she slew Pallas. Palladium, platinum, rhodium, ruthenium, iridium and osmium form a group of elements referred to as the platinum group metals (PGMs). These have similar chemical properties, but palladium has the lowest melting point and is the least dense of them. Over half of the supply of palladium and its congener platinum goes into catalytic converters, which convert up to 90% of harmful gases from auto exhaust (hydrocarbons, carbon monoxide, and nitrogen dioxide) into less-harmful substances (nitrogen, carbon dioxide and water vapor). Palladium is also used in electronics, dentistry, medicine, hydrogen purification, chemical applications, groundwater treatment and jewelry. Palladium plays a key role in the technology used for fuel cells, which combine hydrogen and oxygen to produce electricity, heat, and water.

palmitic[pælmítik]: French. *palme*, palm. 야자유에서 추출한, 팔미트산의.

palmitic acid: hexadecanoic acid in IUPAC nomenclature, one of the most common saturated fatty acids found in animals and plants. Its molecular formula is $CH_3(CH_2)_{14}CO_2H$. As its name indicates, it is a major component of the oil from palm trees (palm oil, palm kernel oil, and coconut oil). Palmitate is a term for the salts and esters of palmitic acid. The palmitate anion is the observed form of palmitic acid at basic pH.

pantothenic acid: G. *pas, pasa, pan,* all, *pantothen,* from all(every) sides. Also called vitamin B_5 (a B vitamin), a water-soluble vitamin required to sustain life (essential nutrient). Pantothenic acid is needed to form coenzyme-A (CoA), and is critical in the metabolism and synthesis of carbohydrates, proteins, and fats.

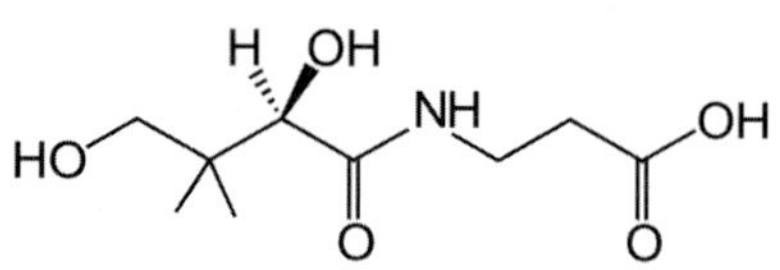

para-[pǽrə]: G. *para-,* 모음 앞에서는 *par(a)-, akin to,* similar, beside, beyoned, aside, amiss. ~으로부터, ~옆에, 이성체(異性體)의. See also: Arene Substitution Pattern

paraffin(e)[pǽrəfin]: L. *parum*, little + *affinis*, closely related, neutral quality and low chemical reactivity. 메탄계 탄화수소, 파라핀.

paraffin oil: 등유.

paralysis[pərǽləsis]: G. *paralusis*, *paraluesthai*, *para*, beside + *luein*, loosen, be disabled at the side, disabling of nerves. 마비, 중풍, 무기력. Loss of muscle function for one or more muscles.

Paralysis can be accompanied by a loss of feeling (sensory loss) in the affected area if there is sensory damage as well as motor.

penicillin[pénəsilin]: L, *penicillus*, *penicillium*, paintbrush (because of the brush-like fruiting bodies). 페니실린(항생물질의 하나). A group of antibiotics derived from Penicillium fungi. They include penicillin G, procaine penicillin, benzathine penicillin, and penicillin V. Penicillin antibiotics are historically significant because they are the first drugs that were effective against many previously serious diseases, such as syphilis(매독), and infections caused by staphylococci(포도상구균) and streptococci (연쇄구균). Penicillins are still widely used today, though many types of bacteria are now resistant. All penicillins are β-lactam antibiotics and are used in the treatment of bacterial infections caused by susceptible, usually Gram-positive, organisms.

Fig. Penicillin core structure, where "R" is the variable group.

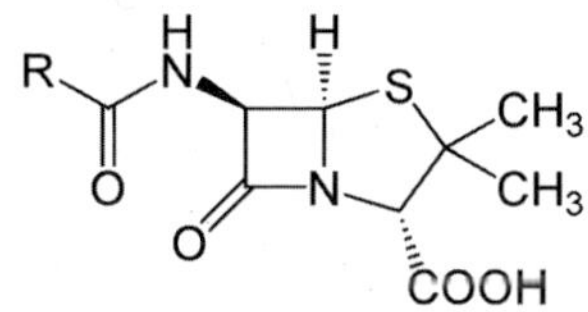

pentane[péntein]: G. *pente*, five + *-ane*, suffix. 펜탄(메탄계 탄화수소의 하나). An organic compound with the formula C_5H_{12} — that is, an alkane with five carbon atoms. The term may refer to any of three structural isomers, or to a mixture of them: in the IUPAC nomenclature, however, pentane means exclusively the *n*-pentane isomer; the other two being called "methylbutane" and "dimethylpropane"

peptide[péptaid]: G. *peptone,* neuter of *peptos*, cooked, digestid, + *ide*, oxide. 펩티드. Short polymers of amino acids linked by peptide bonds. They have the same chemical structure as proteins, but only shorter in length.

peptide bond: a covalent chemical bond formed between two molecules when the carboxyl group of one molecule reacts with the amine group of the other molecule, thereby releasing a molecule of water (H_2O). This is a dehydration synthesis reaction (also known as a condensation reaction), and usually occurs between amino acids. The resulting C(O)NH bond is called a peptide bond, and the resulting molecule is an amide. The four-atom functional group -C(=O)NH- is called a peptide link. Polypeptides and proteins are chains of amino acids held together by peptide bonds, as is the backbone of peptide nucleic acid (=PNA).

Fig. Dehydration synthesis (condensation) reaction forming an amide

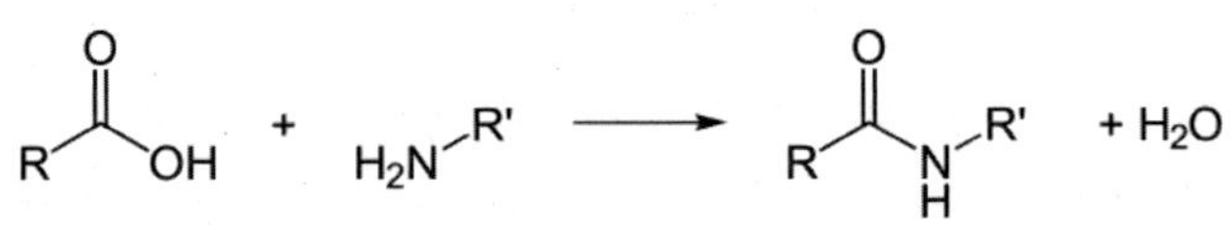

per-[pə(:)r]: L. *per-,* through, completely, very, the maximum (or supposed maximum) proportion of the specified element or group in a chemical compound. 과(過)의.

perfume[pə́:rfju:m, pə:rfjú:m / pə:fju:m]: L. *per-,* through, throughly + *fumare,* to smoke, a fragrant liquid. 향수, 향료, 향수를 뿌리다.

peri-[péri]: G > L. *peri,* around, about, surrounding, near, beyond. 주위의, 근처의, 넘어선.

<u>pericyclic reaction</u>: 고리형 협동 반응. A type of organic reaction with multiple carbon-carbon bond making and breaking wherein the transition state of the molecule has a cyclic geometry, and the reaction progresses in a concerted fashion. Pericyclic reactions are usually rearrangement reactions. The major classes of pericyclic reactions are: Electrocyclic reactions, Cycloadditions, Sigmatropic reactions, and Claisen rearrangement.

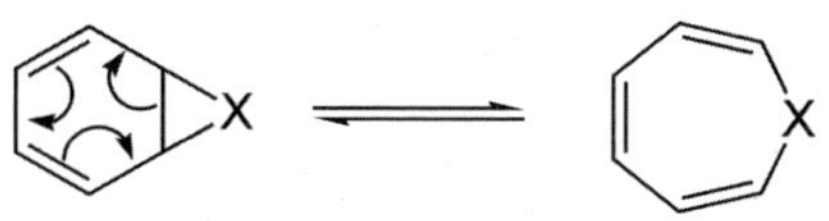

X = CR_2, NR, O

Fig. A conversion of Isochorismate into salicylate and Pyruvate in a catalyzed, true pericyclic reaction.

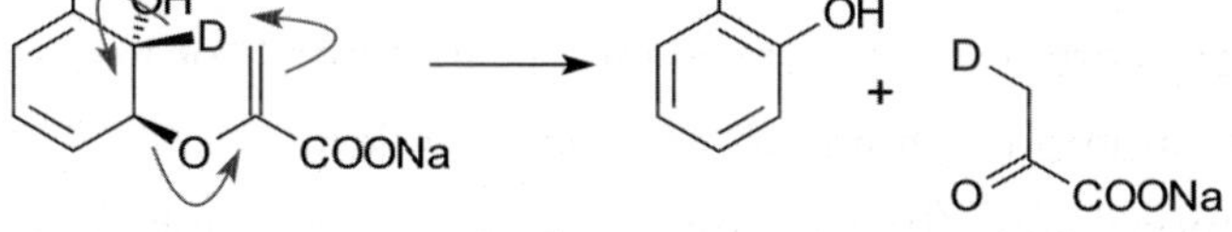

Sigmatropic reactions:

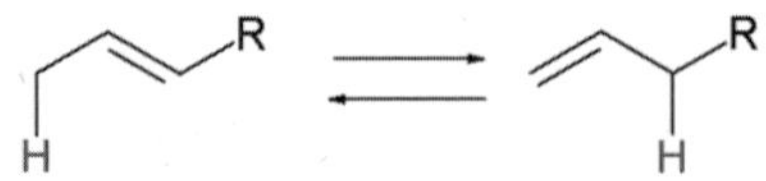

[1,3]

[1,5]

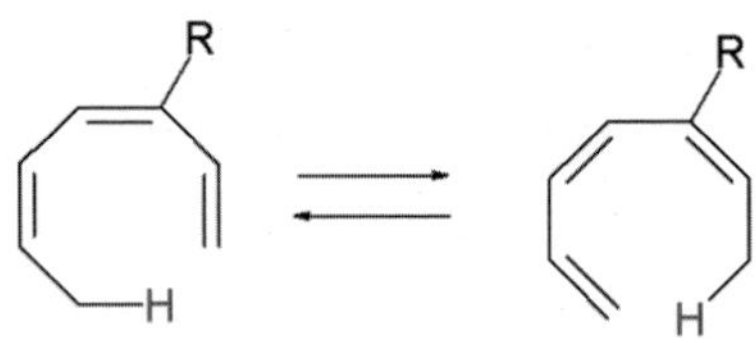

[1,7]

Claisen rearrangement:

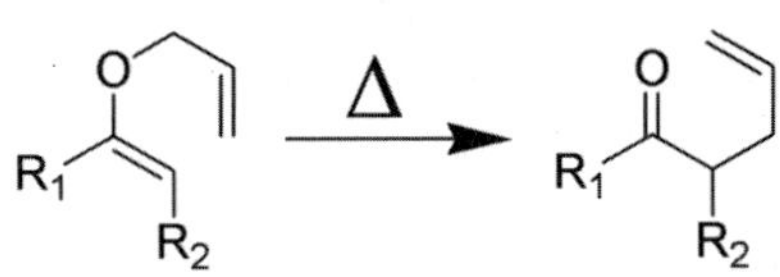

periodic[piəriάdik / -ɔd-]: L. *periodikos,* coming round at intervals. 주기적인, 정기적으로 출현하는.

periodic law: 주기율

periodic table: 주기율표. A tabular arrangement of the chemical elements, organized on the basis of their atomic numbers, electron configurations, and recurring chemical properties. Elements are presented in order of increasing atomic number (number of protons). The standard form of the table comprises an 18-column-by-7-row main grid of elements, with a double row of elements below. The table can also be deconstructed into four rectangular blocks: the s-block to the left, the p-block to the right, the d-block in the middle, and the f-block below that. The rows of the table are called periods; the columns are called groups, with some of these having names such as halogens or noble gases. Since, by definition, a periodic table incorporates recurring trends, any such table can be used to derive relationships between the properties of the elements and predict the properties of new, yet to be discovered or synthesized, elements. As a result, a periodic table — whether in the standard form or some other variant — provides a useful framework for analyzing chemical behavior, and such tables are widely used in chemistry and other sciences. Although precursors exist, Dmitri Mendeleev is generally credited with the publication, in 1869, of the first widely recognized periodic table. He developed his table to illustrate periodic trends in the properties of the then-known elements. Mendeleev also predicted some properties of then-unknown elements that would be expected to fill gaps in this table. Most of his predictions were proved correct when the elements in question were subsequently discovered. Mendeleev's periodic table has since been expanded and refined with the discovery or synthesis of further new elements and the development of new theoretical models to explain chemical behavior. All elements from atomic numbers 1 (hydrogen) to 118 (ununoctium) have been discovered or synthesized. The elements from 1 to 98 (californium) have been found to exist naturally, although some are found only in trace amounts and were initially discovered by synthesis in laboratories. The elements after 98 have only been synthesized in laboratories. Production of elements beyond ununoctium is being pursued, with the question of how the periodic table may need to be modified to accommodate any such additions being a matter of ongoing debate. Numerous synthetic radionuclides(방사성 핵종) of naturally occurring elements have also been produced in laboratories.

Group→ / ↓Period	1	2	3	4	5	6	7	8	9	10	11	12	13	14	15	16	17	18
1	1 H																	2 He
2	3 Li	4 Be											5 B	6 C	7 N	8 O	9 F	10 Ne
3	11 Na	12 Mg											13 Al	14 Si	15 P	16 S	17 Cl	18 Ar
4	19 K	20 Ca	21 Sc	22 Ti	23 V	24 Cr	25 Mn	26 Fe	27 Co	28 Ni	29 Cu	30 Zn	31 Ga	32 Ge	33 As	34 Se	35 Br	36 Kr
5	37 Rb	38 Sr	39 Y	40 Zr	41 Nb	42 Mo	43 Tc	44 Ru	45 Rh	46 Pd	47 Ag	48 Cd	49 In	50 Sn	51 Sb	52 Te	53 I	54 Xe
6	55 Cs	56 Ba		72 Hf	73 Ta	74 W	75 Re	76 Os	77 Ir	78 Pt	79 Au	80 Hg	81 Tl	82 Pb	83 Bi	84 Po	85 At	86 Rn
7	87 Fr	88 Ra		104 Rf	105 Db	106 Sg	107 Bh	108 Hs	109 Mt	110 Ds	111 Rg	112 Cn	113 Uut	114 Fl	115 Uup	116 Lv	117 Uus	118 Uuo

Lanthanides	57 La	58 Ce	59 Pr	60 Nd	61 Pm	62 Sm	63 Eu	64 Gd	65 Tb	66 Dy	67 Ho	68 Er	69 Tm	70 Yb	71 Lu
Actinides	89 Ac	90 Th	91 Pa	92 U	93 Np	94 Pu	95 Am	96 Cm	97 Bk	98 Cf	99 Es	100 Fm	101 Md	102 No	103 Lr

Standard form of the periodic table. The colors represent different categories of elements explained below.

pernicious[pərníʃəs]: L. *pernicious, pernicies*, destructive, ruin base on death. 해로운, 유독한, 파괴적인, 악성의.

pernicious anemia: 악성빈혈. Also known as Biermer's anemia, Addison's anemia, or Addison–Biermer anemia, one of many types of the larger family of megaloblastic anemias(급성빈혈). It is caused by loss of gastric parietal(체벽의) cells which are responsible, in part, for the secretion of Intrinsic Factor, a protein essential for subsequent absorption of vitamin B_{12} in the ileum[장골(腸骨)].

peroxide[pərάksaid]: *per-*, very + *oxide*, oxygen. 과산화물. A compound containing an oxygen–oxygen single bond or the peroxide anion ($[O-O]^{2-}$). The O-O group is called the peroxide group or peroxo group. In contrast to oxide ions, the oxygen atoms in the peroxide ion have an oxidation state of −1. The simplest stable peroxide is hydrogen peroxide. Superoxides, dioxygenyls, ozones and ozonides compound are considered separately. Peroxide compounds can be roughly classified into organic and inorganic. Whereas the inorganic peroxides have an ionic, salt-like character, the organic peroxides are dominated by the covalent bonds. The oxygen-oxygen chemical bond of peroxide is unstable and easily split into reactive radicals via homolytic cleavage. For this reason, peroxides are found in nature only in small quantities, in water, atmosphere, plants, and animals. Peroxides have a bleaching effect on organic substances and therefore are added to some detergents and hair colorants.

Fig. Types of peroxides, from top to bottom: peroxide ion, organic peroxide, organic hydroperoxide, peracid. The peroxide group is marked in blue. R, R^1 and R^2 mark hydrocarbon radicals.

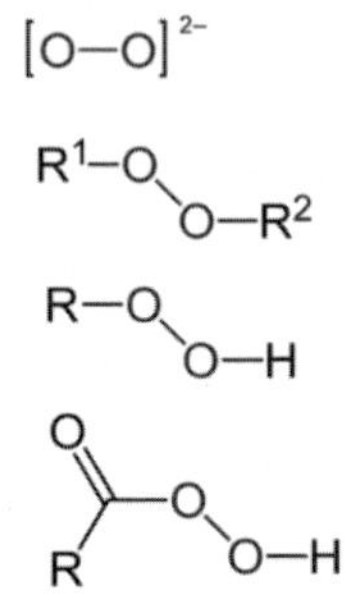

perspective[pərspéktiv]: L. *perspective*, from the verb *perspicere*, *per-*, through + *specere*, to look. 투시도, 투영도, 원근화(遠近畵).

perturbation[pə:rtərbéiʃən]: L. from the verb *perturbare*, disturb greatly, agitation, distress, physical disturbance, disorder, confusion. 섭동(攝動), 교란.

pest[pest]: L. *pestis*, plague. 해충, 역병, 페스트.

petroleum[pətróuliəm]: L. *petra*, rock + *oleum*, oil. 석유. Crude oil, naturally occurring, flammable liquid consisting of a complex mixture of hydrocarbons of various molecular weights and other liquid organic compounds, that are found in geologic formations beneath the Earth's surface. Petroleum is recovered mostly through oil drilling. This latter stage comes after the studies of structural geology (at the reservoir scale), sedimentary basin analysis, reservoir characterization (mainly in terms of porosity and permeable structures). It is refined and separated, most easily by boiling point, into a large number of consumer products, from petrol and kerosene to asphalt and chemical reagents used to make plastics and pharmaceuticals. Petroleum is used in manufacturing a wide variety of materials.

phase[feiz]: G > F. *phase* < *phasis,* appearance. A physically distinct and homogeneous form of matter characterized by its state (as gas, liquid, or solid) and composition and separated by a bounding surface from other forms. 상(相), 위상(位相).

phase-transfer catalyst (= PTC): 상전이 촉매(相轉移 觸媒). A catalyst that facilitates the migration of a reactant from one phase into another phase where reaction occurs. PTC is a special form of heterogeneous catalysis. Ionic reactants are often soluble in an aqueous phase but insoluble in an organic phase in the absence of the phase transfer catalyst. The catalyst functions like a detergent for solubilizing the salts into the organic phase. Phase transfer catalysis refers to the acceleration of the reaction upon the addition of the phase transfer catalyst. By using a PTC process, one can achieve faster reactions, obtain higher conversions or yields, make fewer byproducts, eliminate the need for expensive or dangerous solvents that will dissolve all the reactants in one phase, eliminate the need for expensive raw materials and/or minimize waste problems. Phase transfer catalysts are especially useful in green chemistry — by allowing the use of water, the need for organic solvents is reduced. For example, the nucleophilic aliphatic substitution reaction of an aqueous sodium cyanide solution with an etherial solution of 1-bromooctane does not readily occur. The 1-bromooctane

is poorly soluble in the aqueous cyanide solution, and the sodium cyanide does not dissolve well in the ether. Upon the addition of small amounts of hexadecyltributylphosphonium bromide, a rapid reaction ensues to give nonyl nitrile:
$C_8H_{17}Br(org) + NaCN(aq) \rightarrow C_8H_{17}CN(org) + NaBr(aq)$
(catalyzed by a $R_4P^+Cl^-$ PTC)

phenol[fí:noul, -nɑl / -nɔl]: French. *phenole*, based on *phene*, benzene. carbolic acid. 페놀. An organic compound with the chemical formula C_6H_5OH. It is a white, crystalline solid. This functional group consists of a phenyl, bonded to a hydroxyl (-OH). It is a mildly acidic compound that requires careful handling.

phenolphthalein[fi:nɔ:lfeǽli:n / -nɔ:lfeǽliin]: F. *phene,* to bring to light + *-ol*, suffix + *phthal,* < naphthalic + *-ein,* in, suffix, the formula $C_{20}H_{14}O_4$ and is often written as "HIn" or "phph" in shorthand notation. Often used in titrations, it turns colorless in acidic solutions and pink in basic solutions.

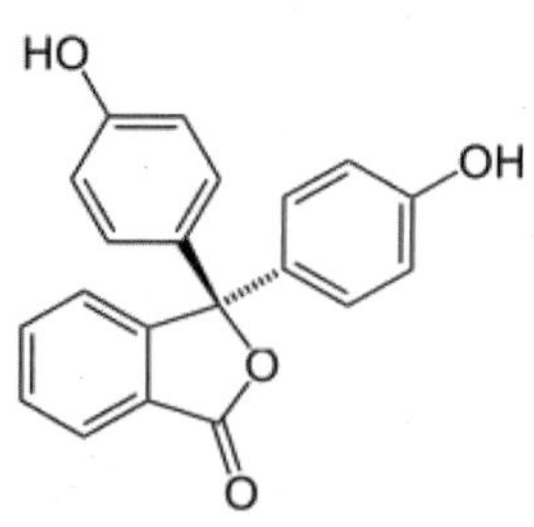

phenyl[fénl, fíːnl / fíːnail]: G. *phaino*, shinning. 페닐기.

Pheromones[férəmòun]: G. *pherein*, to bear, convey, to carry + *hormone*, impetus. 페로몬, 유인물질. A secreted or excreted chemical factor that triggers a social response in members of the same species. Pheromones are chemicals capable of acting outside the body of the secreting individual to impact the behavior of the receiving individual. There are alarm *pheromones*, *food trail pheromones*, *sex pheromones*, and many others that affect behavior or physiology. Their use among insects has been particularly well documented. In addition, some vertebrates and plants communicate by using pheromones.

phil-[fil]: G > L. *philla*, philo-, like, love, affection. 좋아하는, 사랑하는. philosophy (철학).

-phil[fil]: G > L. *philla*, phile, like, love. 좋아하는, 사랑하는. electrophile (친전자체), nucleophile (친핵체).

phosgene[fásdʒiːn / fɔ́z-]: G. *phōs*, light -*genesis,* birth. 포스겐 (무색 독가스). The chemical compound with the formula $COCl_2$. This colorless gas gained infamy as a chemical weapon during World War I. It is also a valued industrial reagent and building block in synthesis of pharmaceuticals and other organic compounds. In low concentrations, its odor resembles freshly cut hay or grass. In addition to its industrial production, small amounts occur naturally from the breakdown and the combustion of organochlorine compounds, such as those used in refrigeration systems. The name, sounding similar to "phosphine", does not mean it has any phosphorus.

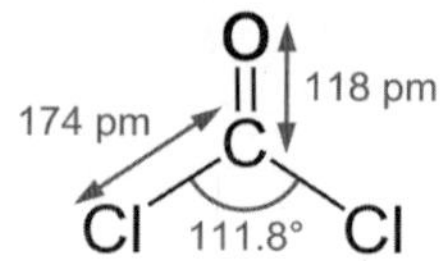

phosphine[fɑ́sfi:n/fɔ́s-]: L. *phospho-, phosphorus* + *-ine.* phosphine. 포스핀, 기체상의 인화수소. A colourless, foul-smelling, poisonous gas, PH_3, which is analogous to ammonia and when impure is often spontaneously flammable in air.

phosphorus[fɑ́sfərəs / fɔ́s]: G. *phōsphoros*, *phōs*, light + *-phoros*, bringing. 인(비금속 원소의 하나). The chemical element that has the symbol **P** and atomic number 15. A multivalent nonmetal of the nitrogen group, phosphorus as a mineral is almost always present in its maximally oxidized state, as inorganic phosphate rocks. Elemental phosphorus exists in two major forms — white phosphorus and red phosphorus — but due to its high reactivity, phosphorus is never found as a free element on Earth.

phosphate: French, *phosphore*, phosphorus. 인산염.

photo-[fóutou, tə]: G. *phōs, phōt-os*, light. 빛.

photon: 광자(光子).

photosynthesis: L. *photosynthesis: phot(o)*, light + *syn-*, with together + *tithenai,* to put. 광합성(光合成).

photochemistry: 광화학(光化學). A sub-discipline of chemistry, is the study of the interactions between light and atoms or molecules. Photochemistry describes chemical reactions that proceed with the absorption of light. Everyday examples include photosynthesis, the degradation of plastics and the formation of vitamin D with sunlight.

physics[fíziks]: L. < G. *phusika* > *physica*, natural things, from *phusis*, nature. 물리학, 물리적 현상. A natural science that involves the study of matter and its motion through spacetime, along with related concepts such as energy and force. More broadly, it is the general analysis of nature, conducted in order to understand how the universe behaves.

physiology[fiziálədʒi / -ɔ́l-]: L > G. *physiologia*, > *phusiologia*, *physio-*, natural + *-logy*, philosophy. 생리학. The science of the function of living systems. This includes how organisms, organ systems, organs, cells, and bio-molecules carry out the chemical or physical functions that exist in a living system.

pi[pai]: 파이 [그리스 알파벳의 열 여섯째 자(Π, π)의 명칭; 로마자의 P, p에 해당. <수학> 파이 (π), 원주율 (약 3.1416). A mathematical constant that is the ratio of any Euclidean circle's circumference to its diameter. π is approximately equal to 3.14. Many formulae in mathematics, science, and engineering involve π, which makes it one of the most important mathematical constants. For instance, the area of a circle is equal to π times the square of the radius of the circle. π is an irrational number, which means that its value cannot be expressed exactly as a fraction having integers in both the numerator and denominator (unlike 22/7). Consequently, its decimal representation never ends and never repeats.

Fig. The circumference of a circle is slightly more than three times as long as its diameter. The exact ratio is called π.

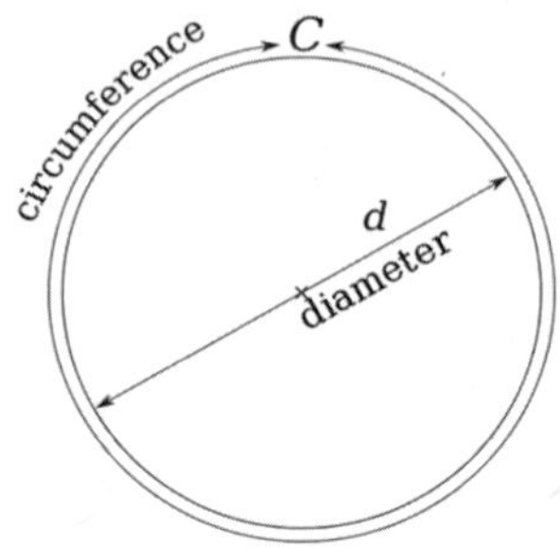

pico-[pí:kou, -kə, pái-]: L. *beccus,* little bit. 10^{-12}(1조분의 1).

pineal[píniəl, páin-, painí:əl]: L. *pinea*, pine cone, the anatomical term refers to the shape of the gland. 송과체의, 솔방울 같은.

piperidine[pipérədi:n, -din, pai-]: L. *piper*, pepper + *-ide*, suffix + *-ine*, suffix, forming names of amines. A peppery-smelling liquid formed by the reduction of pyridine. 피페리딘(무색, 수용성 액체; 용제로 쓰임). An organic compound with the molecular formula $(CH_2)_5NH$. This heterocyclicamine consists of a six-membered ring containing five methylene units and one nitrogen atom. It is a colorless fuming liquid with an odor described as ammoniacal, pepper-like. Piperidine is a widely used building block and chemical reagent in the synthesis of organic compounds, including pharmaceuticals.

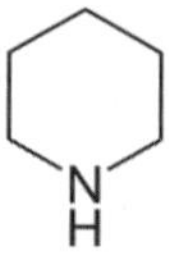

piperine[pípəri:n, -in]: L. *piper*, pepper + *-ine*, suffix, forming names of alkaloids. 피페린(후추의 매운 맛의 주성분). The alkaloid responsible for the pungency of black pepper and long pepper, along with chavicine (an isomer of piperine). It has also been used in some forms of traditional medicine and as an insecticide. Piperine forms monoclinic needles, is slightly soluble in water and more so in alcohol, ether or chloroform: the solution in alcohol has a pepper-like taste. It yields salts only with strong acids.

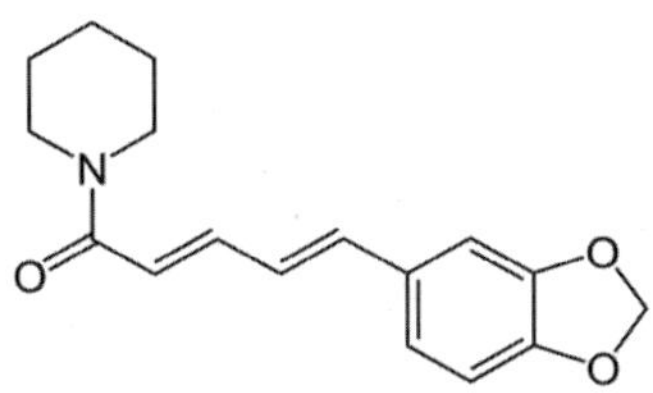

plane[plein]: L. *planum, planus,* flat surface, plain. 평면, 면, 평지, 평원, 명료한.

plane of symmetry: 대칭면. An imaginary plane that bisects a molecule into mirror images. Reflection symmetry, line symmetry, mirror symmetry, mirror-image symmetry, or bilateral symmetry is symmetry with respect to reflection. That is, a figure which does not change upon undergoing a reflection has reflectional symmetry. In 2D there is a line of symmetry, in 3D a plane of symmetry. An object or figure which is indistinguishable from its transformed image is called mirror symmetric.

Fig. The axes of symmetry drawn in, and no axes is asymmetric.

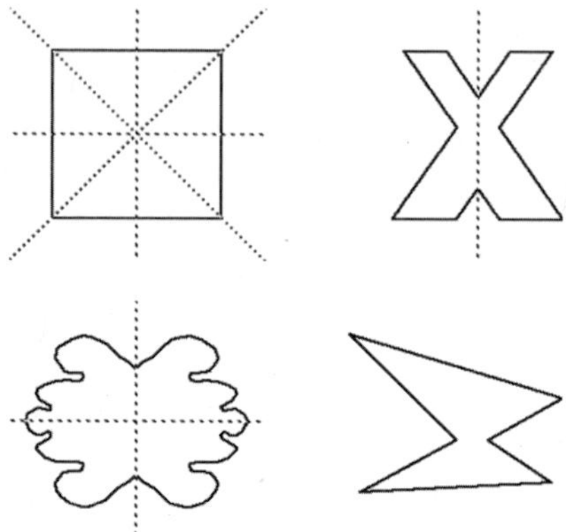

plastic[plǽstik]: L. < G. *plastikos, plassein* > *plasticus*, to mould. 플라스틱. Any of a wide range of synthetic or semi-synthetic organic solids that are moldable. Plastics are typically organic polymers of high molecular mass, but they often contain other substances. They are usually synthetic, most commonly derived from petrochemicals, but many are partially natural.

Fig. Household items made of various types of plastic.

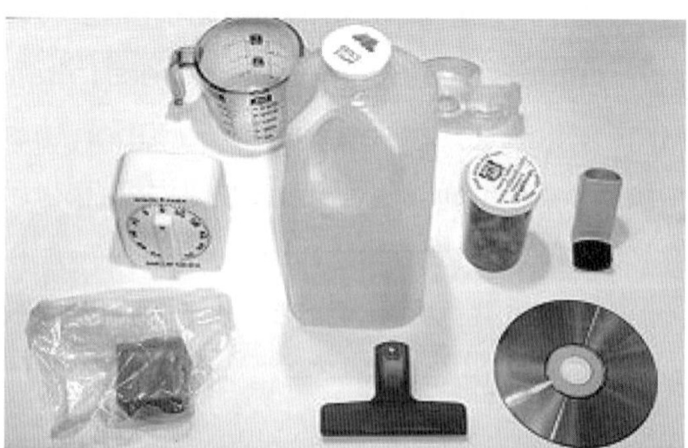

polar[póulər]: L. *polus*, end of axis. 극성, 극선, 극면.

polarity: 극성. A separation of electric charge leading to a molecule orits chemical groups having an electric dipole or multipole moment. Polar molecules interact through dipole–dipole intermolecular forces and hydrogen bonds. Molecular polarity is dependent on the difference in electronegativity between atoms in a compound and the asymmetry of the compound's structure. For example, a molecule of water is polar because of the unequal sharing of its electrons between oxygen and hydrogen in which the former has larger electronegativity than the latter, resulting in a "bent" structure, whereas methane is considered nonpolar because the carbon shares the electrons with the hydrogen atoms almost uniformly. Polarity underlies a number of physical properties including surface tension, solubility, and melting- and boiling-points.

polarizability: 편극[도,률,성], 분극성. The relative tendency of a charge distribution, like the electron cloud of an atom or molecule, to be distorted from its normal shape by an external electric field, which may be caused by the presence of a nearby ion or dipole.

polarization (also polarisation): 편광, 편극, 편극 세기, 분극. A property of certain types of waves that describes the orientation of their oscillations. Electromagnetic waves such as light exhibit polarization; acoustic waves (sound waves) in a gas or liquid do not have polarization because the direction of vibration and direction of propagation are the same.

Fig. Animation of a circularly polarized wave as a sum of two components.

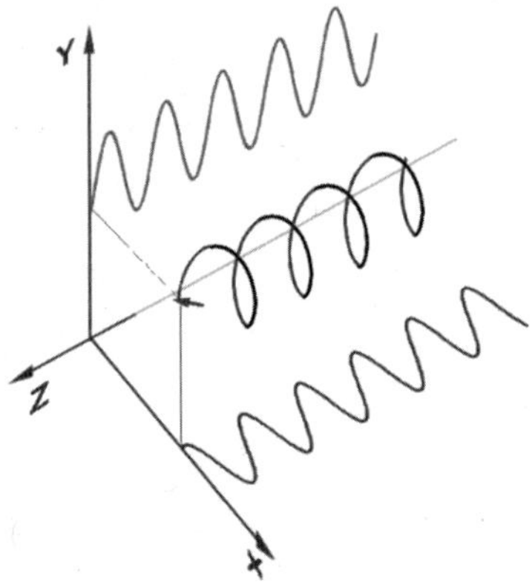

polarize[pouləraiz]: restrict the vibrations of (a wave, especially light) wholly or partially to one direction. (광학)[빛]을 편광시키다. Cause (something) to acquire polarity, -에 극성(極性)을 부여하다, 분극하다.

polarimeter[poulərímətər]: L. *polaris*, polar + -meter, *mete*, measure + *-er*, person, thing. 편광계. A scientific instrument used to measure the angle of rotation caused by passing polarized light through an optically active substance. Some chemical substances are optically active, and polarized light will rotate either to the left (counter-clockwise) or right (clockwise) when passed through these substances. The amount by which the light is rotated is known as the angle of rotation.

pole[poul]: German. *pole*, field. 극(極).

pollution[pəlúːʃən]: L. *pollution,* from the verb *polluere*, *pol*, forth + *lutum*, mud. 오염, 공해.

polymer[pάlimər / pɔl-]: G. *polumeros*, *polu-*, many + *meros*, a share. having many parts, 중합체(重合體), 고분자(高分子). A large molecule (macromolecule) composed of repeating structural units. These subunits are typically connected by covalent chemical bonds. Although the term polymer is sometimes taken to refer to plastics, it actually encompasses a large class comprising both natural and synthetic materials with a wide variety of properties. Because of the extraordinary range of properties of polymeric materials, they play an essential and ubiquitous role in everyday life. This role ranges from familiar synthetic plastics and elastomers to natural biopolymers such as nucleic acids and proteins that are essential for life. Natural polymeric materials such as shellac, amber, and natural rubber have been used for centuries. A variety of other natural polymers exist, such as cellulose, which is the main constituent of wood and paper. The list of synthetic polymers includes synthetic rubber, Bakelite, neoprene, nylon, PVC, polystyrene, polyethylene, polypropylene, polyacrylonitrile, PVB, silicone, and many more. Most commonly, the continuously linked backbone of a polymer used for the preparation of plastics consists mainly of carbon atoms. A simple example is polyethylene, whose repeating unit is based on ethylenemonomer.

<u>polymerization</u>: 중합 (반응). A process of reacting monomer molecules together in a chemical reaction to form

three-dimensional networks or polymer chains. There are many forms of polymerization and different systems exist to categorize them.

Fig. An example of alkene polymerization, in which each Styrene monomer unit's double bond reforms as a single bond with another styrene monomer and forms polystyrene.

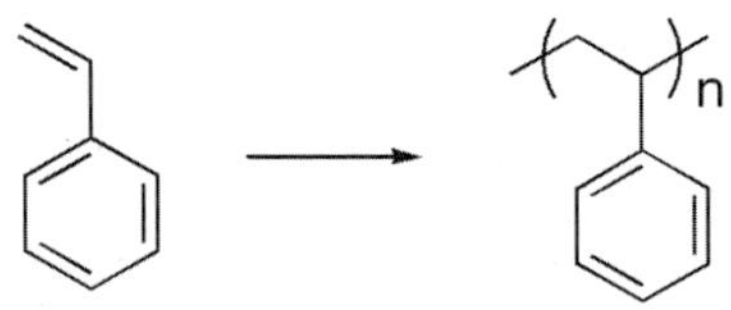

porphyrin[pɔ́:rfirin]: G. *porphyra, porphyras,* shellfish yielding purple dye, purple, purple cloth, deep-red or purple fluorescent crystalline pigments. 포르피린. A group of organic compounds of which many occur in nature, most well-known as the pigment in red blood cells. They are heterocyclic macrocycles characterised by the presence of four modified pyrrole subunits interconnected at their α carbon atoms via methine bridges (=CH-). Porphyrins are aromatic, and they obey Hückel's rule for aromaticity in that they possess 4n+2 π electrons (n=4 for the shortest cyclic path) that are delocalized over the macrocycle. The macrocycles, therefore, are highly-conjugated systems and, as a consequence, have very intense absorption in the visible region and therefore are deeply colored. The macrocycle has 26 pi electrons in total. The parent porphyrin is porphine, and substituted porphines are called porphyrins.

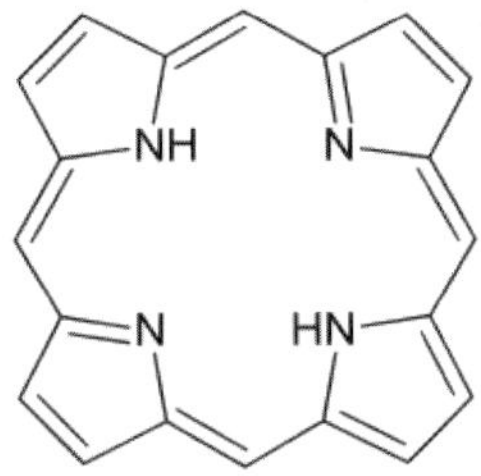

potential[pətén∫əl]: L. *potentialis, potentia, potent-,* power, being able. 전위(電位), 퍼텐셜

potential energy: 퍼텐셜 에너지, 위치 에너지. The energy of a body or a system with respect to the position of the body or the arrangement of the particles of the system. The SI unit of measure for energy and work is the Joule (symbol J). Chemical potential energy is a form of potential energy related to the structural arrangement of atoms or molecules. This arrangement may be the result of chemical bonds within a molecule or otherwise. Chemical energy of a chemical substance can be transformed to other forms of energy by a chemical reaction. As an example, when a fuel is burned the chemical energy is converted to heat, same is the case with digestion of food metabolized in a biological organism. Green plants transform solar energy to chemical energy through the process known as photosynthesis, and electrical energy can be converted to chemical energy through electrochemical reactions. The similar term chemical potential is used to indicate the potential of a substance to undergo a change of configuration, be it in the form of a chemical reaction, spatial transport, particle exchange with a reservoir, etc.

precursor[prikə́:rsər, prí:kə:r-]: L. *praecursor*, from *praecurs*, *prae-*, pre, before, beforehand + *currere*, to run, forerunner. 선구물질, 전구체(前驅體), 선행물. A compound that participates in the chemical reaction that produces another compound. In biochemistry, the term "precursor" is used more specifically to refer to a chemical compound preceding another in a metabolic pathway.

preservative[prizə́:rvətiv]: L. *praeservativus*, kept, from the verb *praeservare, prae-,* before, in advance + *servare,* to keep. 방부제, 예방약.

primary[práimeri, -məri / -məri]: L. *primarius, primus*, first. 일차의, 주요한, 초기의, 기본적인.

pro[prə, prou, prɑ]: L. *pro*, in front of, on behalf of, instead of, on account of. 전에, 앞에, 전방에, 대신에.

<u>prochirality</u>: 선구 손대칭성, 선구 키랄성. In stereochemistry, prochiral molecules are those that can be converted from achiral to chiral in a single step. If two identical substituents are attached to a sp^3-hybridized atom, the descriptors *pro*-R and *pro*-S are used to distinguish between the two. Promoting the *pro*-R substituent to higher priority than the identical alternative would result in an *R* chirality center at the original sp^3-hybridized atom, and vice versa. A trigonal planar sp^2-hybridized atom can be converted to a chiral center when a substituent is added to the *re* or *si* face of the molecule. These faces are labelled

re if the substituents at the trigonal atom are disposed in decreasing Cahn-Ingold-Prelog priority order in a clockwise order, and *si* if the priorities decrease in counter-clockwise order; but the designation of the resulting chiral centre as *s* or *r* depends on the priority of the incoming group.

Fig. An sp^2-hybridized carbon atom, with *re* and *si* faces.

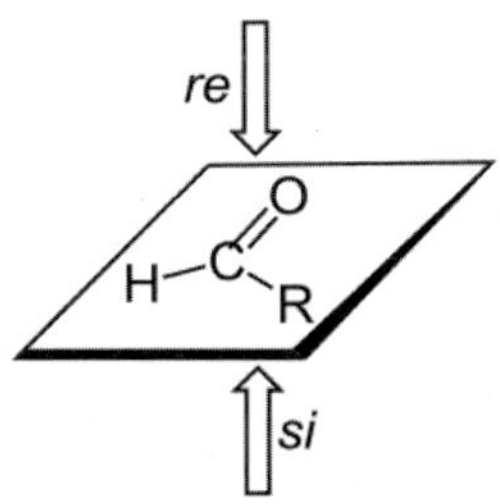

proenzyme(선구 효소, 효소 전구체): a zymogen (or proenzyme) is an inactive enzyme precursor. A zymogen requires a biochemical change (such as a hydrolysis reaction revealing the active site, or changing the configuration to reveal the active site) for it to become an active enzyme. The amino acid chain that is released upon activation is called the activation peptide.

progesterone[proudʒéstəroun]: L. *pro-*, in front of, on behalf of, instead of + *gestation*, carry in the womb + *-ine*, suffix. 프로게스테론, 황체(黃體)호르몬. 여성sex 호르몬. A C-21 steroid hormone involved in the female menstrual cycle, pregnancy (supports gestation) and embryogenesis of humans and other species. Progesterone belongs to a class of hormones called progestogens, and is the major naturally occurring human progestogen.

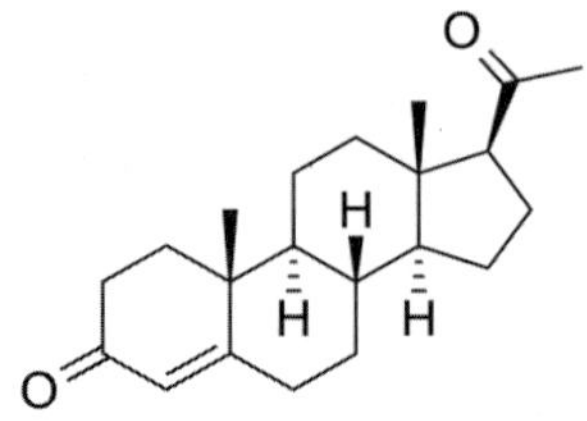

projection[prədʒékʃən]: L. *projection, proicere*, throw forth. 투영도(投影圖).

Fisher projection: devised by Hermann Emil Fischer in 1891, a two-dimensional representation of a three-dimensional organic molecule by projection. Fischer projections were originally proposed for the depiction of carbohydrates and used by chemists, particularly in organic chemistry and biochemistry. The use of Fischer projections in non-carbohydrates is discouraged, as such drawings are ambiguous when confused with other types of drawing.

Fig. Projection of a tetrahedral molecule onto a planar surface.

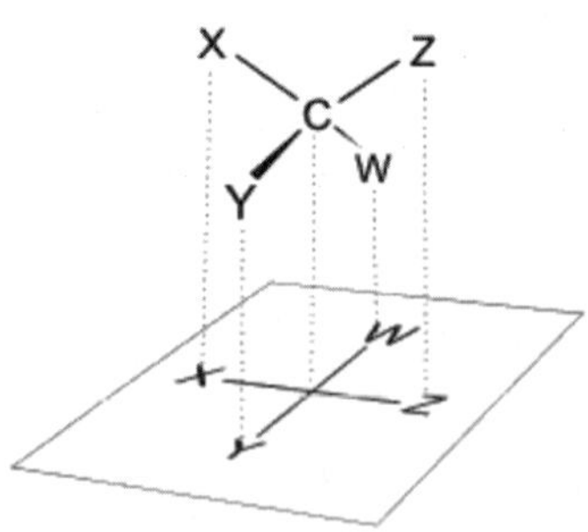

Fig. Visualizing a Fischer projection.

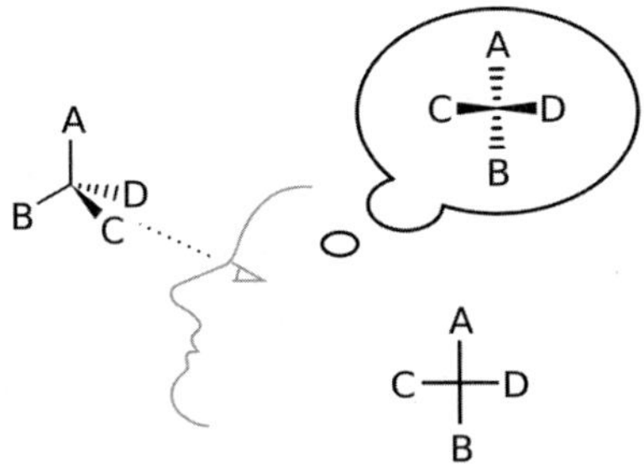

propionic acid[proupiɑ́nik, -óun-, ǽsid] (= propanoic acid): G. *pro-*, before, in front + *pión*, fat, first fatty acid. 프로피온산. Being the first in the order of the carboxylic acid series to form fatty compounds [formic (methanoic) and acetic (ethanoic) acids, which precede it in the series, lack this property]. A naturally-occurring carboxylic acid with chemical formula CH_3CH_2COOH. It is a clear liquid with a pungent odor. The anion $CH_3CH_2COO^-$ as well as the salts and esters of propionic acid are known as propionates (or propanoates).

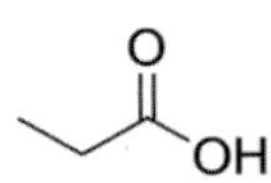

propagation[prɑpəgéiʃən / prɔp-]: L. *propagation*, *propagare*, multiplied from layers. 전파, 전개.

propagation reaction: 전파 반응. A process in which a reactive intermediate is continuously regenerated during the course of a chemical reaction. In polymerization reaction, the reactive end-groups of a polymer chain react in each propagation step with a new monomer molecule transferring the reactive group to that last unit.

prostaglandin[prɑstəglǽndin / prɔs-]: 1930s. prostate, L. *prostata*, *pro*, before + *statos*, standing, one that stands before + *gland*, L. *glandulae*, throat glands + -in, suffix. 프로스타글란딘(일군의 불포화 지방산의 총칭). Any member of a group of lipid compounds that are derived enzymatically from fatty acids and have important functions in the animal body. Every prostaglandin contains 20 carbon atoms, including a 5-carbon ring. They are mediators and have a variety of strong physiological effects, such as regulating the contraction and relaxation of smooth

Fig. E_1 - Alprostadil

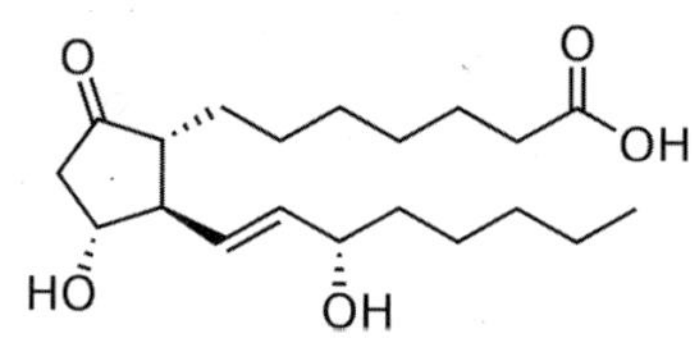

I_2 - Prostacyclin

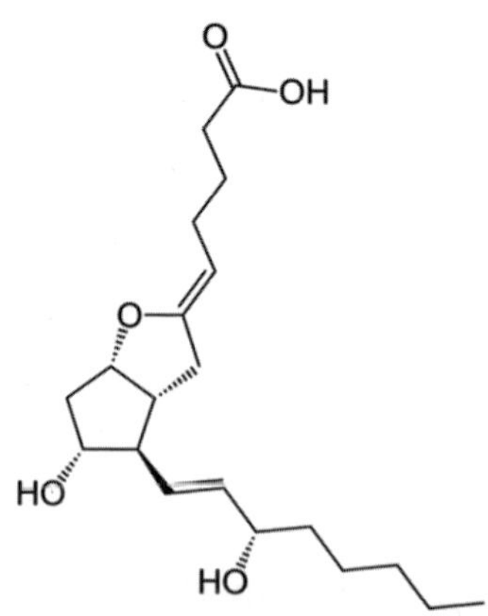

muscle tissue. Prostaglandins are not hormones, but autocrine or paracrine, which are locally acting messenger molecules. They differ from hormones in that they are not produced at a discrete site but in many places throughout the human body. Also, their target cells are present in the immediate vicinity of the site of their secretion (of which there are many).

protect[prətékt]: L. *protect*, covered in front. - 을 보호하다, 지키다.

<u>protecting group or protective group</u>: 보호 원자단, 보호기. Introduced into a molecule by chemical modification of a functional group in order to obtain chemoselectivity in a subsequent chemical reaction. It plays an important role in multistep organic synthesis.

Fig. Acetal protection of a ketone during reduction of an ester, vs. reduction to a dialcohol when unprotected.

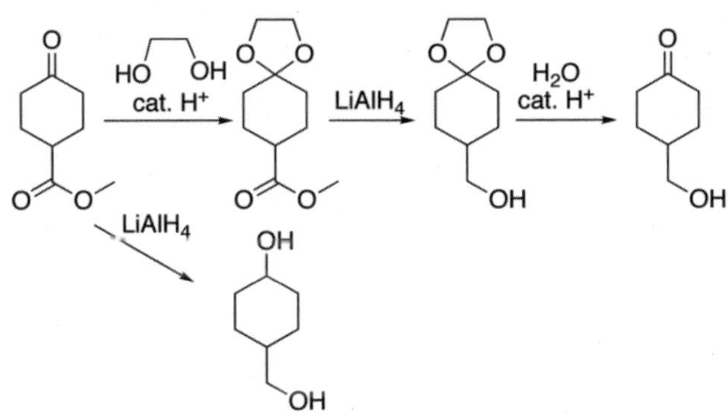

In many preparations of delicate organic compounds, some specific parts of their molecules cannot survive the required reagents or chemical environments. Then, these parts, or groups, must be protected. For example, lithium aluminium hydride is a highly reactive but useful reagent capable of reducing esters to alcohols. It will always react with carbonyl groups, and this cannot be discouraged by any means. When a reduction of an ester is required in the presence of a carbonyl, the attack of the hydride on the carbonyl has to be prevented. For example, the carbonyl is converted into an acetal, which does not react with hydrides. The acetal is then called a protecting group for the carbonyl. After the step involving the hydride is complete, the acetal is removed (by reacting it with an aqueous acid), giving back the original carbonyl. This step is called deprotection. Protecting groups are more commonly used in small-scale laboratory work and initial development than in industrial production processes because their use adds additional steps and material costs to the process. However, the availability of a cheap chiral building block can overcome these additional costs.

protein[próuti:n, -tin]: G. *prōteios*, *prōtos*, primary, first. 단백질. Biochemical compounds consisting of one or more polypeptides typically folded into a globular or fibrous form, facilitating a biological function. A polypeptide is a single linear polymer chain of amino acids bonded together by peptide bonds between the carboxyl and amino groups of adjacent amino acid residues. The sequence of amino acids in a protein is defined by the sequence of a gene, which is encoded in the genetic code. In general, the genetic code specifies 20 standard amino acids. Sometimes proteins have non-peptide groups attached, which can be called prosthetic groups(보결 원자단, 보결 분자단) or cofactors(보조인자). Proteins can also work together to achieve a particular function, and they often associate to form stable protein complexes

proton[próutɑn / -tɔn]: G. *prōtos*. first. 양성자. A subatomic particle with the symbol **p** or **p+** and a positive electric charge of 1 elementary charge. One or more protons are present in the nucleus of each atom, along with neutrons. The number of protons in each atom is its atomic number. In the standard model of particle physics, the proton is a hadron, composed of quarks. Prior to that model becoming a consensus in the physics community, the proton was considered a fundamental particle. A proton is composed of two up quarks and one down quark, and is about 1.6–1.7 fm in diameter. The free proton is stable and is found naturally in a number of situations. Free protons exist in plasmas in which temperatures are too high to allow them to combine with electrons. Free protons of high energy and velocity make up 90% of cosmic rays, which propagate in vacuum for interstellar distances. Free protons are emitted directly from atomic nuclei in some rare types of radioactive decay, and also result from the decay of free neutrons, which are unstable. In all such cases, protons must lose sufficient velocity and (kinetic energy) to allow them to become associated with electrons, since this is a relatively low-energy interaction. However, in such an association, the character of the bound proton is not changed, and it remains a proton. The attraction of low-energy protons to electrons, either free electrons or electrons as present in normal matter, causes such protons to soon form chemical bonds with atoms. This happens at sufficiently "cold" temperatures (comparable to temperatures at the surface of the Sun). In interaction with normal (non plasma) matter, low-velocity free protons are attracted to electrons in any atom or molecule with which they come in contact, causing them to combine. In vacuum, a sufficiently slow proton may pick up a free electron, becoming a neutral hydrogen atom, which then will then react chemically with other atoms if they are available and sufficiently cold.

pseud(o)-[sú:dou,-də/sjú:-]: 모음 앞에서는 *pseud-,* G. *pseudein,* to deceive, false. 유사의, 거짓의, 가짜의.

pseudo-first-order reaction: 유사 일차반응. Measuring a second-order reaction rate with reactants A and B can be problematic: The concentrations of the two reactants must be followed simultaneously, which is more difficult; or measure one of them and calculate the other as a difference, which is less precise. A common solution for that problem is the pseudo-first-order approximation. If either [A] or [B] remains constant as the reaction proceeds, then the reaction can be considered pseudo-first-order because, in fact, it depends on the concentration of only one reactant. If, for example, [B] remains constant, then: $r = k[A][B] = k'[A]$ where $k' = k[B]_0$ (k' or k_{obs} with units s^{-1}) and an expression is obtained identical to the first order expression above. One way to obtain a pseudo-first-order reaction is to use a large excess of one of the reactants ([B]》[A] would work for the previous example) so that, as the reaction progresses, only a small amount of the reactant is consumed, and its concentration can be considered to stay constant. By collecting *k′* for many reactions with different (but excess) concentrations of [B], a plot of *k′* versus [B] gives *k* (the regular second order rate constant) as the slope. Example: The hydrolysis of esters by dilute mineral acids follows pseudo-first-order kinetics where the concentration of water is present in large excess.

$CH_3COOCH_3 + H_2O \rightarrow CH_3COOH + CH_3OH$

purine[pjúəri(:)n]: L. *purus*, pure + *uricum*, uric acid + *-ine*, suffix. 푸린(무색의 침상 결정); 요산 화합물의 원질. A heterocyclic aromatic organic compound, consisting of a pyrimidine ring fused to an imidazole ring. Purines, including substituted purines and their tautomers, are the most widely distributed kind of nitrogen-containing heterocycle in nature. Purines and pyrimidines make up the two groups of nitrogenous bases, including the two groups of nucleotide bases. Two of the four deoxyribonucleotides and two of the four ribonucleotides, the respective building-blocks of DNA and RNA, are purines.

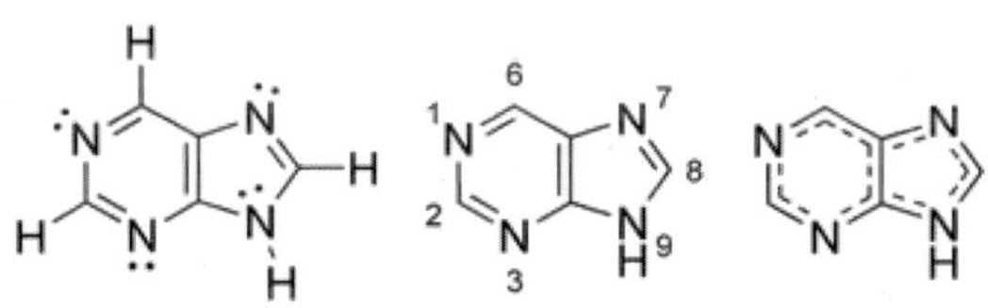

pyridine[pírədi:n]: G. *pur*, fire + *-ide*, suffix + *-ine*, feminine form. 피리딘. A basic heterocyclic organic compound with the chemical formula C_5H_5N. It is structurally related to benzene, with one C-H group replaced by a nitrogen atom. The pyridine ring occurs in many important compounds, including azines and the vitamins, nicotinamide and pyridoxal.

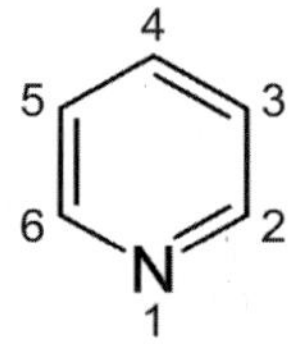

pyridoxine[pirədάksi:n / -dɔk-]: 1930s. *pyrid(ine)* + *ox-*, oxygen + *-ine*, suffix. 피리독신. One of the compounds that can be called vitamin B_6, along with pyridoxal and pyridoxamine. IUPAC name: 4,5-Bis(hydroxymethyl)-2-methylpyridine-3-ol.

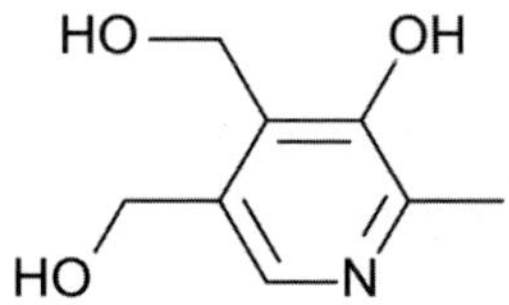

pyrimidine[pairímədi:n]: German. *pyrimidine*, from pyridine with the insertion of -im from imde. 피리미딘. A heterocyclic aromatic organic compound similar to benzene and pyridine, containing two nitrogen atoms at positions 1 and 3 of the six-member ring. It is isomeric with two other forms of diazine.

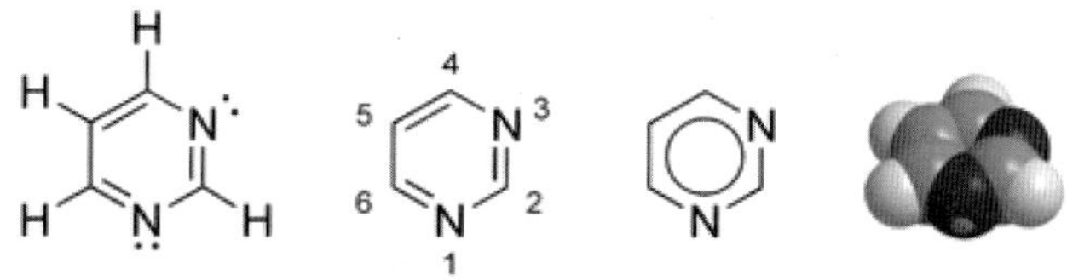

pyro-[páirou, -rə]: G. *pur*, fire, h 또는 모음 앞에서는 pyr-, fire, heat. 열.

pyrolysis[pairɑ́ləsis / paiərɔ́l-]: G. *pyr-*, fire + *lysis*, separating. 열분해. A thermochemical decomposition of organic material at elevated temperatures without the participation of oxygen. It involves the simultaneous change of chemical composition and physical phase, and is irreversible. Pyrolysis is a special case of thermolysis, and is most commonly used for organic materials, being, therefore, one of the processes involved in charring(탄소화). The pyrolysis of wood, which starts at 200–300 °C (390–570 °F), occurs for example in fires where solid fuels are burning or when vegetation comes into contact with lava (용암) in volcanic eruptions. In general, pyrolysis of organic substances produces gas and liquid products and leaves a solid residue richer in carbon content, char. Extreme pyrolysis, which leaves mostly carbon as the residue, is called carbonization.

pyrophosphate: any salt or ester containing two phosphate groups is called a diphosphate. The anion, the salts, and the esters of pyrophosphoric acid are diphosphates that are known as pyrophosphates.

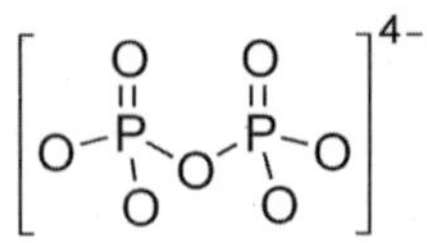

pyruvic[pairú:vik]: L. *pyruvicum*, G. *pyro-*, *pur*, fire + L. *uva*, grape. 피루브산, 초성(焦性)포도산.

pyruvic acid: the simplest of the alpha-keto acids. The carboxylate (COOH) ion (anion) of pyruvic acid, CH_3COCOO^-, is known as pyruvate, and is a key intersection in several metabolic pathways. It can be made from glucose through glycolysis, converted to carbohydrates (such as glycogen) via gluconeogenesis, or to fatty acids through acetyl-CoA. It can also be used to construct the amino acid, alanine and be converted into ethanol. It supplies energy to living cells through the citric acid cycle (also known as the Krebs cycle) when oxygen is present (aerobic respiration), and alternatively ferments to produce lactate when oxygen is lacking (anaerobic respiration).

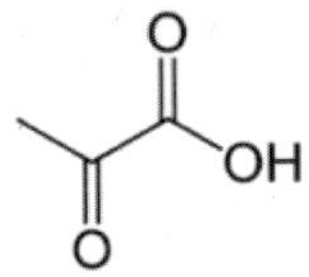

quantity[kwántəti / kwén-]: L. *quantitas, quantus*, how great, how much. 량(量), 수량, 분량.

quantum[kwántəm / kwɔ́n-]: L. *quantus*, how great, how much. 양자(量子). The minimum amount of any physical entity involved in an interaction. Behind this, one finds the fundamental notion that a physical property may be "quantized," referred to as "the hypothesis of quantization". This means that the magnitude can take on only certain discrete values. There is a related term of quantum number. An example of an entity that is quantized is the energy transfer of elementary particles of matter (called fermions) and of photons and other bosons. A photon is a single quantum of light, and is referred to as a "light quantum". The energy of an electron bound to an atom (at rest) is said to be quantized, which results in the stability of atoms, and of matter in general. As incorporated into the theory of quantum mechanics, this is regarded by physicists as part of the fundamental framework for understanding and describing nature at the infinitesimal level. Normally quanta are considered to be discrete packets with energy stored in them. Max Planck considered these quanta to be particles that can change their form (meaning that they can be absorbed and released). This phenomenon can be observed in the case of black body radiation, when it is being heated and cooled.

quantum mechanics (=quantum physics or quantum theory): a branch of physics providing a mathematical description of much of the dual particle-like and wave-like behavior and interactions of energy and matter. It departs from classical mechanics primarily at the atomic and subatomic scales, the so-called quantum realm. In advanced topics of quantum mechanics, some of these behaviors are macroscopic and only emerge at very low or very high energies or temperatures. The name "quantum mechanics", coined by Max Planck, derives from the observation that some physical quantities can change only by discrete amounts, or quanta. For example, the angular momentum of an electron bound to an atom or molecule is quantized. In the context of quantum mechanics, the wave–particle duality of energy and matter and the uncertainty principle provide a unified view of the behavior of photons, electrons and other atomic-scale objects.

quartet[kwɔ:rtét]: L. *quartus*, fourth. 사중선, 사중항 (상태), 사전자계.

quinine[kwáinain/kwiní:n]: Spanish. *quina,* cinchona* bark + *-ine,* suffix. 키니네, 퀴닌(말라리아 특효약).

cinchona*: 기나(幾那)나무, 페루의 총독 부인 chinchon이 1638년에 열병치료에 사용. A natural white crystalline alkaloid having antipyretic (fever-reducing), antimalarial, analgesic (painkilling), anti-inflammatory properties and a bitter taste. It is a stereoisomer of quinidine which, unlike quinine, is an anti-arrhythmic. Quinine contains two major fused-ring systems: the aromatic quinoline and the bicyclic quinuclidine. Though it has been synthesized in the lab, quinine occurs naturally in the bark of the cinchona tree. The medicinal properties of the cinchona tree were originally discovered by the Quechua Indians of Peru and Bolivia; later, the Jesuits were the first to bring the cinchona to Europe. Quinine was the first effective treatment for malaria caused by Plasmodium falciparum, appearing in therapeutics in the 17th century.

It remained the antimalarial drug of choice until the 1940s, when other drugs replaced it. Since then, many effective antimalarials have been introduced, although quinine is still used to treat the disease in certain critical situations, i.e., as in impoverished regions. Quinine is available with a prescription in the United States and over-the-counter, in very small quantities, in tonic water. Quinine is also used to treat lupus and arthritis. Quinine was also frequently prescribed in the U.S. as an "off-label" treatment for nocturnal leg cramps, but this has since been prohibited, in effect, by an FDA statement warning against the practice. Quinine is very sensitive to ultraviolet light (UV) and will fluoresce in direct sunlight, due to its highly conjugated resonance structure (see Quinoline).

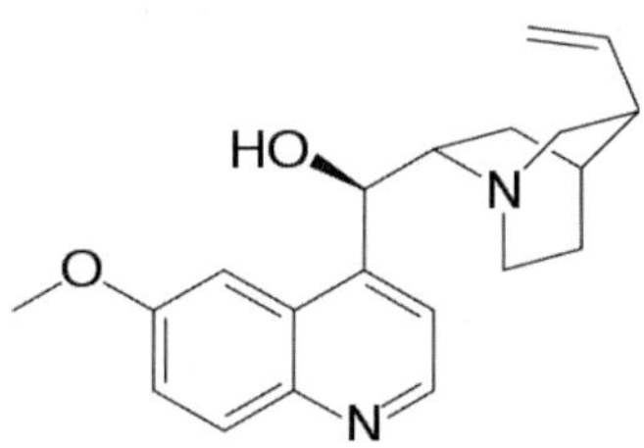

quinoline[kwinəlin, -lin]: Spanish, *quina,* cinchona* bark + *-ol,* oil, + *-ine,* suffix. 퀴놀린(식품 방부제 및 염료 제조용). A heterocyclic aromatic organic compound with the chemical formula C_9H_7N. It is a colorless hygroscopic liquid with a strong odor. Aged samples, if exposed to light, become yellow and later brown. Quinoline is only slightly soluble in cold water but dissolves readily in hot water and most organic solvents.

Quinoline itself has few applications, but many of its derivatives are useful in diverse applications. A prominent example is quinine, which is found naturally in plants as alkaloids. 4-Hydroxy-2-alkylquinolines (HAQs) are involved in antibiotic resistance.

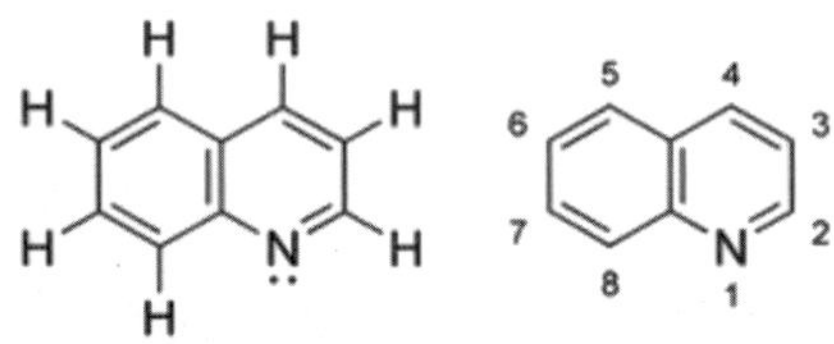

chloroquine: a 4-aminoquinoline drug used in the treatment or prevention of malaria.

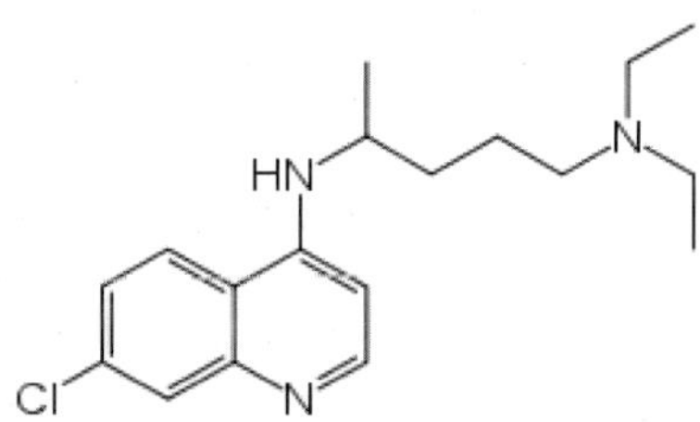

quinone[kwinóun, kwínoun]: Spanish. *quina* cinchona* bark + *-one,* suffix. 퀴논(염료 따위의 원료). A class of organic compounds that are formally derived from aromatic compounds (such as benzene or naphthalene) by exchanging an even number of –CH= groups by –C(=O)– groups, with any necessary rearrangement of double bonds, resulting in a fully conjugated cyclic dione structure. The class includes derivatives of heterocyclic aromatic compounds.

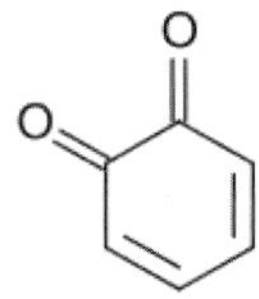

1,2-benzoquinone

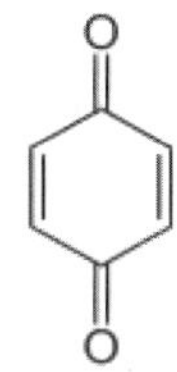

1,4-benzoquinone

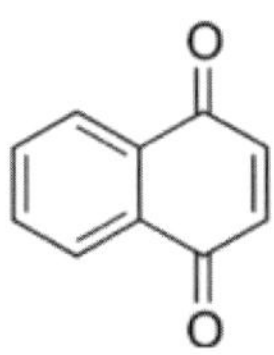

1,4-naphthoquinone

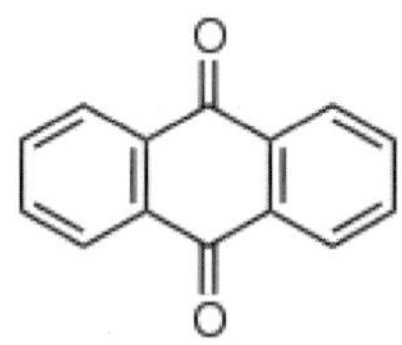

9,10-anthraquinone

hydroquinone: benzene-1,4-diol or quinol, an aromatic organic compound which is a type of phenol, having the chemical formula $C_6H_4(OH)_2$. A crystalline compound made by the reduction of benzoquinone. 하이드로퀴논(사진현상약).

racemic[reisí:mik, -sém- / rə-]: French. *racemique*, derived from grape juice, bunch of grapes, (originally referring to tartaric acid). 라세미의.

racemic mixture, or racemate: 라세미 혼합물. One that has equal amounts of left- and right-handed enantiomers of a chiral molecule. The first known racemic mixture was "racemic acid", which Louis Pasteur found to be a mixture of the two enantiomeric isomers of tartaric acid.

racemization[ræsəməzéiʃən, resi:mə- / -mai-]: French. *racemique*, derived from grape juice (originally refering to tartaric acid). 라세미화. Refers to the converting of an enantiomerically pure mixture (one where only one enantiomer is present) into a mixture where more than one of the enantiomers are present. If the racemization results in a mixture where the enantiomers are present in equal quantities, the resulting sample is described as racemic or a racemate. Chiral molecules have two forms (at each point of asymmetry), which differ in their optical characteristics: The *levorotatory form* [the *(−)-form*] will rotate the plane of polarization of a beam of light to the left, whereas the *dextrorotatory* form [the *(+)-form*] will rotate the plane of polarization of a beam of light to the right. The two forms, which are non-superimposable when rotated in 3-dimensional space, are said to be *enantiomers*. Racemization occurs when one pure form of an enantiomer is converted into equal proportion of both enantiomers, forming a racemate. When there are both equal numbers of dextrorotating and levorotating molecules, the net optical rotation of a racemate is zero.

radiation[reidiéiʃən]: L. *radiation* from *radiare*, emit rays. 복사(선), 방사(선). A process in which energetic particles or energetic waves travel through a medium or space. Two types of *radiation* are commonly differentiated in the way they interact with normal chemical matter: **ionizing** and **non-ionizing** radiation. The word *radiation* is often colloquially used in reference to ionizing radiation (i.e., radiation having sufficient energy to ionize an atom), but the term radiation may correctly also refer to non-ionizing radiation (e.g., radio waves, heat or visible light). The particles or waves *radiate* (i.e., travel outward in all directions) from a source. This aspect leads to a system of measurements and physical units that are applicable to all types of radiation. Both ionizing and non-ionizing radiation can be harmful to organisms and can result in changes to the natural environment. In general, however, ionizing radiation is far more harmful to living organisms per unit of energy deposited than non-ionizing radiation, since the ions that are produced by ionizing radiation, even at low radiation powers, have the potential to cause DNA damage. By contrast, most non-ionizing radiation is harmful to organisms only in proportion to the thermal energy deposited, and is conventionally considered harmless at low powers which do not produce significant temperature rise. Ultraviolet radiation in some aspects occupies a middle ground, in having some

Fig. Illustration of the relative abilities of three different types of ionizing radiation to penetrate solid matter. Alpha particles (α) are stopped by a sheet of paper while beta particles (β) are stopped by an aluminium plate. Gamma radiation (γ) is dampened when it penetrates matter.

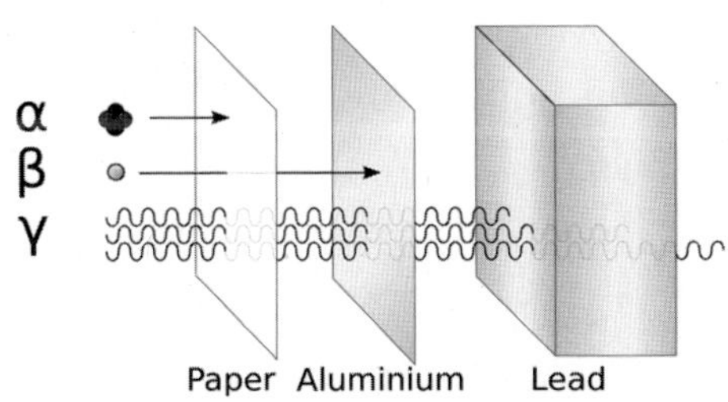

features of both ionizing and non-ionizing radiation. Although nearly all of the ultraviolet spectrum of radiation is non-ionizing, at the same time ultraviolet radiation does far more damage to many molecules in biological systems than is accounted for by heating effects (an example is sunburn). These properties derive from ultraviolet's power to alter chemical bonds, even without having quite enough energy to ionize atoms.

radical[rǽdikəl]: L. *radicalis, radix, radic-,* root. 라디칼. Atoms, molecules, or ions with unpaired electrons on an open shell configuration. Free radicals may have positive, negative or zero charge. Even though they have unpaired electrons, by convention, metals and their ions or complexes with unpaired electrons are not radicals. With some exceptions, the unpaired electrons cause radicals to be highly chemically reactive. Radicals are believed to be involved in degenerative diseases and cancers. Free radicals play an important role in combustion, atmospheric chemistry, polymerization, plasma chemistry, biochemistry, and many other chemical processes, including human physiology. "Radical" and "free radical" are frequently used interchangeably, although a radical may be trapped within a solvent cage or be otherwise bound. In chemical equations, free radicals are frequently denoted by a dot placed immediately to the right of the atomic symbol or molecular formula as follows:

$$Cl_2 \xrightarrow{UV} Cl\cdot + Cl\cdot$$

Radical reaction mechanisms use single-headed arrows to

depict the movement of single electrons:

radical ion: a free radical species that carries a charge. Radical ions are encountered in organic chemistry as reactive intermediates and in mass spectrometry as gas phase ions. Positive radical ions are called radical cations whereas negative radical ions are called radical anions. In organic chemistry, a radical ion is typically indicated by a superscript dot followed by the sign of the charge: $M^{\bullet+}$ and $M^{\bullet-}$. In mass spectrometry, the sign is written first, followed by the superscripted dot: $M^{+\bullet}$ and $M^{-\bullet}$.

radical anions: many aromatic compounds can undergo one-electron reduction by alkali metals. For example the reaction of naphthalene with sodium in an aprotic solvent yields the naphthalene radical anion - sodium ion salt. In a ESR spectrum this compound shows up as a quintet of quintets (25 lines). In the presence of a proton source the radical anion is protonated and effectively hydrogenated like in the Birch reduction. The electron is transferred from the alkali metal ion to an unoccupied antibonding p-p π* orbital of the aromatic molecule. This transfer is usually only energetically favorable if the aprotic solvent efficiently solvates the alkali metal ion. Effectiveness for this is in the order diethyl ether < THF < 1,2-dimethoxyethane < HMPA.

In principle any unsaturated molecule can form a radical anion, but the antibonding orbitals are only energetically accessible in more extensive conjugated systems. Ease of formation is in the order benzene < naphthalene < anthracene < pyrene, etc. On addition of a proton source, the structure of the resulting hydrogenated molecule is defined by the charge distribution of the radical anion. For instance, the anthracene radical anion forms mainly (but not exclusively) 9,10-dihydroanthracene. An example of a non-carbon radical anion is the superoxide anion, formed by transfer of one electron to an oxygen molecule.

radical cations: cationic radical species do also exist but are much less stable. They appear prominently in mass spectrometry. When a gas-phase molecule is subjected to electron ionization one electron is abstracted by an electron in the electron beam to create a radical cation M^{+}. This species represents the molecular ion or parent ion and will tell the precise molecular weight. On a typical mass spectrum more signals show up because the molecular ion fragments into a complex mixture of ions and uncharged radical species. For example the methanol radical cation fragments into a methyl cation CH_3^+ and a hydroxyl radical. In naphthalene the unfragmented radical cation is by far the most prominent peak in the mass spectrum. Secondary species are generated from proton gain (M+1) and proton loss (M-1).

radical initiators: 라디칼 개시체. Substances that can produce

radical species under mild conditions and promote radical reactions. These substances generally possess weak bonds — bonds that have small bond dissociation energies. Radical initiators are utilized in industrial processes such as polymer synthesis. Typical examples are halogen molecules, azo compounds, and organic peroxides. Like all diatomic-molecules, halogens can generate two free radicals resulting from the homolysis of the bond, but halogens undergo the homolytic fission relatively easily. Chlorine, for example, gives two chlorine radicals (Cl•) by irradiation with ultraviolet light. This process is used for chlorination of alkanes. Azo compounds (R-N=N-R') can be the precursor of two carbon-centered radicals (R• and R'•) and nitrogen gas upon heating and/or by irradiation. For example, AIBN (= azobisisobutyronitrile) and ABCN (= 1,1'-Azobis (cyclohexanecarbonitrile) yield isobutyronitrile and cyclohexanecarbonitrile radicals, respectively. Organic peroxides each have a peroxide bond (-O-O-), which is

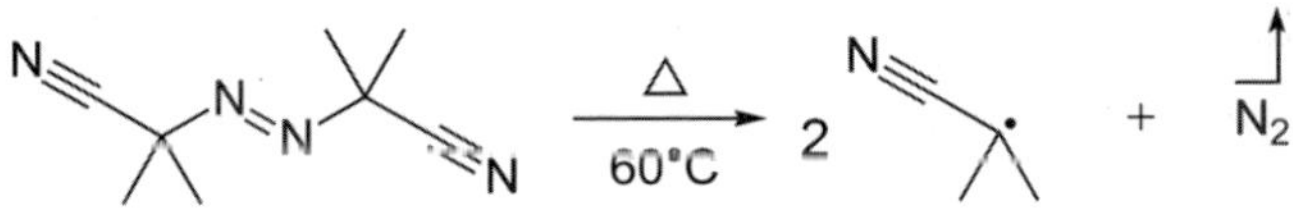

readily cleaved to give two oxygen -centered radicals. The oxyl radicals are rather unstable and believed to be transformed into relatively stable carbon-centered radicals. For example, di-*tert*-butyl peroxide (^{t}BuOO^{t}Bu) gives two *t*-butanoyl radicals (^{t}BuO•) and the radicals become methyl radicals (CH_3•) with the loss of acetone. Benzoyl peroxide [$(PhCOO)_2$] generates benzoyloxyl radicals (PhC OO•), each of which loses carbon dioxide to be converted into a phenyl radical (Ph•). Methyl ethyl ketone peroxide is also common, and acetone peroxide is on rare occasions used as a radical initiator, too. Radical initiators, especially azo compounds and organic peroxides, are inherently unstable. They must be kept in a cool place or refrigerated. Care should be taken with the handling of the compounds or an explosion may occur.

rate[reit]: L. *rata, ratus, reri*, reckoned. 속도.

rate law (= rate equation): 속도법칙. For a chemical reaction, an equation that links the reaction rate with concentrations or pressures of reactants and constant parameters (normally rate coefficients and partial reaction orders). To determine the rate equation for a particular system one combines the reaction rate with a mass balance for the system. For a generic reaction $aA + bB \rightarrow C$ with no intermediate steps in its reaction mechanism (that is, an elementary reaction), the rate is given by $r = k[A]^x[B]^y$ where [A] and [B] express the concentration of the species A and B, respectively [usually in moles per liter (molarity, M)]; x and y are not the respective stoichiometric coefficients of the balanced equation; they must be determined experimentally. k is the *rate coefficient* or *rate constant* of the reaction. The value of this coefficient k depends on conditions such as temperature, ionic strength, surface area of the adsorbent or light irradiation. For elementary reactions, the rate equation can be derived from first principles using collision theory. The equation may involve fractional exponential coefficients, or it may depend on the concentration of an intermediate species. The rate equation is a differential equation, and it can be integrated to obtain an integrated rate equation that links concentrations of reactants or products with time. If the concentration of one of the reactants remains constant (because it is a catalyst or it is in great excess with respect to the other reactants), its concentration can be grouped with the rate constant, obtaining a pseudo constant: If B is the reactant whose concentration is constant, then $r = k[A][B] = k'[A]$. The second-order rate equation has been reduced to a pseudo-first-order rate equation. This makes the treatment to obtain an integrated rate equation much easier.

rate-determining step (RDS): 속도 결정 단계. A chemistry term for the slowest step in a chemical reaction. The rate-determining step is often compared to the neck of a funnel; the rate at which water flows through the funnel is determined by the width of the neck, not by the speed at which water is poured in. In similar manner, the rate of reaction depends on the rate of the slowest step. However it is not clear what 'slowest step' means. For example does it refer to the step with the smallest rate constant; if the reaction is reversible, does it refer to the ratio of the forward and reverse rate constants or does it simply refer to the step that has the smallest flux? If the latter, then at steady state all steps carry the same flux and therefore there is no slowest step. In metabolic pathways the rate limiting step is better defined and a measure, the flux control coefficient is given to a step to signify how rate-limiting the step is. Theory and experiment also suggests that there is no single rate limiting step but a range of rate limitingness across the entire reaction network.

For example, the reaction $NO_{2(g)} + CO_{(g)} \rightarrow NO_{(g)} + CO_{2(g)}$ can be thought of as occurring in two elementary steps:

$NO_2 + NO_2 \rightarrow NO + NO_3$ *(slow step)*

$NO_3 + CO \rightarrow NO_2 + CO_2$ *(fast step)*

As the second step consumes the NO_3 produced in the slow first step, it is limited by the rate of the first step. For this reason, the rate-determining step is reflected in the rate equation of a reaction. Another example of a rate-determining step is the formation of a carbocation from a haloalkane during the S_N1 reaction of tertiary haloalkanes with sodium hydroxide. In a reaction coordinate, the transition state with the highest energy is the rate-determining step of a given reaction.

react[riǽkt]: L. *react*, from the verb *reagere*, done again. 반응하다.

reaction coordinate: 반응 좌표. An abstract one-dimensional coordinate which represents progress along a reaction pathway. It is usually a geometric parameter that changes during the conversion of one or more molecular entities. Reaction coordinates are often plotted against free energy to demonstrate in some schematic form the potential energy profile associated to the reaction. In the formalism of transition-state theory the reaction coordinate is that coordinate in set of curvilinear coordinates obtained from the conventional ones for the reactants which, for each reaction step, leads smoothly from the configuration of the reactants through that of the transition state to the configuration of the products. The reaction coordinate is typically chosen to follow the path along the gradient (path of shallowest ascent/deepest descent) of potential energy from reactants to products. For example, in the homolytic dissociation of molecular hydrogen, an apt coordinate system to choose would be the coordinate corresponding to the bond length.

Fig. Diagram of a catalytic reaction, showing the energy depending on the reaction coordinate. For a catalysed reaction, the activation energy is lower.

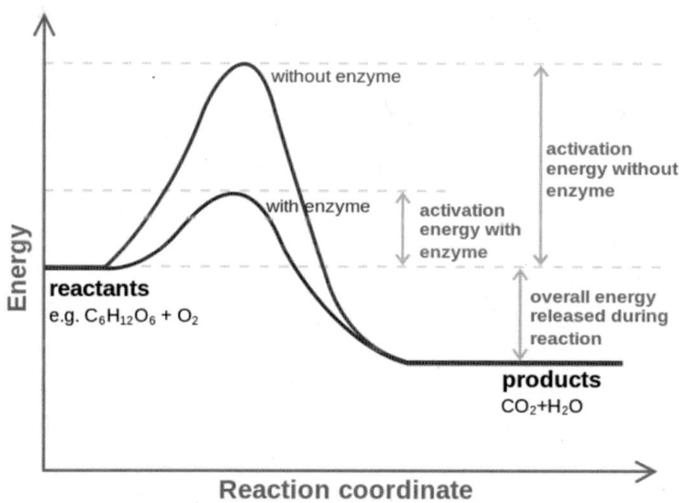

reaction mechanism: 반응 메카니즘. 반응 과정. The step by step sequence of elementary reactions by which overall chemical change occurs. Although only the net chemical change is directly observable for most chemical reactions, experiments can often be designed that suggest the possible sequence of steps in a reaction mechanism. Recently, electrospray ionization mass spectrometry has been used to corroborate the mechanism of several organic reaction proposals. A chemical mechanism describes in detail exactly what takes place at each stage of an overall chemical reaction (transformation). It also describes each reaction intermediate, activated complex, and transition state, and which bonds are broken (and in what order), and which bonds are formed (and in what order). A complete mechanism must also account for all reactants used, the function of a catalyst, stereochemistry, all products formed and the amount of each, and what the relative rates of the steps are. Reaction intermediates are chemical species, often unstable and short-lived, which are not reactants or products of the overall chemical reaction, but are temporary products and reactants in the mechanism's reaction steps. Reaction intermediates are often free radicals or ions. Transition states can be unstable intermediate molecular states even in the elementary reactions. Transition states are commonly molecular entities involving an unstable number of bonds and/or unstable geometry which may be at chemical potential maxima.

Fig. S_N2 reaction mechanism. Note the negatively-charged transition state in brackets in which the central carbon atom in question shows five bonds, an unstable condition.

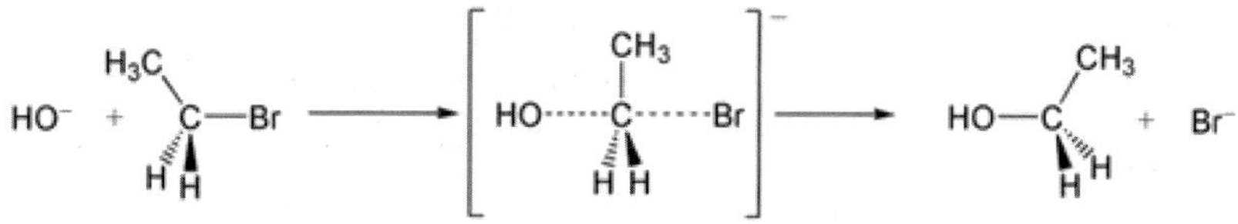

<u>reaction rate constant</u>: 반응속도 상수. *k* or λ, quantifies the speed of a chemical reaction. For a chemical reaction where substance A and B are reacting to produce C, the reaction rate has the form:

Reaction: A + B → C

$$\frac{d[C]}{dt} = k(T)[A]^m[B]^n$$

k(T) is the reaction rate constant that depends on temperature. [C] is the concentration of substance C in moles per volume of solution assuming the reaction is taking place throughout the volume of the solution (for a reaction taking place at a boundary it would denote something like moles of C per area). The exponents *m* and *n* are called orders and depend on the reaction mechanism. They can be determined experimentally. A single-step reaction can also be written as

$$\frac{d[C]}{dt} = Ae^{-E_a/RT}[A]^m[B]^n$$

E_a is the activation energy and R is the Gas constant. Since at temperature T the molecules have energies according to a Boltzmann distribution, one can expect the proportion of collisions with energy greater than E_a to vary with $e^{-E_a/RT}$. A is the pre-exponential factor or frequency factor. The Arrhenius equation gives the quantitative basis of the relationship between the activation energy and the reaction rate at which a reaction proceeds.

rearrange[riəréindʒ]: Old French. *arangier*, from *re-* + *a-* + *rangier*, put in order again. 자리옮김. 재배열하다.

<u>rearrangement reaction</u>: 자리옮김 반응. A broad class of organic reactions where the carbon skeleton of a molecule is rearranged to give a structural isomer of the original molecule. Often a substituent moves from one atom to another atom in the same molecule. In the example below the substituent R moves from carbon atom 1 to carbon atom 2:

Intermolecular rearrangements also take place. Three key rearrangement reactions are 1,2-rearrangements, pericyclic reactions and olefin metathesis. 1,2-rearrangement: an organic reaction where a substituent moves from one atom to another atom in a chemical compound. In a 1,2 shift the movement involves two adjacent atoms but moves over larger distances are possible.

Examples: the Wagner-Meerwein rearrangement:

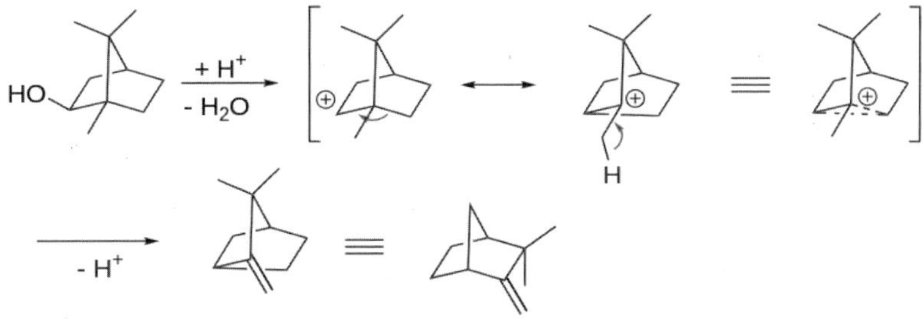

the Beckmann rearrangement:

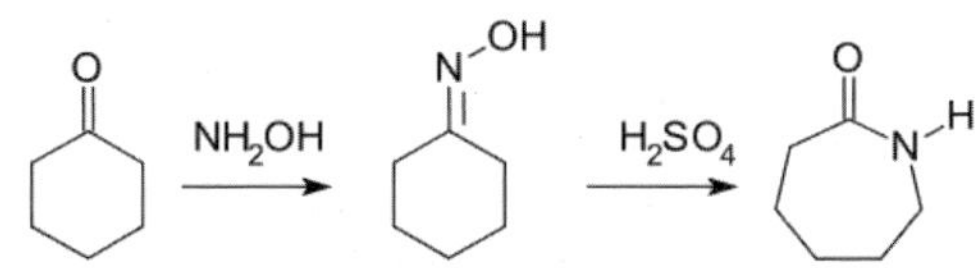

receptor[riséptər]: L. *receptor, recept,* from the verb *recipere*, taken back. 수용체(기). A protein molecule that receives and responds to a neurotransmitter, or other substance.

reduction[ridʌkʃən]: L. *reduction, reducere*, bring back, restore. 환원. Part of a reduction-oxidation (redox) reaction in which atoms have their oxidation number (oxidation state) changed.

relative[rélətiv]: L. *ralativus*, having reference or relation. 상대적인, 관련되어 있는.

<u>relative configuration</u>: 상대(적)배열(배치). The configuration of a compound relative to the configuration of another compound.

renal[rí:nl]: L. *renalis, renes*, kidneys. 신장(腎臟)의.

replication[repləkéiʃən]: L. *replication, replicare*, hold back, repear, make a reply. 복제, 반복.

residue[rézəd*j*u:/-dju:]: L. *residuum*, something remaining. 찌꺼기, 잔기(殘基). 유수(留數).

resin[rézin]: G > L. *resina* < *rhētinē* pine resin. 수지. A hydrocarbon secretion of many plants, particularly coniferous (침엽수의) trees. It is valued for its chemical properties and associated uses, such as the production of varnishes, adhesives, and food glazing agents; as an important source of raw materials for organic synthesis; and as constituents of incense and perfume. In perfumery such products are often termed "ambered", based on Fossilized resins being the source of the gemstone amber. Resins are also a material in nail polish.

resolution[rezəlú:ʃən]: L. *resolution, resolvere*, loosen, release. 분리, 분별, 분할.

resolution of racemic mixtures: 라세미 혼합물의 분리. The separation of a racemate into its components, the pure enantiomers, is called a chiral resolution. There are various methods, including crystallization, chromatography, and the use of enzymes. The first successful resolution of a racemate was performed by Louis Pasteur, who manually separated the crystals of a conglomerate (고체 라세미 혼합물).

resonance[rézənəns]: L. *resonare*, *re*, back, again + *sonare*, to sound, resound. 공명(共鳴). Mesomerism (old-fashioned term for resonance). A way of describing delocalized electrons within certain molecules or polyatomic ions where the bonding cannot be expressed by one single Lewis formula. A molecule or ion with such delocalized electrons is represented by several contributing structures (also called resonance structures or canonical forms). Each contributing structure can be represented by a Lewis structure, with only an integer number of covalent bonds between each pair of atoms within the structure. Several Lewis structures are used collectively to describe the actual molecular structure. However these individual contributors cannot be observed in the actual resonance-stabilized molecule; the molecule does not oscillate back and forth between the contributing structures, as might be assumed from the word "resonance". The actual structure is an approximate intermediate between the canonical forms, but its overall energy is lower than each of the contributors. This intermediate form between different contributing structures is called a resonance hybrid. Contributing structures differ only in the position of electrons, not in the position of nuclei. Resonance is a key component of valence bond theory. Electron delocalization lowers the potential energy of the substance and thus makes it more stable than any of the contributing structures. The difference between the potential energy of the actual structure and that of the contributing structure with the lowest potential energy is called the resonance energy or delocalization energy. Resonance is distinguished from tautomerism and conformational isomerism, which involve the formation of isomers, thus the rearrangement of the nuclear positions. Resonance instead considers such molecules to be an intermediate or an average (called a resonance hybrid) between several Lewis structures that differ only in the placement of the π electrons.

Fig. Resonance structure of benzene

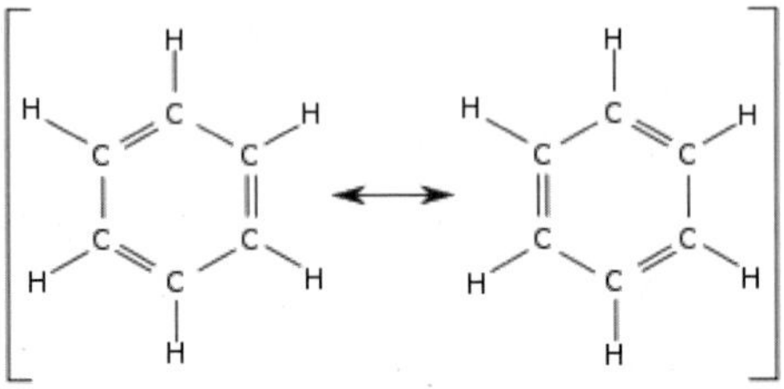

retina[rétənə]: L. *rete*, net. (눈의) 망막.

retinal (=retinaldehyde or vitamin A aldehyde): one of the many forms of vitamin A (the number of which varies from species to species). Retinal is a polyene chromophore, and bound to proteins called opsins, is the chemical basis of animal vision. Bound to proteins called type 1 rhodopsins, retinal allows certain microorganisms to convert light into metabolic energy. Vertebrate animals ingest retinal directly from meat, or produce retinal from one of four carotenoids (beta-carotene, alpha-carotene, gamma-carotene, and beta-cryptoxanthin), which they must obtain from plants or other photosynthetic organisms (no other carotenoids can be converted by animals to retinal, and some carnivores cannot convert any carotenoids at all). The other main forms of vitamin A, retinol, and a partially active form retinoic acid, may both be produced from retinal. Invertebrates such as insects and squid use hydroxylated forms of retinal in their visual systems, which derive from conversion from other xanthophylls.

All-trans-retinal

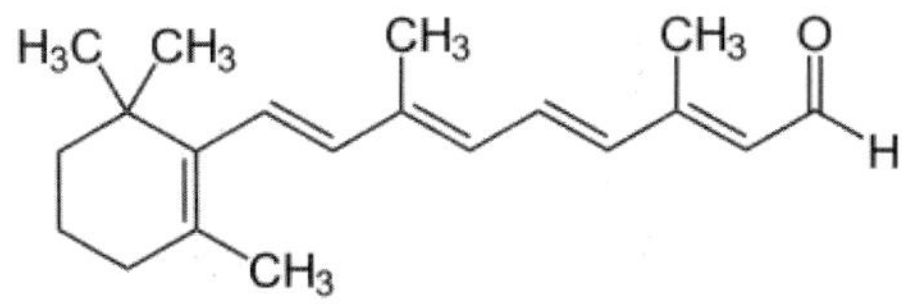

retro-[rétrou, -rə]: French. *rétro, rétrograde,* retrograde. backward, behind. 되돌림, 역행하는, 퇴화하는.

retro Diels-Alder reaction (rDA): reverse Diels-Alder reaction. The microscopic reverse of the Diels-Alder reaction — the formation of a diene and dienophile from a cyclohexene. It can be accomplished spontaneously with heat, or with acid or base mediation.

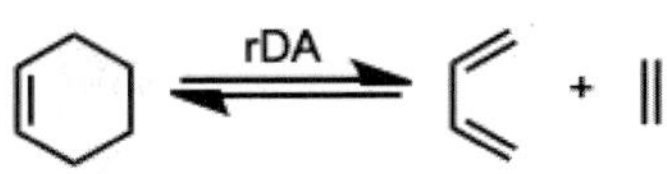

retrovirus[retrəváiərəs]: L. from the initial letters of *reverse transcriptase* + *virus.* 레트로 바이러스, RNA 종양바이러스. A virus whose genetic information is stored in its RNA.

reverse[rivə́:rs]: L. *reversus*, turned back, past participle of *revertere*, *re-*, back + *vertere*, to turn. 역의, 반대의, 거꾸로의, 뒤집힌.

reversible reaction: 가역반응. A chemical reaction that results in an equilibrium mixture of reactants and products. For a reaction involving two reactants and two products this can be expressed symbolically as

$$aA + bB \rightleftharpoons cC + dD$$

A and B can react to form C and D or, in the reverse reaction, C and D can react to form A and B. This is distinct from reversible process in thermodynamics.

rheumatic[ru:mǽtik, ru-]: L. *rheumaticos,* suffering from abnormal fluid discharge. 류머티즘의, 류머티즘 환자.

rheumatism: 류머티즘.

rhinocerous[rainάsərəs / -nɔ́s-]: G. *rhinokeros, rhin*, nose + *keras*, horn. A large, heavily built plant-eating mammal with one or two horns on the nose and thick folded skin, native to Africa and South Asia. 꼬뿔소, 무소.

riboflavin[raibouflέivin, ráibouflеivin]: 1930s. *ribose*, arabinose + L. *flavus*, yellow + *-in*, suffix. 리보플라빈(vitamin B_2). An easily absorbed micronutrient with a key role in maintaining health in humans and animals. It is the central component of the cofactors flavin adenine dinucleotide (FAD) and Flavin mononucleotide (FMN), and is therefore required by all flavoproteins. As such, vitamin B_2 is required for a wide variety of cellular processes. It plays a key role in energy metabolism, and for the metabolism of fats, ketone bodies, carbohydrates, and proteins. It is also used as an orange-red food colour additive, designated in Europe as the E number E101, Milk, cheese, leafy green vegetables, liver, kidneys, legumes, tomatoes, yeast, mushrooms, and almonds are good sources of vitamin B_2, but exposure to light destroys riboflavin. The name "riboflavin" comes from "ribose" (the sugar whose reduced form, ribitol, forms part of its structure) and "flavin", the ring-moiety which imparts the yellow color to the oxidized molecule. The reduced form, which occurs in metabolism along with the oxidized form, is colorless. Riboflavin is best known visually as the vitamin which imparts the orange color to solid B-vitamin preparations, the yellow color to vitamin supplement solutions, and the unusual fluorescent-yellow color to the urine of persons who supplement with high-dose B-complex preparations (no other vitamin imparts any color to urine).

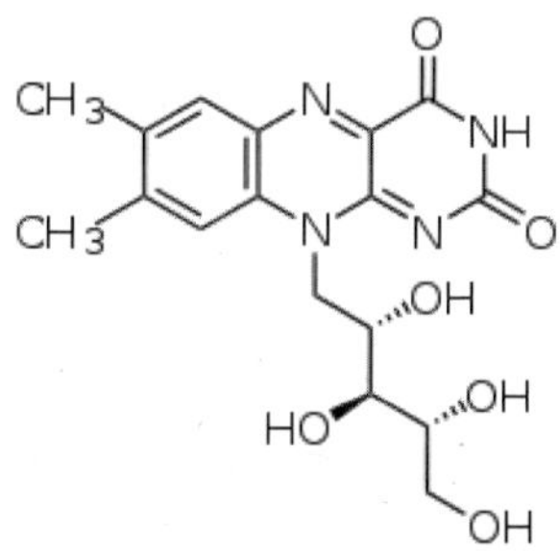

saccharide[sǽkəraid, -rid]: L. *saccharum*, sugar + ide. 당류(糖類).

saccharide(s), mono-; G. *monos,* alone, single.
di-; G. *dis-,* two, double.
oligo-; G. *oligos,* small.
poly-; G. *polus*, much, *polloi,* many.

saccharin[sǽkərin]: L. *saccharum, sacchari,* sugar + *in,* not, without. 사카린. An artificial sweetener. The basic substance, benzoic sulfilimine, has effectively no food energy and is much sweeter than sucrose, but has a bitter or metallic aftertaste, especially at high concentrations. It is used to sweeten products such as drinks, candies, cookies, medicines, and toothpaste.

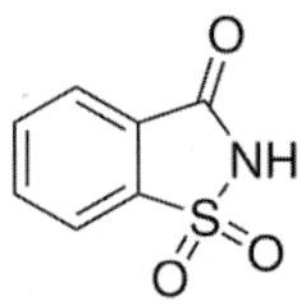

salt[sɔ:lt]: G > L. *sal* < *hals*, salt. 염(塩), 소금. Any chemical compound formed from the reaction of an acid with a base, with all or part of the hydrogen of the acid replaced by a metal or other cation.

saponification[səpanəfikéiʃən /-pə-]: L. *sapo, sapon-*, soap + *fication*, from verbs ending in*-ficare*, turn a fat, or an oil into soap by reaction with an alkali. 비누화 (반응), 에스테르 가수분해 (반응). Sodium hydroxide (NaOH) is a caustic[가성(苛性)의, 부식성의] base. If NaOH is used a hard soap is formed, whereas when potassium hydroxide (KOH) is used, a soft soap is formed. Vegetable oils and animal fats are fatty esters in the form of triglycerides. The alkali breaks the ester bond and releases the fatty acid salt and glycerol. If necessary, soaps may be precipitated by salting it out with saturated sodium chloride. The saponification value is the amount of base required to saponify a fat sample.

Fig. Saponification of a triglyceride with sodium hydroxide

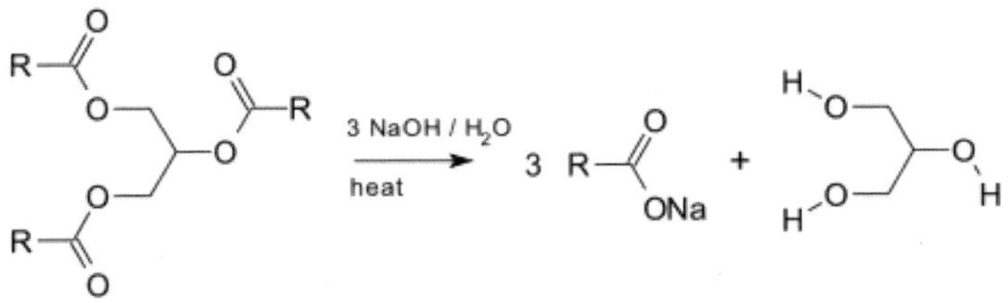

saponin[sǽpənin]: L. *sapo, sapon-*, soap. 사포닌, (식물에 분포하는 배당체; 수용액에서 거품이 잘 일어난다). A class of chemical compounds, one of many secondary metabolites found in natural sources, with saponins found in particular abundance in various plant species. More specifically, they are amphipathic glycosides grouped, in terms of phenomenology (empirical observations of phenomena, 현상학), by the soap-like foaming they produce when shaken in aqueous solutions, and, in terms of structure, by their composition of one or more hydrophilic glycoside moieties combined with a lipophilic triterpene derivative. A ready and therapeutically relevant example is the cardio-active agent digoxin, from common foxglove.

Fig. Chemical structure of the saponin solanine

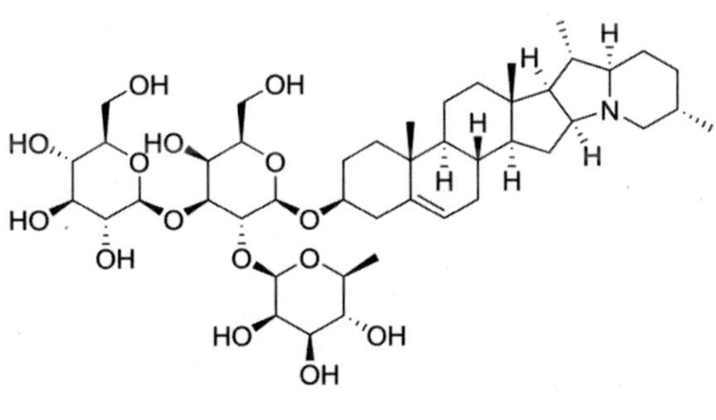

saturate[sǽtʃəreit]: L. *saturat-*, *satur*, filled, glutted, full, from the verb *saturare*. - 을 포화시키다.

saturated solution: 포화용액

scheme[ski:m]: L. *schema, schematis,* form, figure. 도표, 분류, 계획, 설계.

science[sáiəns]: L. *scientia,* knowledge. 체계적 지식, 과학(연구), 학술, 자연과학, 술(術).

scurvy[skə́:rvi]: L. *scorbutus*, to break belly. 괴혈병. A disease resulting from a deficiency of vitamin C, which is required for the synthesis of collagen in humans. The chemical name for vitamin C, ascorbic acid. Scurvy often presents itself initially as symptoms of malaise(불안, 권태) and lethargy(혼수), followed by formation of spots on the skin, spongy gums, and bleeding from the mucous membranes (점막, 粘膜). Spots are most abundant on the thighs(가랑이, 넓적다리) and legs, and a person with the ailment(병) looks pale, feels depressed, and is partially immobilized. As scurvy advances, there can be open, suppurating(화농) wounds, loss of teeth, jaundice(황달), fever, neuropathy(신경장애) and death.

secondary[sékənderi/-dəri]: L. *secundarius, secundaria, secundarium,* from *secundus,* of the second class or quality. 2차적인.

secondary carbocation: 2차 탄소양이온.

Fig. Order of stability of examples of tertiary (III), secondary (II), and primary (I) alkyl carbocations.

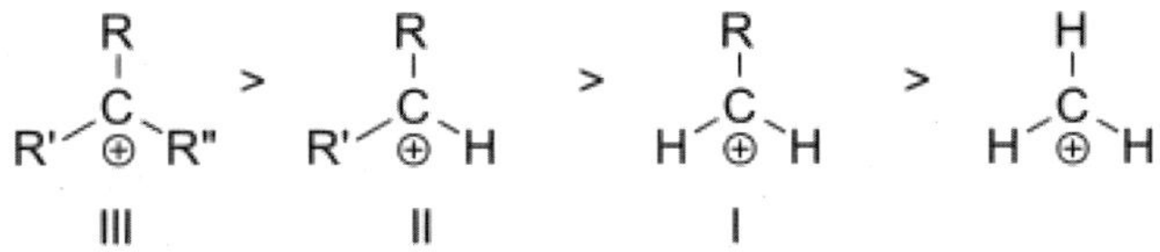

secretion[sikríːʃən]: L. *secretion*, separation, from *secret-*, moved apart, from the verb *secernere*. 분비(작용), 분비액(물).

sedative[sédətiv]: L. *sedativus, sedat-, sedare*, settled, promoting calm or inducing sleep. 진정시키는, 진정제.

sedimentation[sedəməntéiʃən]: L. *sedimentum*, settling, from *sedere*, sit. 퇴적(작용).

selenium[silíːniəm]: G. *selēnē*, moon. 셀렌, 셀레늄. A chemical element with atomic number 34, chemical symbol **Se**, and an atomic mass of 78.96. It is a nonmetal, whose properties are intermediate between those of adjacent chalcogen elements (산소족 원소) sulfur and tellurium. It rarely occurs in its elemental state in nature, but instead is obtained as a side-product in the refining of other elements. Selenium is found in sulfide ores such as pyrite, where it partially replaces the sulfur Minerals that are selenide or selenate compounds are also known, but are rare. The chief commercial uses for selenium today are in glassmaking and in pigments. Uses in electronics, once important, have been supplanted by silicon semiconductor devices. It is a semiconductor with the unusual property of conducting electricity better in the light than in the dark, and is used in photocells. Selenium is a trace mineral that is essential to good health but required only in small amounts. Selenium is incorporated into proteins to make selenoproteins, which are important antioxidant enzymes. The antioxidant properties of selenoproteins help prevent cellular damage from free radicals. Free radicals are natural by-products of oxygen metabolism that may contribute to the development of chronic diseases such as cancer and heart disease. Other selenoproteins help regulate thyroid function and play a role in the immune system.

serine[séri(:)n, siər-]: L. *sericum*, silk + *-ine*, suffix. 세린(인체 내에서 합성 가능한 아미노산의 일종). Abbreviated as Ser or S, an amino acid with the formula, $HO_2CCH(NH_2)CH_2OH$. It is one of the proteinogenic amino acids. By virtue of the hydroxyl group, serine is classified as a polar amino acid.

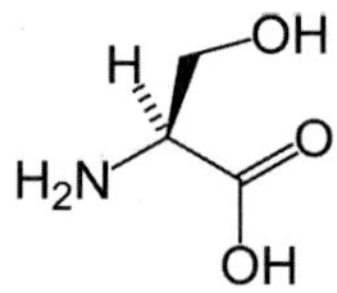

serotonin[serətóunin, siərt-]: 1940s. *serum*, L. whey + *tonic*, G. *tonikos*, of or for stretching + *-in*, suffix. 세로토닌(혈관 수축 작용을 하는 호르몬). 5-Hydroxytryptamine(5-HT) is a monoamine neurotransmitter. Biochemically derived from tryptophan, serotonin is primarily found in the gastrointestinal(GI) tract, platelets, and in the central nervous system(CNS) of animals including humans. It is a well-known contributor to feelings of well-being; therefore it is also known as a "happiness hormone" despite not being a hormone.

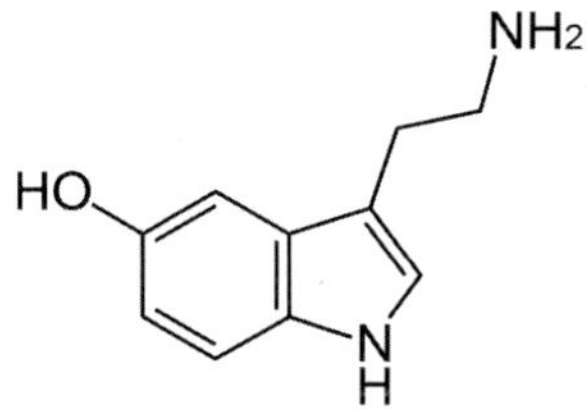

serum[síərəm]: L. *serum, seri,* whey, watery fluid, protein-rich liquid which separates out when blood coagulates. 혈청(血淸, blood ~), 유장(乳漿, 치즈를 만들 때 우유가 응고한 뒤 분리되는 액체).

sesqui[séskwi, -kwə]: L. *semi-*, half + *que*, one. one and half (1배반), (원소의 화합 비율이 3:2의).

<u>sesquiterpenes</u>: a class of terpenes that consist of three isoprene units and have the molecular formula $C_{15}H_{24}$. Like monoterpenes, sesquiterpenes may be acyclic or contain rings, including many unique combinations.

Fig. δ-Cadinene, a sesquiterpene

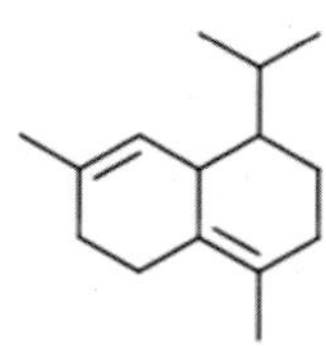

shield[ʃi:ld]: German. *schild*, divide, separate. 가려막이.

shielding: 가려막기, 가리움. Phenomenon caused by electron donation to the environment of a proton. The electron shield the proton from the full effect of the applied magnetic field. The more a proton is shielded, the farther to the right its signal appears in an NMR spectrum.

sigma[sígmə]: the eighteenth letter of the Greek alphabet (Σ, σ), transliterated as 's'.

sigma (σ) bond: 시그마 결합. The strongest type of covalent chemical bond. Sigma bonding is most clearly defined for diatomic molecules using the language and tools of symmetry groups. In this formal approach, a σ-bond is symmetrical with respect to rotation about the bond axis. By this definition, common forms of sigma bonds are s+s, p_z+p_z, s+p_z and $d_z^2+d_z^2$ (where z is defined as the axis of the bond).

Fig. *σ bond* between two atoms : localisation of electron density.

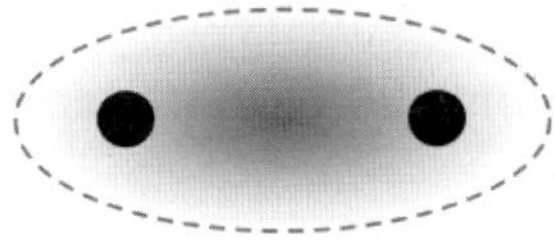

silicon[sílikən]: L. *silex, silicis*, flint. 규소. A chemical element with the symbol **Si** and atomic number 14. A tetravalent metalloid, it is less reactive than its chemical analog carbon, but more reactive than germanium(Ge), Controversy about silicon's character dates to its discovery: silicon was first prepared and characterized in pure form in 1824, and given the name silicium with an -ium word-ending to suggest a metal. However, its final name, suggested in 1831, reflects the more physically similar elements carbon and boron. Silicon is the eighth most common element in the universe by mass, but very rarely occurs as the pure free element in nature. It is most widely distributed in dusts, sands, planetoids, and planets as various forms of silicon dioxide (silica) or silicates. Over 90% of the Earth's crust is composed of silicate minerals, making silicon the second most abundant element in the earth's crust (about 28% by mass) after oxygen. Most silicon is used commercially without being separated, and indeed often with little processing of compounds from nature. These include direct industrial building-use of clays, silica sand and stone. Silica is used in ceramic brick. More modern silicon compounds such as silicon carbide form abrasives and high-strength ceramics. Silicon is the basis of the ubiquitous synthetic silicon-based polymers called silicones.

skeleton[skélətn]: L. *skeleton,* basic structure of molecule, dried body, mummy. 골격, 분자의 골격구조, 뼈대, 해골.

skeletal formula of an organic compound: a shorthand representation of its molecular structure. Skeletal formulae are ubiquitous in organic chemistry because they show complicated structures clearly and they are quick and simple to draw. The skeletal formula of the antidepressant drug escitalopram, featuring skeletal representations of heteroatoms, a triple bond, benzene rings and stereochemistry.

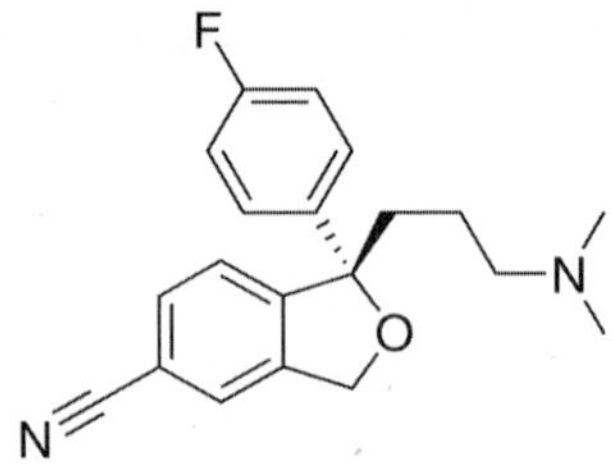

sodium[sóudiəm]: Arabic. *suda*, headache, as the headache-alleviating properties of sodium carbonate or soda were well known in early times. 나트륨, 소듐. A chemical element with the symbol **Na** (from Latin: *natrium*) and atomic number 11. It is a soft, silvery-white, highly reactive metal and is a member of the alkali metals; its only stable isotope is ^{23}Na. It is an abundant element that exists in numerous minerals such as feldspars(장석), sodalite and rock salt. Many salts of sodium are highly soluble in water and are thus present in significant quantities in the Earth's bodies of water, most abundantly in the oceans as sodium chloride.

solid[sάlid /sɔ́l-]: L. *solidus, salvus, sollus*, safe, entire. 고체. One of the three classical states of matter (the others being gas and liquid). It is characterized by structural rigidity and resistance to changes of shape or volume. Unlike a liquid, a solid object does not flow to take on the shape of its container, nor does it expand to fill the entire volume available to it like a gas does. The atoms in a solid are tightly bound to each other, either in a regular geometric lattice (crystalline solids, which include metals and ordinary water ice) or irregularly (an amorphous solid such as common window glass).

soluble[sάljubl /sɔ́l-]: L. *solubilis, solvere*, loosen, unfasten. 녹는, 가용성.

<u>solubility</u>: 용해도. The property of a solid, liquid, or gaseous chemical substance called solute to dissolve in a solid, liquid, or gaseous solvent to form a homogeneous solution of the solute in the solvent. The solubility of a substance fundamentally depends on the used solvent as well as on temperature and pressure. The extent of the solubility of a substance in a specific solvent is measured as the saturation concentration where adding more solute does not increase the concentration of the solution. Most often, the solvent is a liquid, which can be a pure substance or a mixture. One may also speak of solid solution, but rarely of solution in a gas (see vapor-liquid equilibrium instead). The extent of solubility ranges widely, from infinitely soluble (fully miscible) such as ethanol in water, to poorly soluble, such as silver chloride in water. The term insoluble is often applied to poorly or very poorly soluble compounds. Under certain conditions, the equilibrium solubility can be exceeded to give a so-called supersaturated solution, which is metastable. Solubility is not to be confused with the ability to dissolve or liquefy a substance, because the solution might occur not only because of dissolution but also because of a chemical reaction. For example, zinc is insoluble in hydrochloric acid, but does dissolve in it by chemical reaction into zinc chloride and hydrogen, where zinc chloride is then soluble in hydrochloric acid. Solubility does not also depend on particle size or other

kinetic factors; given enough time, even large particles will eventually dissolve.

solvent[sάlvənt / sɔ́l-]: L. *solvent*, *solvere,* loosening, unfastening, paying. 용매. A liquid, solid, or gas that dissolves another solid, liquid, or gaseous solute, resulting in a solution that is soluble in a certain volume of solvent at a specified temperature. Common uses for organic solvents are in dry cleaning (e.g. tetrachloroethylene), as paint thinners (e.g. toluene, turpentine), as nail polish removers and glue solvents (acetone, methyl acetate, ethyl acetate), in spot removers (e.g. hexane, petrol ether), in detergents (citrus terpenes), in perfumes (ethanol), and in chemical synthesis. The use of inorganic solvents (other than water) is typically limited to research chemistry and some technological processes.

sonic[sάnik/sɔ́n-]: L. *sonus, soni,* sound + *ic,* suffix, relating to or using sound waves. 음파의, 소리의, 음향의.

<u>ultrasonic</u>: (*ultra*, beyond + sonic) 초음파의, 초음속의.

species[spi:ʃi:z, -si:z]: L. *specere*, to look, appearance, form. 종(種) (분자, 이온 등의 다른 화학적 실체).

specific[spisífik]: L. *species*, appearance, generally means per unit of something (often mass). (단위량 또는 물리량에 대한) 비(比)의, 고유의, 특이의.

<u>specific rotation[α]</u>: 고유 광회전도. Defined as the observed angle of optical rotation *α* when plane-polarized light is

passed through a sample with a path length of 1 decimeter and a sample concentration of 1 gram per 1 millilitre in stereochemistry. It is the main property used to quantify the chirality of a molecular species or a mineral. The specific rotation of a pure material is an intrinsic property of that material at a given wavelength and temperature. Values should always be accompanied by the temperature at which the measurement was performed and the solvent in which the material was dissolved. Often the temperature is not specified; in these cases it is assumed to be room temperature. The formal unit for specific rotation values is deg $dm^{-1}cm^{3}g^{-1}$ but scientific literature uses just degrees. A negative value means levorotatory rotation and a positive value means dextrorotatory rotation. Optical rotation is measured with an instrument called a polarimeter. There is a linear relationship between the observed rotation and the concentration of optically active compound in the sample. There is a non-linear relationship between the observed rotation and the wavelength of light used. Specific rotation is calculated using either of two equations, depending on the sample you are measuring:
For pure liquids:

$$[\alpha]_{\lambda}^{T} = \frac{\alpha}{l \times \rho}$$

In this equation, l is the path length in decimeters, and ρ (Greek letter "rho") is the density of the liquid in g/mL, for a sample at a temperature T (given in degrees Celsius) and wavelength λ (in nanometers). If the wavelength of the light used is 589 nanometer (the sodium D line), the symbol "D" is used. The sign of the rotation (+ or -) is always given.

spectrum[spéktrəm]: L. *specere*, to look, literally image, apparition. 스펙트럼, 분광(分光). Applies to any signal that can be decomposed into frequency components. A spectrum is a usually 2-dimensional plot, of a compound signal, depicting the components by another measure. As the wavelengths get bigger below the red visible light they become infrared, microwave, and radio. As the wavelengths get smaller above violet light, they become ultra-violet, x-ray, and gamma ray.

spectroscopy: 분광(법,학). The study of the interaction between matter and radiated energy. Historically, spectroscopy originated through the study of visible light dispersed according to its wavelength, e.g., by a prism. Later the concept was expanded greatly to comprise any interaction with radiative energy as a function of its wavelength or frequency. Spectroscopic data is often represented by a spectrum, a plot of the response of interest as a function of wavelength or frequency. Spectrometry and spectrography are terms used to refer to the measurement of radiation intensity as a function of wavelength and are often used to describe experimental spectroscopic methods. Spectral measurement devices are referred to as spectrometers, spectrophotometers, spectrographs or spectral analyzers.

sphere[sfiər]: G. *sphaira*, ball. 공, 구(球), 구체, 구면, 지구의.

spherical[sférikəl, sfíər-]: G. *sphairikos*, from *sphaira*, sphere. 구형의, 구면의, 천체의.

spin[spin]: Old English. *spinnan*, draw out and twist(fibre). 스핀, 자전.

spin-spin coupling: 스핀-스핀 상호작용. 스핀-스핀 짝지음. The splitting of a signal in an NMR spectrum described by the *N*+1 rule.

spiro-[spáiərou, -rə]: L. < G. spira < speira, a coil. 나선형의, 코일.

squalene[skwéili:n]: L. *squalamine*, *squalus*, shark + amine + *-ene*, suffix. 스콸렌(상어의 간유에 함유된 탄화수소). A natural organic compound originally obtained for commercial purposes primarily from shark liver oil, though plant sources (primarily vegetable oils) are used as well, including amaranth seed, rice bran, wheat germ, and olives. All plants and animals produce squalene, including humans. Squalene has been proposed to be an important part of the Mediterranean diet as it may be a chemopreventative substance that protects people from cancer. Squalene is a hydrocarbon and a triterpene, and is a natural and vital part of the synthesis of cholesterol, steroid hormones, and vitamin D in the human body. Squalene is used in cosmetics, and more recently as an immunologic adjuvant (보조의) in vaccines.

Fig. Two molecules of farnesyl pyrophosphate condense with reduction by NADPH to form squalene - by squalene synthase.

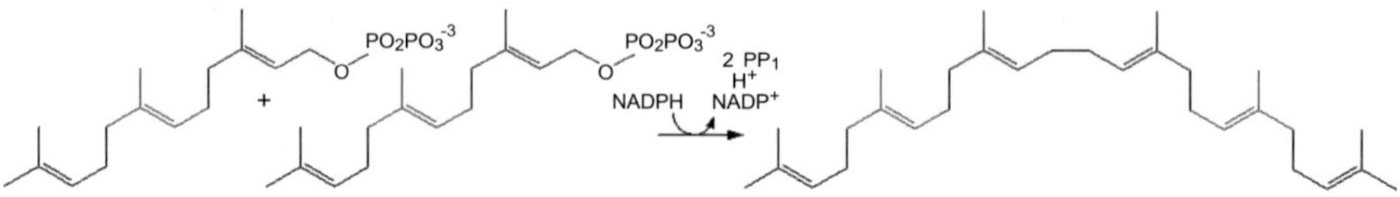

staphylococcus[stæ̀fəloukɑkəs / -kɔk-]: G. *staphulē*, bunch of grapes + *kokkos*, berry. 포도상구균. 복수형: -cocci[-kaksai].

starch[stɑ:rtʃ]: German. *stärken*, strengthen. 녹말, 전분, 정력, 활기, 원기. A carbohydrate consisting of a large number of glucose units joined together by glycosidic bonds. This polysaccharide is produced by all green plants as an energy store. It is the most common carbohydrate in the human diet and is contained in large amounts in such staple foods as potatoes, wheat, maize (corn), rice, and cassava. Pure starch is a white, tasteless and odorless powder that is insoluble in cold water or alcohol. It consists of two types of molecules: the linear and helical amylose and the branched amylopectin. Depending on the plant, starch generally contains 20 to 25% amylose and 75 to 80% amylopectin. Glycogen, the glucose store of animals, is a more branched version of amylopectin. Starch is processed to produce many of the sugars in processed foods. Dissolving starch in warm water gives wheatpaste that can be used as a thickening, stiffening or gluing agent. The biggest industrial non-food use of starch is as adhesive in the papermaking process.

-static[stǽtik]: L. *staticus*, causing to stand. 정적인.

bacteriostat: 세균발육 억제제.

bacteriostatic: 세균발육 억제의.

stearic[stiǽrik, stíər-]: G. *stear*, tallow[수지(獸脂), 쇠기름, 양기름]. 스테아린의.

stearic acid: 스테아린산. The saturated fatty acid with an 18 carbon chain and has the IUPAC name octadecanoic acid. It is a waxy solid, and its chemical formula is $CH_3(CH_2)_{16}CO_2H$. The salts and esters of stearic acid are called stearates. Stearic acid is noteworthy as the most common saturated fatty acid, another common one being the C16 compound palmitic acid.

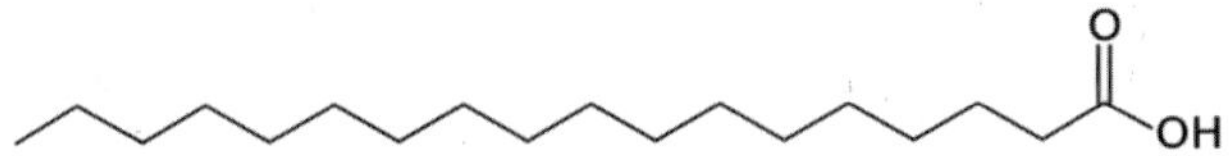

steric[stérik, stiər-]: G. *stereos*, solid + -ic, suffix. relating to the spatial arrangement of atoms in a molecule, especially as it affects chemical reactions. 입체(구조)의.

stereocenters (=stereogenic center): 입체 중심. An atom, bearing groups such that an interchanging of any two groups leads to a stereoisomer. A chirality center is a stereocenter consisting of an atom holding a set of ligands (atoms or groups of atoms) in a spatial arrangement which is not superposable on its mirror image. A chiral center is a generalized extension of an *asymmetric carbon atom*, which is a carbon atom bonded to four different entities, such that an interchanging of any two groups gives rise to an enantiomer. In organic chemistry a chirality center usually refers to a carbon, phosphorus, or sulfur atom, though it is also possible for other atoms to be chirality centers in organic and inorganic chemistry. A molecule can have multiple stereocenters, giving it many stereoisomers. In compounds whose stereoisomerism is due to tetrahedral stereogenic centers, the total number of hypothetically possible stereoisomers will not exceed 2^n, where *n* is the number of tetrahedral stereocenters. Molecules with symmetry frequently have fewer than the maximum possible number of stereoisomers. Having two chiral centers may give a meso compound which is achiral. Certain configurations may not exist due to steric reasons. Cyclic compounds with chiral centers may not exhibit chirality due to the presence of a two-fold rotation axis. Planar chirality may also provide for chirality without having an actual chiral center present.

stereochemistry: 입체 화학. A subdiscipline of chemistry, involves the study of the relative spatial arrangement of atoms within molecules. An important branch of stereochemistry is the study of chiral molecules. Stereochemistry is also known as 3D chemistry because the prefix "stereo-" means "three-dimensionality". The study of stereochemical problems spans the entire range of organic, inorganic, biological, physical and supramolecular chemistries. Stereochemistry includes methods for determining and describing these relationships; the effect on the physical or biological properties these relationships impart upon the molecules in question, and the manner in which these relationships influence the reactivity of the molecules in question (dynamic stereochemistry).

stereoisomers: 입체 이성질체. Isomeric molecules that have the same molecular formula and sequence of bonded atoms (constitution), but that differ *only* in the three-dimensional orientations of their atoms in space. This contrasts with structural isomers, which share the same molecular formula, but the bond connections and/or their order differ(s) between different atoms/groups. In stereoisomers, the order and bond connections of the constituent atoms remain the same, but their orientation in space differ.

stereoselectivity: 입체 선택성. The property of a chemical reaction in which a single reactant forms an unequal mixture of stereoisomers during the non-stereospecific creation of a new stereocenter or during the non-stereospecific transformation of a pre-existing one. The selectivity arises from differences in steric effects and electronic effects in the mechanistic pathways leading to the different products. Stereoselectivity can vary in degree but it can never be total since the activation energy difference between the two pathways is finite. However, in favorable cases, the minor stereoisomer may not be detectable by the analytic methods used.

stereospecificity: 입체 특이성. The property of a reaction mechanism that leads to different stereoisomeric reaction products from different stereoisomeric reactants, or which operates on only one (or a subset) of the stereoisomers. In contrast, stereoselectivity is the property of a reactant mixture where a non-stereospecific mechanism allows for the formation of multiple products, but where one (or a subset) of the products is favored by factors, such as steric access, that are independent of the mechanism. A stereospecific mechanism specifies the stereochemical outcome of a given reactant, whereas a stereoselective reaction selects products from those made available by the same, non-specific mechanism acting on a given reactant. Given a single, stereoisomerically pure starting material, a stereospecific mechanism will give 100% of a particular stereoisomer, although loss of stereochemical integrity can easily occur through competing mechanisms with different stereochemical outcomes. A stereoselective process will normally give multiple products even if only one mechanism is operating on an isomerically pure starting material. The term stereospecific reaction is ambiguous, since the term reaction itself can mean a single-mechanism transformation (such as the Diels-Alder reaction), which could be stereospecific, or the outcome of a reactant mixture that may proceed through multiple competing mechanisms, specific and non-specific. In the latter sense, the term stereospecific reaction is commonly misused to mean 'highly stereoselective reaction'. Chiral synthesis is built on a combination of stereospecific transformations (for the interconversion of existing stereocenters) and stereoselective ones (for the creation of new stereocenters), where also the optical activity of a chemical compound is preserved. The quality of stereospecificity is focused on the reactants and their stereochemistry; it is concerned with the products too, but only as they provide evidence of a difference in behavior between reactants. Of stereoisomeric reactants, each behaves in its own specific way.

steric effects: 입체효과. Arise from the fact that each atom within a molecule occupies a certain amount of space. If atoms are brought too close together, there is an associated cost in energy due to overlapping electron clouds (Pauli or Born repulsion), and this may affect the molecule's preferred shape (conformation) and reactivity.

steric hindrance (= steric resistance): 입체 장애. Occurs when the size of groups within a molecule prevents chemical reactions that are observed in related smaller molecules. Although steric hindrance is sometimes a problem, it can also be a very useful tool, and is often exploited by chemists to change the reactivity pattern of a molecule by stopping unwanted side-reactions (steric protection). Steric hindrance between adjacent groups can also restrict torsional bond angles.

steroid[stíərɔid, stér-]: 1930s. *sterol,* steroid alcohol + *-oid,* form. 스테로이드(스테롤, 담즙산, 성호르몬 따위 지방 용해성 화합물). A type of organic compound that contains a specific arrangement of four rings that are joined to each other. Examples of steroids include cholesterol, the sex hormones estradiol and testosterone, and the anti-inflammatory drug dexamethasone.

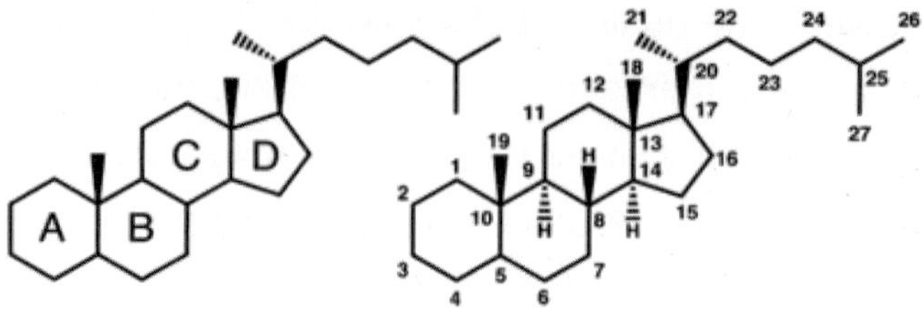

stoichiometry[stɔikiɑ́mətri]: G. *stoicheion, stoicheiou*, element + *metry,* measure, quantitative measuring of the elements in chemical reaction. 화학양론(化學量論).

strain[strein]: L. *stringere*, draw tight. 스트레인, 변형, 균주(생물), 계통(생물), 긴장, 무리. A molecule experiences strain when its chemical structure undergoes some stress which raises its internal energy in comparison to a strain-free reference compound. The internal energy of a molecule consists of all the energy stored within it. A strained molecule has an additional amount of internal energy which an unstrained molecule does not. This extra internal energy, or strain energy, can be likened to a compressed spring. Much like a compressed spring must be held in place to prevent release of its potential energy, a molecule can be held in an energetically unfavorable conformation by the bonds within that molecule. Without the bonds holding the conformation in place, the strain energy would be released.

streptococus[stréptəkɑkəs/-kɔk-]: G. *streptos,* chain + *kokkos,* berry, A form of bacterial organism in which the cocci are arranged in chains or chaplets. 연쇄상구균(連鎖狀球菌).

streptomycin: G. *streptos,* chain, twisted + *myces,* fungus. 스트렙토마이신 (항생물질의 하나). A bactericidal antibiotic drug, the first of a class of drugs called aminoglycosides to be discovered, and was the first antibiotic remedy for tuberculosis. It is derived from the actinobacterium(광박테리아) Streptomyces griseus. Streptomycin cannot be given orally, but must be administered by regular intramuscular injections. An adverse effect of this medicine is ototoxicity, which can lead to hearing loss.

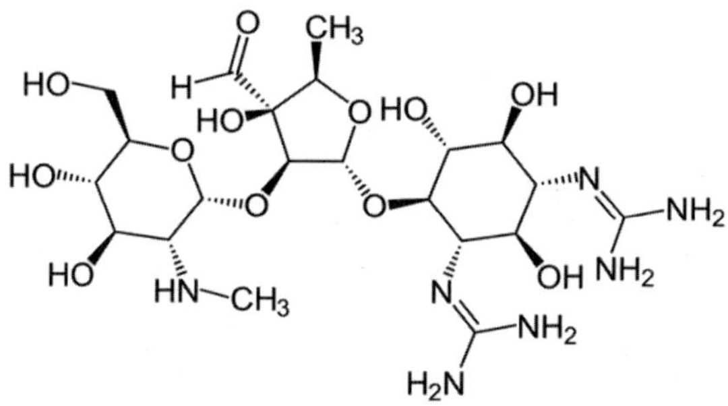

structure[strʌ́ktʃər]: L. *structura, struere,* to build. the arrangement of elements of compound. 화학구조. The chemical structure refers to both molecular geometry and to electronic structure. The structural formula of a chemical compound is a graphical representation of the molecular structure showing how the atoms are arranged. A protein structure is the three dimensional coordinates of the atoms within (macro) molecules made of protein.

sub-[sʌb, səb]: L. *sub*, under, close to, below, slightly.

subacid: 약간 신, (의견 따위가)조금 신랄하고 공격적인.

subshell: (원자의) 각, (전자)껍질.

subunit: 구성하는 기본 단위.

sublimate[sʌ́bləmeit] : L. *sublimate*, *sublimare*, rasised up. - 을 승화시키다.

sublimation: the change from solid to gas without passing through the liquid phase.

substituent[sʌbstitʃúənt, -tjú-]: L. *substituent, substituere*, standing in place of. 치환기(置換基). An atom or group of atoms substituted in place of a hydrogen atom on the parent chain of a hydrocarbon. The terms substituent, side-chain, group, branch, or pendant group are used almost interchangeably to describe branches from a parent structure, though certain distinctions are made in the context of polymer chemistry. In polymers, side chains extend from a backbone structure. In proteins, side chains are attached to the alpha carbon atoms of the amino acid backbone. In addition, when naming hydrocarbons that contain a substituent, positional numbers are used to indicate which carbon atom the substituent is attached to when such information is needed to distinguish between structural isomers. The polar effect exerted by a substituent is a combination of the inductive effect and the mesomeric effect. Additional steric effects result from the volume occupied by a substituent.

substitution[sʌbstətʃú:ʃən / -tjú:-]: L. from *substitutus, substituere, statuere*, put in place of, set up. 치환.

substitution reaction: 치환반응. A functional group in a particular chemical compound is replaced by another group. In organic chemistry, the electrophilic and nucleophilic substitution reactions are of prime importance. Organic substitution reactions are classified in several main organic reaction types depending on whether the reagent that brings about the substitution is considered an electrophile or a nucleophile, whether a reactive intermediate involved in the reaction is a carbocation, a carbanion or a free radical or whether the substrate is aliphatic or aromatic. Detailed understanding of a reaction type helps to predict the product outcome in a reaction. It also is helpful for optimizing a reaction with regard to variables such as temperature and choice of solvent.

nucleophilic substitution: 친핵성 치환. In organic and inorganic chemistry, a fundamental class of reactions in which an electron nucleophile selectively bonds with or attacks the positive or partially positive charge of an atom or a group of atoms called the leaving group; the positive or partially positive atom is referred to as an electrophile. The most general form for the reaction may be given as

Nu: + R-LG → R-Nu + LG:

The electron pair (:) from the nucleophile (Nu) attacks the substrate (R-LG) forming a new bond, while the leaving group (LG) departs with an electron pair. The principal product in this case is R-Nu. The nucleophile may be electrically neutral or negatively charged, whereas the substrate is typically neutral or positively charged. An example of nucleophilic substitution is the hydrolysis of an alkylbromide, R-Br, under alkaline conditions, where the attacking nucleophile is the OH^- and the leaving group is Br^-. $R\text{-}Br + OH^- \rightarrow R\text{-}OH + Br^-$ Nucleophilic substitution reactions are commonplace in organic chemistry, and they can be broadly categorised as taking place at a saturated aliphatic carbon or at (less often) a saturated aromatic or other unsaturated carbon centre. In 1935, Edward D. Hughes and Sir Christopher Ingold studied nucleophilic substitution reactions of alkyl halides and related compounds. They proposed that there were two main mechanisms at work, both of them competing with each other. The two main mechanisms are the S_N1 reaction and the S_N2 reaction. S stands for chemical substitution, N stands for nucleophilic, and the number represents the kinetic order of the reaction. In the S_N2 reaction, the addition of the nucleophile and the elimination of leaving group take place simultaneously. S_N2 occurs where the central carbon atom is easily accessible to the nucleophile. By contrast the S_N1 reaction involves two steps. S_N1 reactions tend to be important when the central carbon atom of the substrate is surrounded by bulky groups, both because such groups interfere sterically with the S_N2 reaction (discussed above) and because a highly substituted carbon forms a stable carbocation. An example of a substitution reaction taking place by a so-called borderline mechanism as originally studied by Hughes and Ingold is the reaction of 1-phenylethyl chloride with sodium methoxide in methanol.

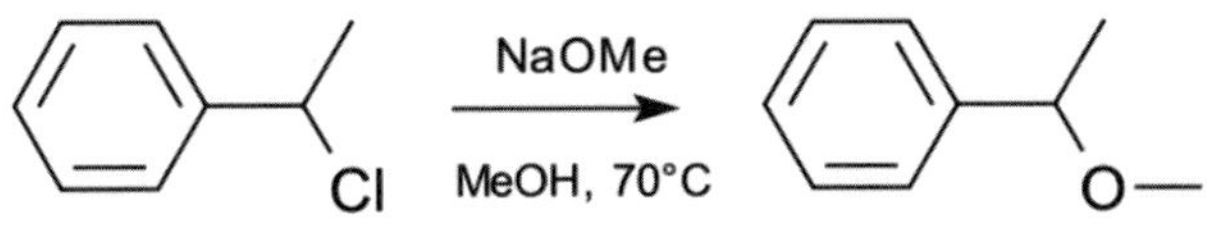

rate = k_1 [$PhCHClCH_3$] + k_2 [$PhCHClCH_3$][NaOMe]

The reaction rate is found to the sum of S_N1 and S_N2 components with 61% (3.5 M, 70°C) taking place by the latter.

Nucleophilic substitution at carbon

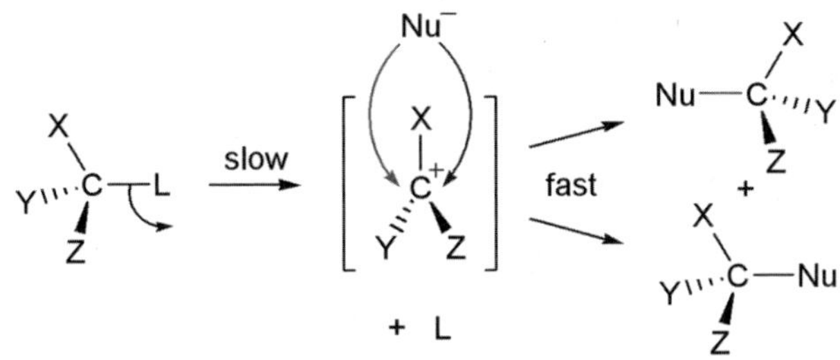

S_N1 mechanism

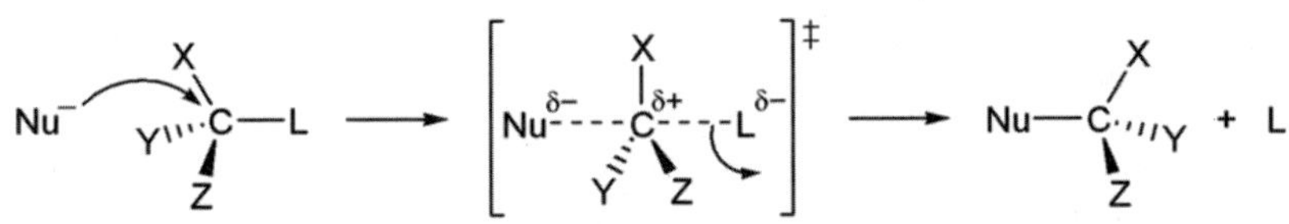

S_N2 mechanism

electrophilic substitution: 친전자성 치환. Chemical reactions in which an electrophile displaces a functional group in a compound, which is typically, but not always, a hydrogen atom. **Electrophilic aromatic substitution reactions** are characteristic of aromatic compounds, and are important ways of introducing functional groups onto benzene rings. The other main type of electrophilic substitution reaction is an **electrophilic aliphatic substitution reaction**. In electrophilic substitution in aromatic compounds, an atom appended to the aromatic ring, usually hydrogen, is replaced by an electrophile. The most important reactions of this type that take place are aromatic nitration, aromatic halogenation, aromatic sulfonation and acylation and alkylating Friedel-Crafts reactions. In electrophilic substitution in aliphatic compounds, an electrophile displaces a functional group. This reaction is similar to nucleophilic aliphatic substitution where the reactant is a nucleophile rather than an electrophile. The four possible electrophilic aliphatic substitution reaction mechanisms are, S_E1, S_E2(front), S_E2(back) and S_Ei (Substitution Electrophilic), are also similar to the nucleophile counterparts S_N1 and S_N2. In the S_E1 course of action the substrate first ionizes into a carbanion and a positively charged organic residue. The carbanion then quickly recombines with the electrophile. The S_E2 reaction mechanism has a single transition state in which the old bond and the newly formed bond are both present.

Electrophilic aliphatic substitution reactions are:

- Nitrosation
- Ketone halogenation
- Keto-enol tautomerism
- aliphatic diazonium coupling
- carbene insertion into C-H bonds
- Carbonyl Alpha-Substitution Reactions

radical substitution: 라디칼 치환. A radical substitution reaction involves radicals. An example is the Hunsdiecker reaction.

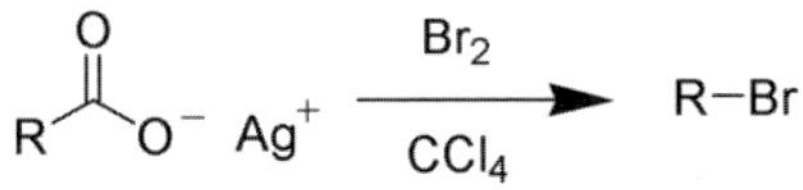

organometallic substitution: 유기금속 치환. Coupling reactions are a class of metal-catalyzed reactions involving an organometallic compound RM and an organic halide R'X that together react to a compound of the type R-R' with formation of a new carbon-carbon bond. Examples are the Heck reaction and the Ullmann reaction. Many variations exist.

Heck reaction

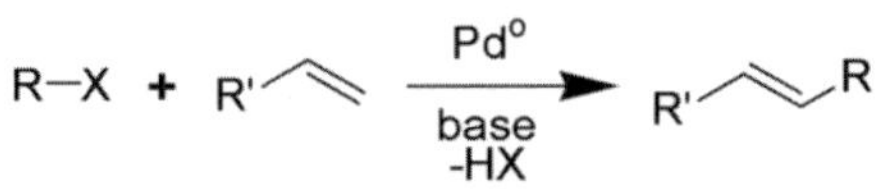

Ullmann reaction

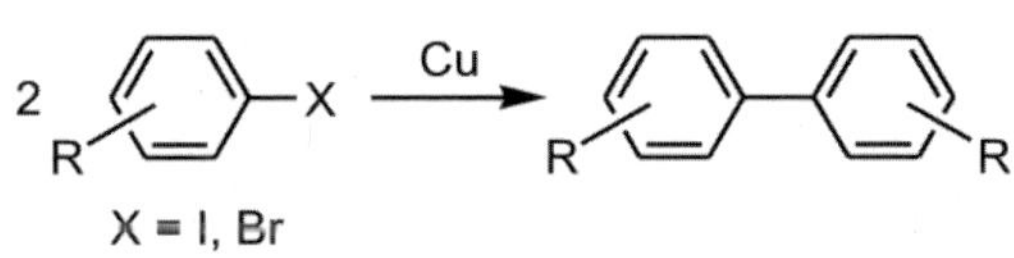

succinic[səksínik]: L. *succinum*, amber (from which it was first derived). 호박의(琥珀의), 호박산(숙신산)의.

succinic acid (= butanedioic acid), historically known as spirit of amber. a dicarboxylic acid. Succinate plays a biochemical role in the citric acid cycle. The carboxylate anion is called succinate and esters of succinic acid are called alkyl succinates.

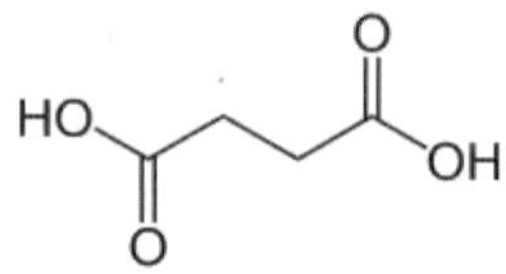

sucrose[sú:krous/sʲu:-]: French. *sucre,* sugar + L. *-osus,* suffix, the organic compound commonly known as table sugar and sometimes called saccharose. The molecule is a disaccharide derived from glucose and fructose with the molecular formula $C_{12}H_{22}O_{11}$. 설탕. 자당, 수크로우스.

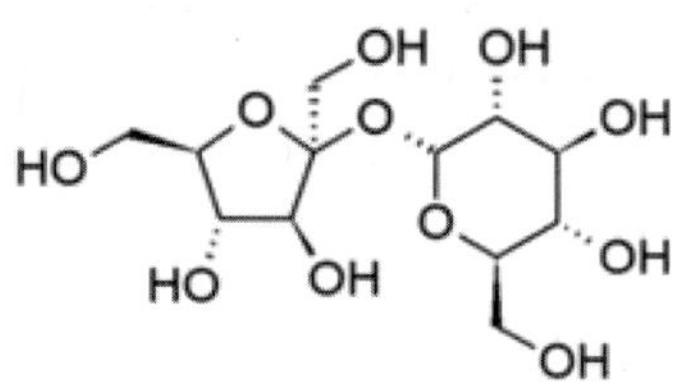

sugar[ʃúgər]: Arabic, *sukkar.* 당(糖). Scientifically, loosely refers to monosaccharide or disaccharides. Monosaccharides are also called "simple sugars," the most important being glucose. Almost all sugars have the formula $C_nH_{2n}O_n$ (n is between 3 and 7). Glucose has the molecular formula $C_6H_{12}O_6$. The names of typical sugars end with "-ose," as in "glucose", "dextrose", and "fructose". Sometimes such words may also refer to any types of carbohydrates soluble in water. The acyclic mono- and disaccharides contain either aldehyde groups or ketone groups. These carbon-oxygen double bonds (C=O) are the reactive centers. All saccharides with more than one ring in their structure result from two or more monosaccharides joined by glycosidic bonds with the resultant loss of a molecule of water (H_2O) per bond. Monosaccharides in a closed-chain form can form glycosidic bonds with other monosaccharides, creating disaccharides (such as sucrose) and polysaccharides (such as starch). Enzymes must hydrolyse or otherwise break these glycosidic bonds before such compounds become metabolised. After digestion and absorption the principal monosaccharides present in the blood and internal tissues include glucose, fructose, and galactose. Many pentoses and hexoses can form ring structures. In these closed-chain forms, the aldehyde or ketone group remains unfree, so many of the reactions typical of these groups cannot occur. Glucose in solution exists mostly in the ring form at equilibrium, with less than 0.1% of the molecules in the open-chain form. Biopolymers of sugars are common in nature. Through photosynthesis plants produce glucose, which has the formula $C_6H_{12}O_6$, and convert it for storage as an energy reserve in the form of other carbohydrates such as starch, or (as in cane and beet) as sucrose (table sugar). Sucrose has the chemical formula $C_{12}H_{22}O_{11}$. Starch, consisting of two different polymers of glucose, is a readily degradable chemical energy stored by cells, convertable to other types of energy. Cellulose is a polymer of glucose used by plants as structural component. DNA and RNA are built up of the sugars ribose and deoxyribose.

Fig. sucrose: a disaccharide of glucose (left) and fructose (right), important molecules in the body.

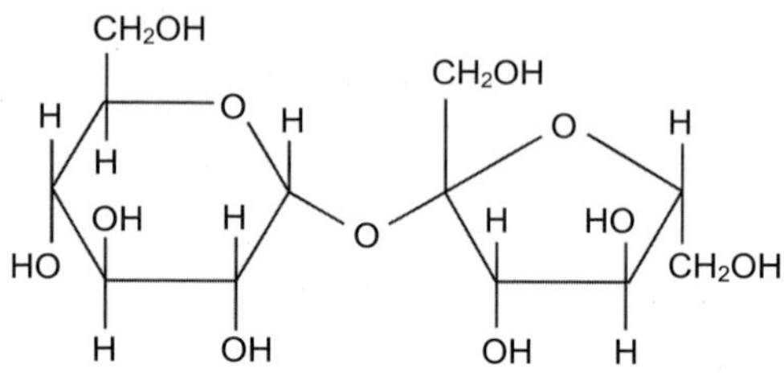

suicide[sú:əsaid / s*j*u:-]: L. *suicida*, *sui*, of oneself + *caedere*, kill, act of suicide. 자살, 자살행위, 자살자.

suicide inhibitor: a compound that inactivates an enzyme by undergoing part of its normal catalytic mechanism.

sulfur (= sulphur)[sʌ́lfər]: L. *sulfur*, sulphur, S. 황, 유황. The chemical element with atomic number 16. In the periodic table it is represented by the symbol **S**. It is an abundant, multivalent non-metal. Under normal conditions, sulfur atoms form cyclic octatomic molecules with chemical formula S_8. Elemental sulfur is a bright yellow crystalline solid when at room temperature. Chemically, sulfur can react as either an oxidant or reducing agent. It oxidizes most metals and several nonmetals, including carbon, which leads to its negative charge in most organosulfur compounds, but it reduces several strong oxidants, such as oxygen and fluorine. It is also the lightest element to easily produce stable exceptions to the octet rule.

sulfide: 황화물. Thioether, the sulfur analog of an ether (RSR).

sulfonamide (= sulphonamide): 설폰아미드. $-S(=O)_2-NH_2$, A sulfonyl group connected to an amine group. A sulfonamide is a compound that contains this group. The general formula is RSO_2NH_2, where R is some organic group. For example, "methanesulfonamide" is $CH_3SO_2NH_2$. Any sulfonamide can be considered as derived from a sulfonic acid by replacing a hydroxyl group with an amine group. In medicine, the term "sulfonamide" is sometimes used as a synonym for sulfa drug, a derivative or variation of sulfanilamide.

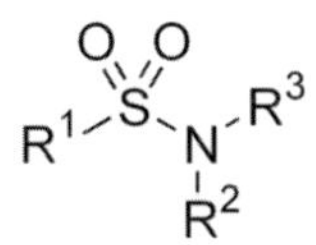

sulfonate ester: the ester of a sulfonic scid (RSO_2OR).

sulfone: 설폰. A chemical compound containing a sulfonyl functional group attached to two carbon atoms. The central hexavalent sulfur atom is double bonded to each of two oxygen atoms and has a single bond to each of two carbon atoms, usually in two separate hydrocarbon substituents.

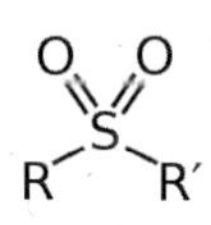

sulfonic acid: 설폰산. Refers to a member of the class of organosulfur compounds with the general formula $RS(=O)_2$–OH, where R is an organic alkyl or aryl group and the $S(=O)_2$–OH group a sulfonyl hydroxide. A sulfonic acid can be thought of as sulfuric acid with one hydroxyl group replaced by an organic substituent. The parent compound (with the organic substituent replaced by hydrogen) is the hypothetical compound sulfurous acid. It is produced by the process of

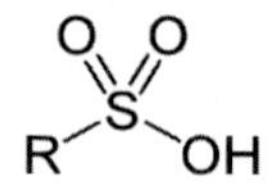

sulfonation. Usually the sulfonating agent is sulfur trioxide. A particularly large scale application of this method is the production of alkylbenzenesulfonic acids:

$$RC_6H_5 + SO_3 \rightarrow RC_6H_4SO_3H$$

In this reaction, sulfur trioxide is an electrophile and the arene undergoes electrophilic aromatic substitution. Thiols can be oxidized to sulfonic acids:

$$RSH + 3/2\ O_2 \rightarrow RSO_3H$$

<u>sulfonyl chlorides (=Sulfonic acid chlorides, or sulfonyl chlorides)</u>: a sulfonyl halide with the general formula $R\text{-}SO_2\text{-}Cl$. These compounds react readily with nucleophiles like alcohols and amines (see Hinsberg reaction). If the nucleophile is an alcohol the product is a sulfonic ester, if it is an amine the product is a sulfonamide. Important sulfonic acid chlorides are tosyl chloride, brosyl chloride, nosyl chloride and mesyl chloride. One synthetic procedure to synthesize sulfonic acid chlorides is the Reed reaction. Sulfonyl chloride is very prone to hydrolysis. An ice/water bath can be used to decrease this effect.

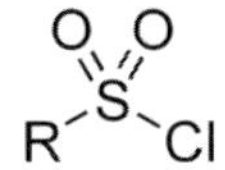

sulfuric acid (=sulphuric acid): 황산. A highly corrosive strong mineral acid with the molecular formula H_2SO_4. The historical name of this acid is oil of vitriol. It is a colorless to slightly yellow viscous liquid and is soluble in water at all concentrations. It is a dibasic acid. The corrosiveness of it is mainly due to its strong acidic nature, strong dehydrating property and if concentrated strong oxidizing property. It has many applications and is a central substance in the chemical industry. Principal uses include lead-acid batteries for cars and other vehicles, mineral processing, fertilizer manufacturing, oil refining, waste water processing, and chemical synthesis.

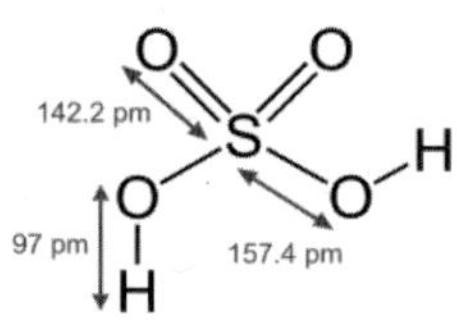

superimpose[su:pərimpóuz /s*j*u:-]: L. super, *supra*, above, beyond, before in time + impose, *imponere*, *in*, In, upon + *ponere*, put. - 을 (- 위에) 놓다, 겹쳐놓다.

suprafacial[su:prəféiʃəl /s*j*u:-]: L. *suprafacial*, *supra*, above, beyond, before in time + *facialis, facies*, face to face, open. 동일면.

suprafacial and antarafacial: 동일면과 반대면. Two topological

concepts in organic chemistry describing the relationship between two simultaneous chemical bond making and/or bond breaking processes in or around a reaction center. The reaction center can be a *p-orbital*, a conjugated system or even a sigma bond.

- This relationship is *antarafacial* when opposite faces are involved (think *anti*).
- It is suprafacial when both changes occur at the same face.

Many sigmatropic reactions and cycloadditions can be either suprafacial or antarafacial, and this determines the stereochemistry. An example is the [1,3]-hydride shift in which the interacting frontier orbitals are the allyl free radical and the hydrogen 1s orbitals. The suprafacial shift is symmetry forbidden because orbitals with opposite algebraic signs overlap. The symmetry allowed antarafacial shift would require a strained transition state and is also unlikely. In contrast a symmetry allowed and suprafacial [1,5]-hydride shift is a common event.

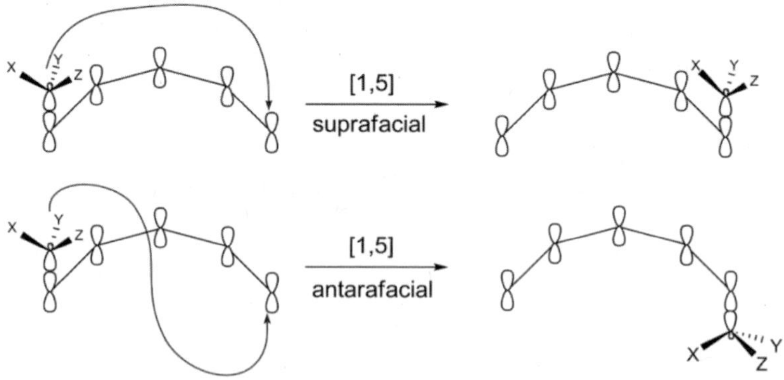

symmetry[sɪ́mətri]: G. *symmetria*, *sun-*, with + *metron*, measure, to measure together, similarity or exact correspondence between different things. 대칭. Exact correspondence in position of the several points or parts of a figure or body with reference to a dividing line, plane, or point, molecular symmetry, axis of symmetry, centre of symmetry, plane of symmetry.

syn-[sin, sən]: G. *syn-,* with, together, similarly, alike. 함께, 똑같이, 양쪽 모두. Syn and anti addition are different ways in which two substituents can be added to a double bond or triple bond.

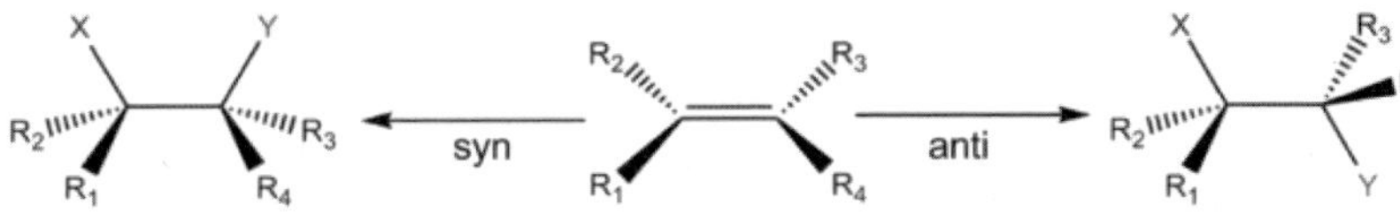

synthesis[sɪ́nθəsis]: G. *sunthesis*, *syn-*, with, together + *thesis*, placing, to place, place together. 합성. A combination of two or more entities that together form something new; alternately, it refers to the creating of something by artificial means.

<u>organic synthesis</u>: 유기합성. The chemical synthesis of organic compounds.

<u>total synthesis</u>: 전합성(全合成). The complete organic synthesis of complex organic compounds, usually without the aid of biological processes.

<u>convergent synthesis or linear synthesis</u>: a strategy to improve the efficiency of multi-step chemical syntheses.

synthon[sínθɔn]: USA. A concept in retrosynthetic analysis. It is defined as a structural unit within a molecule which is related to a possible synthetic operation. The term was coined by E. J. Corey.

Example:

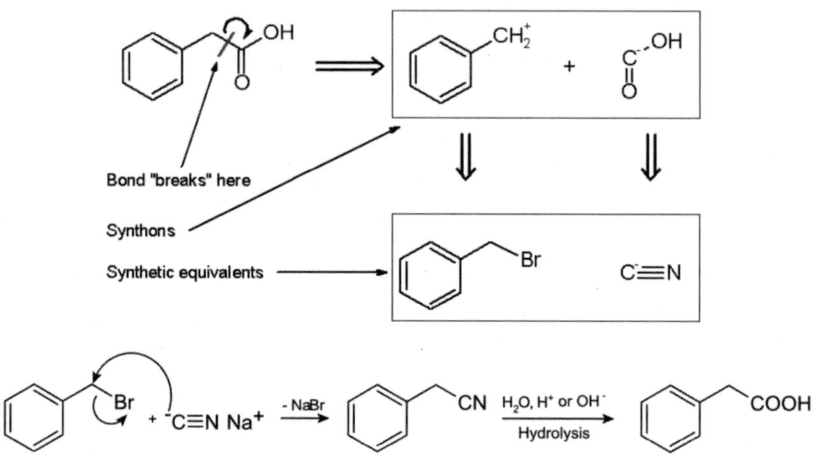

system[sístəm]: G > L. systema < sustēma, sun-, with + histania, set up. 계(系), 체계, 계통.

systematic name: 계통명(系統名), 체계명. A name given in a systematic way to one unique group, organism, object or chemical substance, out of a specific population or collection. Systematic names are usually part of a nomenclature. A semisystematic name or semitrivial name is a name that has at least one systematic part and at least one trivial part. Creating systematic names can be as simple as assigning a prefix or a number to each object (in which case they are a type of numbering scheme), or as complex as encoding the complete structure of the object in the name. Many systems combine some information about the named object with an extra sequence number to make it into a unique identifier. Systematic names often co-exist with earlier common names assigned before the creation of any systematic naming system. For example, many common chemicals are still referred to by their common or trivial names, even by chemists.

tactic[tǽktik]: L. *tactica*, ordered, arranged. 입체 규칙성, 중합체의 구성 단위가 규칙적으로 늘어선.

tactic polymer: 입체규칙성 중합체. A polymer in which the substituents are orderly oriented on the extended carbon chain.

atactic polymer: 혼성배열 중합체. A polymer in which the substituents are randomly oriented on the extended carbon chain.

tartaric[ta:rtǽrik, -tá:r-]: L. *tartarus,* acid potassium tartrate, present in grape juice, deposited in a crude form in the process of fermentation, and adhering to the sides of wine-casks in the form of a hard crust. 주석(酒石)의, 주석에서 나는.

tartaric acid: 타타르산, 주석산. A white crystalline diprotic organic acid. It occurs naturally in many plants, particularly grapes, bananas, and tamarinds, is one of the main acids found in wine. It is added to other foods to give a sour taste, and is used as an antioxidant. Salts of tartaric acid are known as tartrates. It is a dihydroxyl derivative of succinic acid. Tartaric acid played an important role in the discovery of chemical chirality. This property of tartaric acid was first observed in 1832 by Jean Baptiste Biot, who observed its ability to rotate polarized light. Louis Pasteur continued this research in 1847 by investigating the shapes of ammonium sodium tartrate crystals, which he found to be chiral. By manually sorting the differently shaped crystals under magnification, Pasteur was the first to produce a pure sample of levotartaric acid.

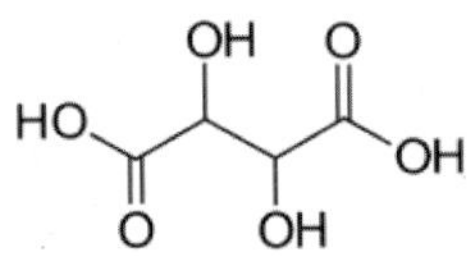

tautomer[tɔ́:təmər]: G. *taut(o)*, the same + *meros*, part, the same part. 토토머. Isomers of organic compounds that readily interconvert by a chemical reaction called tautomerization. It is common that this reaction results in the formal migration of a hydrogen atom or proton, accompanied by a switch of a single bond and adjacent double bond. The exact ratio of the tautomers depends on several factors, including temperature, solvent, and pH. The concept of tautomers that are interconvertible by tautomerizations is called tautomerism. Tautomerism is a special case of structural isomerism and can play an important role in non-canonical base pairing in DNA and especially RNA molecules.

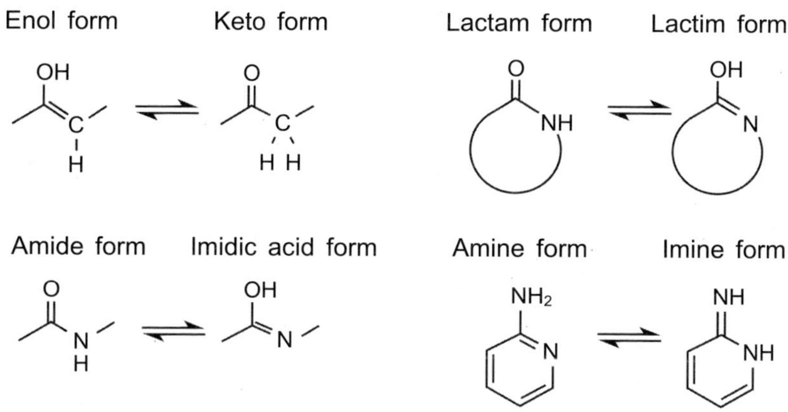

taxol[tǽksɔ:l]: L. *taxus*, yew + *-ol*, (alcoh)ol. 택솔[주목(朱木)에서 얻는 항암제]. When taxol was developed commercially by Bristol-Myers Squibb (BMS) the generic name was changed to 'paclitaxel' and the BMS compound is sold under the trademark 'TAXOL'. Paclitaxel.

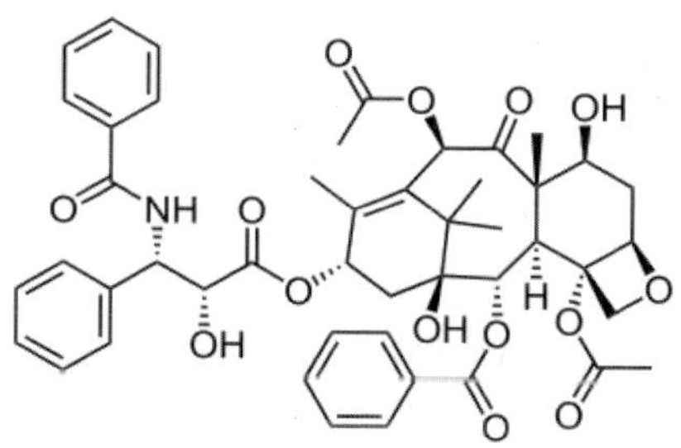

teflon[téflɑn / -lɔn]: 1940s. *tetra-*, four + *fluoro-*, fluorine + -on, suffix. 테플론(합성수지). Polytetrafluoroethylene (PTFE), a synthetic fluoropolymer of tetrafluoroethylene that finds numerous applications. The most well known brand name of PTFE is Teflon by DuPont Co. PTFE is a fluorocarbon solid, as it is a high-molecular-weight compound consisting wholly of carbon and fluorine. PTFE is hydrophobic: neither water nor water-containing substances wet PTFE, as fluorocarbons demonstrate mitigated London dispersion forces due to the high electronegativity of fluorine. PTFE has one of the lowest coefficients of friction against any solid. PTFE is used as a non-stick coating for pans and other cookware. It is very non-reactive, partly because of the strength of carbon–fluorine bonds, and so it is often used in containers and pipework for reactive and corrosive chemicals. Where used as a lubricant, PTFE reduces friction, wear, and energy consumption of machinery.

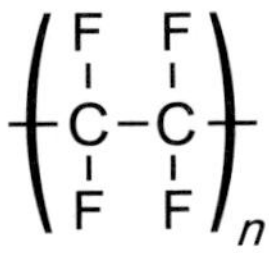

temperature[témpərətʃər]: L. *temperatura, temperare*, restrain. 온도, 기온, 체온. A physical property of matter that quantitatively expresses the common notions of hot and cold. Objects of low temperature are cold, while various degrees of higher temperatures are referred to as warm or hot. Heat spontaneously flows from bodies of a higher temperature to bodies of lower temperature, at a rate that increases with the temperature difference and the thermal conductivity. No heat will be exchanged between bodies of the same temperature; such bodies are said to be in "thermal equilibrium". The temperature of a substance typically varies with the average speed of the particles that it contains, raised to the second power; that is, it is proportional to the mean kinetic energy of its constituent particles. Formally, temperature is defined as the derivative of the internal energy with respect to the entropy. Quantitatively, temperature is measured with thermometers, which may be calibrated to a variety of temperature scales.

tensile[ténsəl / -sail]: L. t*ensilis, tendere*, to stretch. 장력(張力)의. 인장(引長)의.

terminal[tə́:rmənl]: L. *terminalis, terminus*, end, boundary. 말단의, 최종의.

termination[tə:rmənéiʃən]: L. *termination*, *terminare,* to limit, end. 종결(終結).

termination reaction: 종결반응. Any chemical reaction leading to the destruction of a reactive intermediate in a chain propagation step in the course of a polymerization, effectively bringing it to a halt.

terminology[tə:rmənɑ́lədʒi/-nɔ́l-]: L. *terminus, termini,* term + G. *-logia*, study, study for term. 용어론, (총칭)용어. The body of terms used with a particular technical application in a subject of study, theory, profession, etc.

termite[tə́:rmait]: L. t*ermes, termit-*, wood worm. 흰개미.

tertiary[tə́:rʃieri, -ʃəri]: L. *tertiarius, from tertius*, third. 삼차의. Having its functional group located on a carbon atom which is itself bonded to three other carbon atoms.

tertiary carbon: 삼차탄소. Three carbon neighbors.

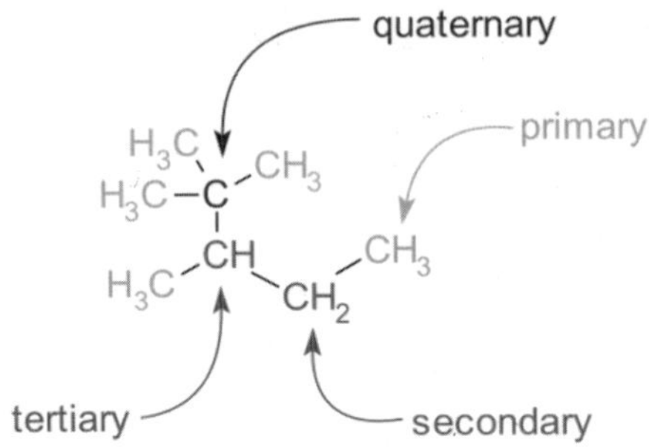

testosterone[testάstəroun]/-tɔ́s-]: G. *testosteron, testis, or testicle,* in man and mammals + *sterone,* formative element in the names of some steroids. 테스토스테론(남성호르몬의 일종). A steroid hormone that stimulates the development of male secondary sexual characteristics and which is produced in the testes(고환), and, in very much smaller quantities, in the ovaries(난소) and adrenal cortex(신장피질).

tether[téðər]: Norwegian. *tjóthr,* fasten, tie with a rope or chain so as to restrict its movement. - 을 속박하다, 묶어두다.

tetra-[tétrə]: G. *tettares*, 모음 앞에서는 tetr-, four.

tetracycline[tetrəsáikli:n]: 1950s. *tetra-*, L. four + *cyclic*, L. *cyclicus*, circle + -ine, suffix. 테트라사이크린(항생물질의 일종). A broad-spectrum polyketide antibiotic produced by the Streptomyces genus of Actinobacteria, indicated for use against many bacterial infections. It is a protein synthesis inhibitor. It is commonly used to treat acne (좌창, 여드름) today, and, more recently, rosacea, and is historically important in reducing the number of deaths from cholera. Tetracycline is marketed under the brand names Sumycin, Tetracyn, and Panmycin, among others. Actisite is a thread-like fiber formulation used in dental applications. It is also used to produce several semisynthetic derivatives, which together are known as the tetracycline antibiotics. The term "tetracycline" is also used to denote the 4-ring system of this compound; "tetracyclines" are related substances that contain the same 4-ring system.

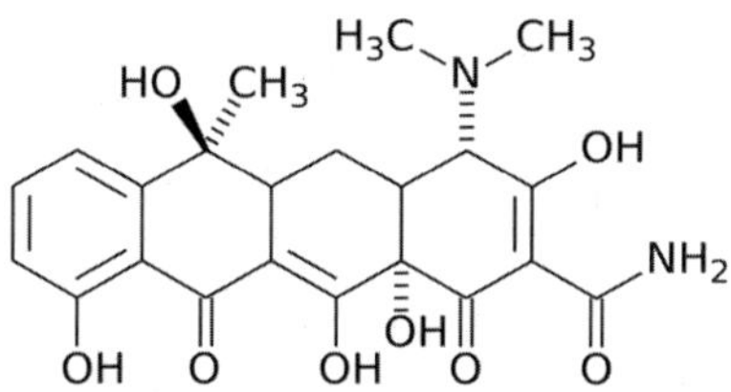

tetrahedron[tetrəhí:drən /-héd-]: G. *tetraedros, tetraedron*, four-sided. 4면체. A polyhedron composed of four triangular faces, three of which meet at each vertex. It has six edges and four vertices. The tetrahedron is the only convex polyhedron that has four faces. The tetrahedron is the three-dimensional case of the more general concept of a Euclidean simplex. The tetrahedron is one kind of pyramid, which is a polyhedron with a flat polygon base and triangular faces connecting the base to a common point. In the case of a tetrahedron the base is a triangle (any of the four faces can be considered the base), so a tetrahedron is also known as a "triangular pyramid".

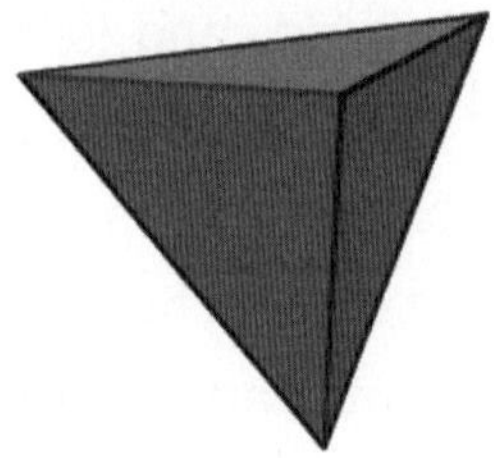

thalidomide[θəlídəmaid]: 1950s. *(ph)thal(ic acid) + (im)ido + (i)mide.* 탈리도마이드. A sedative drug in the late 1950s. In 1961, it was withdrawn due to teratogenicity (기형을 발생하는) and neuropathy (신경장애). There is now a growing clinical interest in thalidomide, and it is introduced as an immunomodulatory agent(면역조절제) used primarily, combined with dexamethasone, to treat multiple myeloma(골수증). The drug is a potent teratogen in zebrafish, chickens, rabbits and primates(영장류) including humans: severe birth defects may result if the drug is taken during pregnancy.

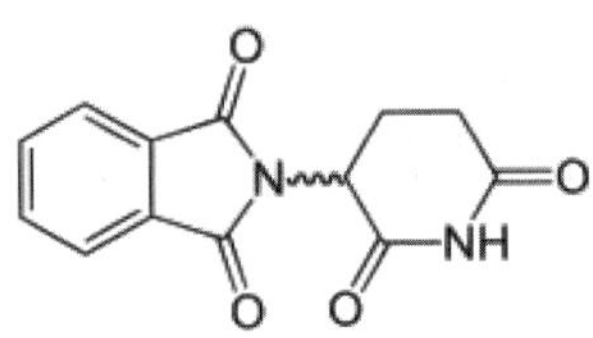

theology[θiάlədʒi/-ɔ́l-]: G. *theos,* god + *-logia, logos*, word, the study of the nature of God and religious belief. (특히 기독교) 신학.

theory[θí:əri, θiəri]: L. *theoria, theoriae,* contemplation, speculation. 이론, 학설, 생각, 고찰.

therapeutic[θerəpjú:tik]: L. *therapeutikos, therapeuein*, minister to, treat medically. 치료(상)의, 치료법의.

thermal[θə́:rməl]: G. *thermē*, heat. 열의.

thermodynamic[θə:rmoudainǽmik]: G. *thermodynamic*, *thermos*, *thermē*, hot, heat + *dunamikos, dunamis*, power. 열역학의.

thermodynamics: 열역학. The science of energy conversion involving heat and other forms of energy, most notably mechanical work. It studies and interrelates the macroscopic variables, such as temperature, volume and pressure, which describe physical, thermodynamic systems.

thermoplastic[θə:rməplǽstik]: G. *thermoplastic*, *thermē*, heat, + L. < G. *plastikos, plassein* > *plasticus*, to mould. 열가소성(熱可塑性) 플라스틱.

thi(o)-[θáiou, θáiə]: G. *theion, theiou,* 모음 앞에서는 *thi-,* brimstone, sulfur. 유황.

thiamine (=thiamin): a water-soluble vitamin of the B complex (vitamin B1), whose phosphate derivatives are involved in many cellular processes.

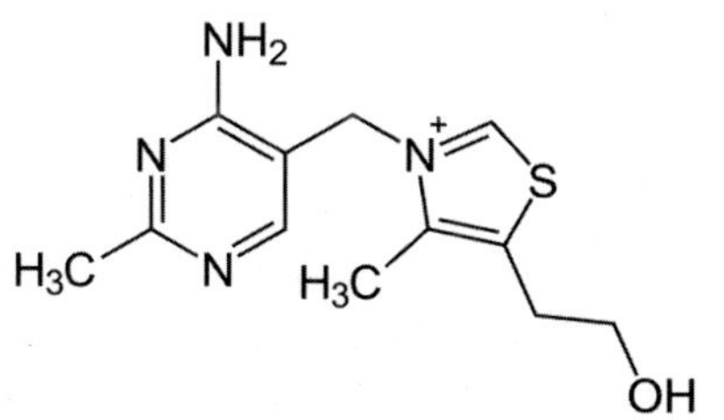

thiol: a compound that contains the functional group composed of a sulfur-hydrogen bond (-SH). Being the sulfur analogue of an alcohol group (-OH), this functional group is referred to either as a thiol group or a sulfhydryl group. More traditionally, thiols are often referred to as mercaptans.

methionine (Met, M): an α-amino acid with the chemical formula $HO_2CCH(NH_2)CH_2CH_2SCH_3$. This essential amino acid is classified as nonpolar.

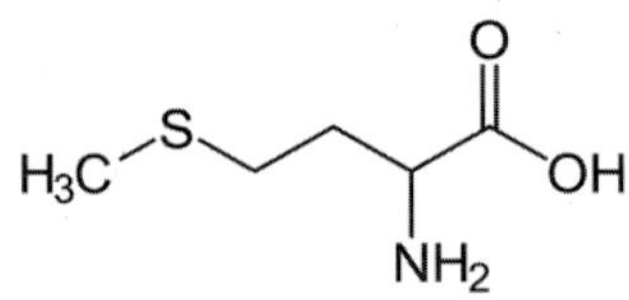

thiophene: heterocyclic compound with the formula C_4H_4S. Consisting of a flat five-membered ring, it is aromatic as indicated by its extensive substitution reactions. Related to thiophene are benzothiophene and dibenzothiophene, containing the thiophene ring fused with one and two benzene rings, respectively. Compounds analogous to thiophene include furan (C_4H_4O) and pyrrole (C_4H_4NH).

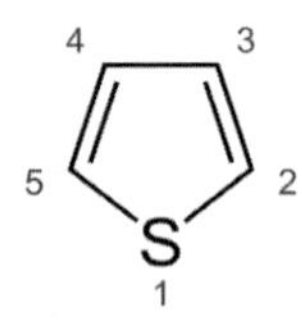

thymine[θáimi(:)n]: L. *thymus,* pertaining to, or connected with the thymus gland + *-ine,* suffix. (가슴샘과 관련이있는) DNA의 주요 성분, abbreviated as T. One of the four nucleobases in the nucleic acid of DNA that are represented by the letters G-C-A-T. The others are guanine, cytosine, and adenine. Thymine (T) almost always pairs with adenine, although thymine dimers also occur due to UV light exposure.

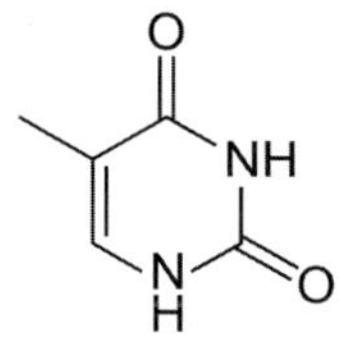

thyroxine[θ*aiərəksi:n*]: 20th cent. thyroid, G. *thureoeidēs, thureos*, shield-shaped (cartilage), oblong shield(장방형 방패) + *ox-*, oxygen + -ine, suffix. 갑상선 호르몬. 3,5,3',5'-Tetraiodothyronine (often abbreviated as T_4), a form of thyroid hormones is the major hormone secreted by the follicular cells of the thyroid gland.

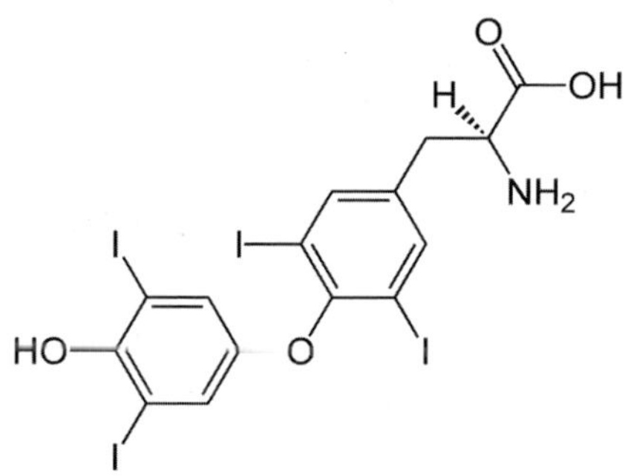

titrate[táitreit]: French. *titrate*, *titrer*, *titre*, fineness of alloyed gold or silver + *-ate*, adjectives ending. - 을 적정(滴定)하다.

titration curve: 적정 곡선. A plaot of pH versus added equivalents of hydroxide ion.

tocopherol[toukάfərɔ:l, -rɑl/-kɔ́fərɔl]: G. *tokos,* offspring + *pherein,* to bear + -ol. Vitamin E. Any or all of a group of closely related fat-soluble compounds that occur esp. in plant oils and are anti-oxidants essential in the diets of many animals and probably of man.

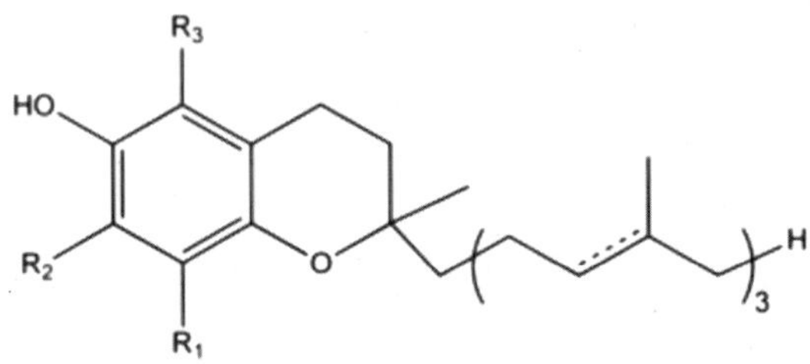

α-tocopherol, $R_1 = R_2 = R_3 = CH_3$
α-tocotrienol, $R_1 = R_2 = R_3 = CH_3$

β-tocopherol, $R_1 = R_3 = CH_3$; $R_2 = H$
β-tocotrienol, $R_1 = R_3 = CH_3$; $R_2 = H$

γ-tocopherol, $R_1 = R_2 = CH_3$ $R_3 = H$
γ-tocotrienol, $R_1 = R_2 = CH_3$ $R_3 = H$

δ-tocopherol, $R_1 = R_2 = R_3 = H$
δ-tocotrienol, $R_1 = R_2 = R_3 = H$

tomography[tɔmάgrəfi / -mɔ́g-]: G. *tomos*, slice, section + -*graphia*, writing. 단층 X선 촬영

Fig. Basic principle of tomography: superposition free tomographic cross sections S_1 and S_2 compared with the projected image P

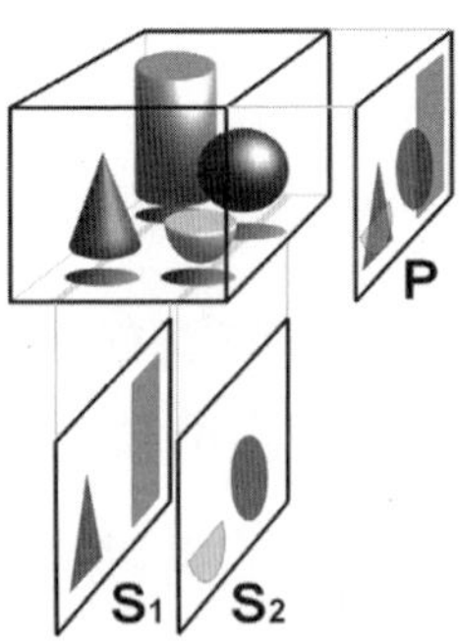

torsion[tɔ́:rʃən]: L. *torsio(n),* variant of *tortio(n)*, twisting, torture, from *torquere*, to twist. 비틀림, 비트는 힘, 염력(捻力).

toxic[tɑ́ksik/ tɔ́k-]: L. *toxicum, toxici,* poisoned, poison. 유독한, 독성의, 유독 화학약품.

trajectory[trədʒéktəri]: L. *trajectoria*, from the verb *traicere*, *trans-*, across + *jacere*, to throw, thrown across. 궤적, 곡선, 직교절선.

trans-[træns / trænz]: L. *trans*, across. 접두어, 다른 편에. on[or to] the other side of, over, across, through, beyond, above.

trans fat: 트랜스 지방. The common name for unsaturated fat with trans-isomer (E-isomer) fatty acid(s). Because the term refers to the configuration of a double carbon-carbon bond, trans fats are sometimes monounsaturated or polyunsaturated, but never saturated. Trans fats are rare in living nature, but can occur in food production process. The consumption of trans fats increases the risk of coronary heart disease by raising levels of LDL cholesterol and lowering levels of "good" HDL cholesterol. Health authorities worldwide recommend that consumption of trans fat be reduced to trace amounts. Trans fats from partially hydrogenated oils are more harmful than naturally occurring oils.

transannular: above ring. 고리횡단.

transannular hydrogen: flagpole hydrogen.

trans isomer: ref. *cis* isomer.

transcribe[trænskráib]: L. *transcribere*, from *trans-*, across + *scribe*, write. 전사(轉寫)하다, 복사하다.

transfer[trænsfə́:r, trǽnsfə:r, trænsfə́:]: L. *transferre*, *trans-*, across + *ferre*, to bear, conveyance of property. - 전달하다, 이동시키다, 이동, 전사(轉寫), 전이.

transition[trænzíʃən, -siʃ-]: L. *transition*, *transire*, go across. 전이(轉移).

transition metal (= transition element): 전이금속. Two possible meanings:

- The IUPAC definition states that a transition metal is "an element whose atom has an incomplete *d* sub-shell, or which can give rise to cations with an incomplete *d* sub-shell".
- Most authors describe a "transition metal" as any element in the *d-block* of the periodic table, which includes groups 3 to 12 on the periodic table. All elements in the *d*-block are metals. In actual practice, the *f-block* is also included in the form of the lanthanide and actinide series.

transition state of a chemical reaction: 화학반응의 전이상태. A particular configuration along the reaction coordinate. It is defined as the state corresponding to the highest energy along this reaction coordinate. At this point, assuming a perfectly irreversible reaction, colliding reactant molecules will always go on to form products.

Fig. The transition state shown below occurs during the S_N2 reaction of bromoethane with a hydroxyl anion.

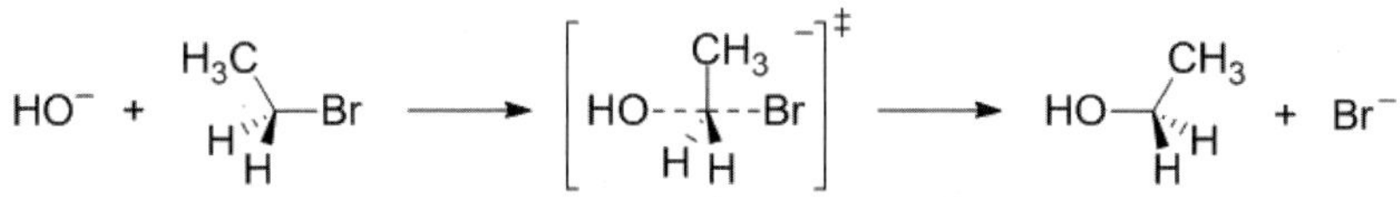

Transition state

translation[trænsléiʃən, trænz-]: L. *translation, translat*, carried across. 병진, 번역, 절단.

translational energy: 병진운동 에너지. It sometimes is convenient to split the total kinetic energy of a body into the sum of the body's center-of-mass translational kinetic energy and the energy of rotation around the center of mass (rotational energy):

$$E_k = E_t + E_r$$

where:

E_k is the total kinetic energy
E_t is the translational kinetic energy
E_r is the *rotational energy* or *angular kinetic energy* in the rest frame

Thus the kinetic energy of a tennis ball in flight is the kinetic energy due to its rotation, plus the kinetic energy due to its translation.

tri-[trai]: G. *treis*, three.

triangle[tráiæŋgl]: L. *tri,* three + *angulus,* angle, a plane figure with three straight sides and three angle. 삼각형, 삼각 석쇠.

trigonal[trigənl]: L. *trigonalis*, characterized by three orbitals lying in a plane and directed to the corners of an equilateral triangle. 삼각(형)의, (결정)삼방정계(三方晶系)의.

triple[tripl]: G. > L. *triplus,* < *triplous,* three times. 3중의, 세 겹의, 3종류의, 3배의.

<u>triple bond</u>: 삼중결합. A chemical bond between two chemical elements involving six bonding electrons instead of the usual two in a covalent single bond. The most common triple bond, that between two carbon atoms, can be found in alkynes. Other functional groups containing triple bond are cyanides and isocyanides. Some diatomic molecules, such as dinitrogen and carbon monoxide are also triple bonded. In skeletal formula the triple bond is drawn as three parallel lines (≡) between the two connected atoms; in typography, this is accomplished with the identity operator. Triple bonds are stronger than single bonds or double bonds and they are also shorter. The bond order is three.

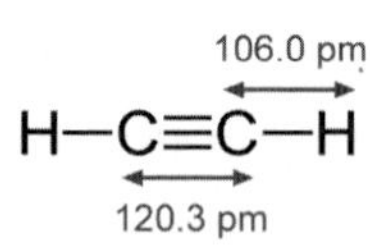

trivial[trívial]: L. *trivium*, *tri-,* three + *via*, road, place where three roads meet. 통속명의.

<u>trivial name</u>: 통속명. A non-systematic name or non-scientific

name. That is, the name is not recognised according to the rules of any formal (*e.g.* IUPAC) system of nomenclature. Many trivial names continue to be used because their sanctioned equivalents are considered too cumbersome for everyday use. For example, "tartaric acid", a compound found in wine, has a systematic name of 2,3-dihydroxybutanedioic acid. Trivial names often have their roots in the common language; in chemistry they may come from historic usages in, for example, alchemy. Frequently, trivial names derive from some notable property of the thing being named.

tropic[trάpik / trɔ́p-]: G. *tropikos, tropē, trepein,* turning, to turn, changing, tending to change. 회귀선, 열대지방(의).

<u>prototropic</u>: 양성자성. The most common form of tautomerism and refers to the relocation of a proton.

trough[trɔ:f, trɑf /trɔf]: Dutch, *trog*, the opposite of a crest, so the minimum or lowest point in a cycle. 고랑, (놀과 놀 사이의) 골

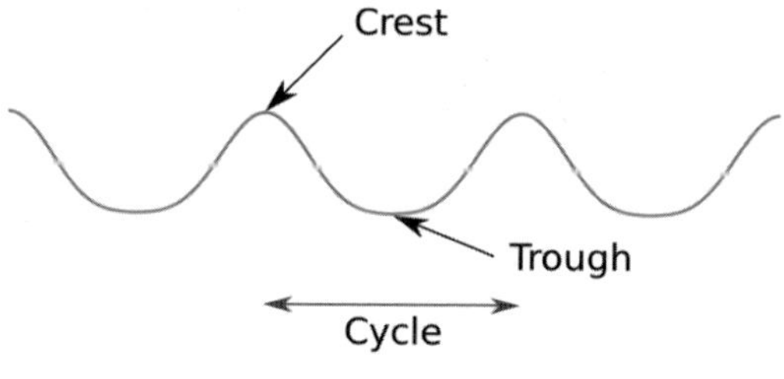

turbid[tə́:rbid]: L. *turbidus, turbida, turbidum*, troubled, muddy. 혼탁, 흐림.

<u>turbidity</u>: 혼탁도, 흐림도.

turpentine[tə́:rpəntain]: L. *ter(e)binthina*(*resina*), *terebinthus*, (resin) of the turpentine tree. 테레빈(송진; 테레빈유). Also called spirit of turpentine, oil of turpentine, and wood turpentine, a fluid obtained by the distillation of resin obtained from trees, mainly pine trees. It is composed of terpenes, mainly the monoterpenes alpha-pinene and beta-pinene. It is sometimes known as *turps*.

typhoid[táifɔid]: 19th cent. *Typhus*, G. *tuphos*, *tuphein*, smoke, stupor(마비, 기절, 혼수상태의), to smoke + *-oid*, form. 티푸스(성)의, 장티푸스의, 장티푸스.

ultra-[ʌ́ltrə]: L. *ultra*, beyond. trans-, super-, hyper-. 초(超).

ultraviolet[ʌltrəváiəlit]: L. *ultraviola*, beyond of violet. 자외선(紫外線)의. Electromagnetic radiation with a wavelength shorter than that of visible light, but longer than X-rays, in the range 10 nm to 400 nm, and energies from 3 eV to 124 eV. It is named because the spectrum consists of electromagnetic waves with frequencies higher than those that humans identify as the colour violet. These frequencies are invisible to humans, but visible to a number of insects and birds. They are also indirectly visible, by causing fluorescent materials to glow with visible light.

urea[juərí:ə, júəriə]: G. *ouron*, urine, 요소(尿素). Carbamide, an organic compound with the chemical formula $CO(NH_2)_2$. The molecule has two $-NH_2$ groups joined by a carbonyl (C=O) functional group. Urea serves an important role in the metabolism of nitrogen-containing compounds by animals and is the main nitrogen-containing substance in the urine of mammals. It is solid, colourless, and odorless (although the ammonia that it gives off in the presence of water, including water vapor in the air, has a strong odor). It is highly soluble in water and non-toxic. Dissolved in water, it is neither acidic nor alkaline. The body uses it in many processes, the most notable one being nitrogen excretion. Urea is widely used in fertilizers as a convenient source of nitrogen. Urea is also an important raw material for the chemical industry. The synthesis of this organic compound by Friedrich Wöhler in 1828 from an inorganic precursor was an important milestone in the development of organic chemistry, as it showed for the first time that a molecule found in living organisms could be synthesized in the lab without biological starting materials (thus contradicting a theory widely prevalent at one time, called vitalism).

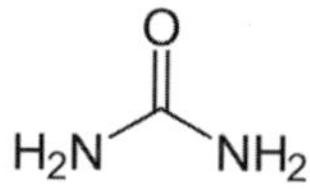

uric acid (= urate): 요산(尿酸). A heterocyclic compound of carbon, nitrogen, oxygen, and hydrogen with the formula $C_5H_4N_4O_3$.

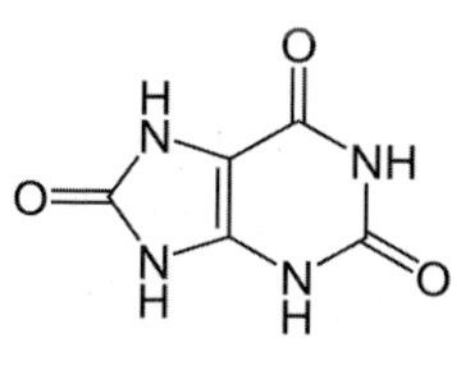

uridine: 우리딘(RNA의 성분). A molecule (known as a nucleoside) that is formed when uracil is attached to a ribose ring (also known as a ribofuranose) via a β-N_1-glycosidic bond.

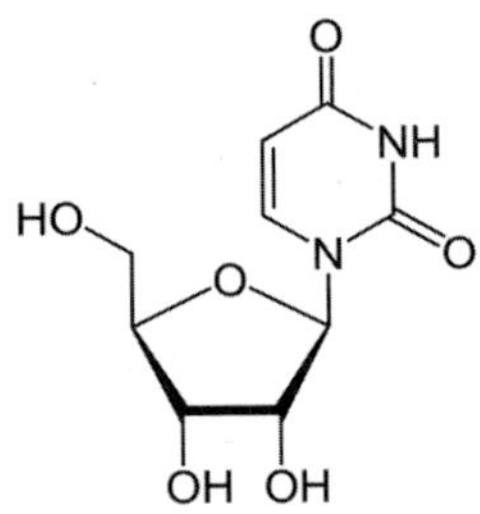

urethane[júərəθein]: French. *urėthane*, urea + ethane. ethyl carbamate. 우레탄(무색 무취의 결정체, 주로 최면제용). A chemical compound with the molecular formula $C_3H_7NO_2$ first prepared in the nineteenth century. Structurally, it is an ester of carbamic acid (NH_2COOH). Despite its common name, it is not a component of polyurethanes.

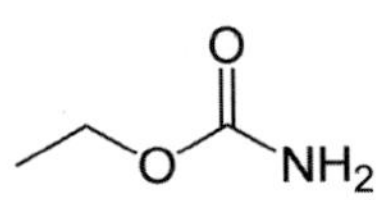

urine[júərin]: L. *urina, urinae,* a watery, typically yellowish fluid stored in the bladder and discharged through the urethra (요도). 오줌, 소변.

valency[véilənsi] = valence: L. > German. *valentia*, *valere*, > *valenz,* power, competence, strength, capacity, be well or strong. The chemical meaning referring to the "combining power of an element" is recorded from 1884. 원자가.

VSEPR(Valence Shell Electron Pair Repulsion) theory: 원자가 껍질 전자쌍 반발 이론. A model in chemistry used to predict the shape of individual molecules based upon the extent of electron-pair electrostatic repulsion. It is also named Gillespie–Nyholm theory after its two main developers. The acronym "VSEPR" is sometimes pronounced "vesper" for ease of pronunciation. The premise of VSEPR is that the valence electron pairs surrounding an atom mutually repel each other, and will therefore adopt an arrangement that minimizes this repulsion, thus determining the molecular geometry. The number of electron pairs surrounding an atom, both bonding and nonbonding, is called its steric number. VSEPR theory is usually compared and contrasted with valence bond theory, which addresses molecular shape through orbitals that are energetically accessible for bonding. Valence bond theory concerns itself with the formation of sigma and pi bonds. Molecular orbital theory is another model for understanding how atoms and electrons are assembled into molecules and polyatomic ions. VSEPR theory has long been criticized for not being quantitative, and therefore limited to the generation of "crude", even though structurally accurate, molecular geometries of covalent molecules.

vector[véktər]: L. *vehere*, convey. 벡터.

vertebrate[və́:rtəbrət, -breit]: L. *vertebratus*, jointed, an animal of a large group distinguished by the possession of a backbone or spiral column, including mammals, birds, reptiles, amphbians [양서류(兩棲類)], and fishes. 척추가 있는, 척추동물의.

vertex[vә́:rteks]: L. *vertere*, to turn, a meeting point of two lines that form an angle. 꼭지점, 정점.

vibration[vaibréiʃən]: L. *vibration,* from the verb *vibrare,* move to and fro. 진동.

<u>normal vibration</u>: 기준 진동

<u>rocking vibration</u>: 좌우 흔듦 진동

<u>scissoring vibration</u>: 가위질 진동

<u>streching vibration</u>: 신축진동

<u>torsional vibration</u>: 비틀림 진동

vicinal[vísənl]: L. *vicinalis, vicinus*, near, neighbour. 이웃자리, 이웃. (결정) 미사면(微斜面)의. Stands for any two functional groups bonded to two adjacent carbon atoms. For example the molecule 2,3-dibromobutane carries two vicinal bromine atoms and 1,3-dibromobutane does not.

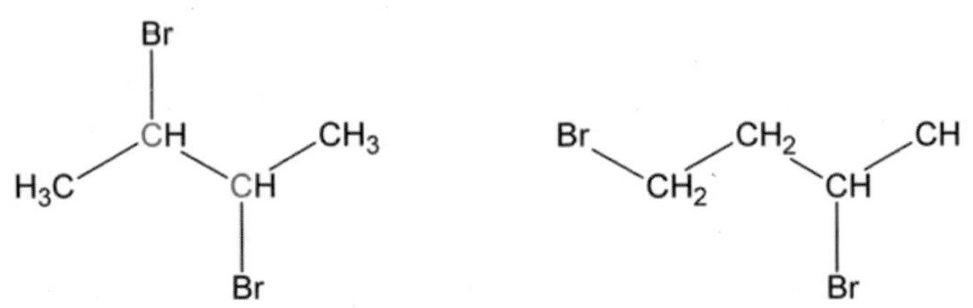

2,3-dibromobutane (at left) and 1,3-dibromobutane (at right).

vinegar[vínigər]: L. *vinum,* wine + *acer,* sour, a sour-tasting liquid containing acetic acid. 식초.

vinyl[váinl]: L. *vinum,* wine + *-yl,* suffix, of or denoting the unsaturated vinyl group (also called ethenyl), $-CH=CH_2$. 비닐기(基).

virus[váiərəs]: L. *virulentia, virulentiae*, slimy liquid, poison. 바이러스, 여과성 병원체.

visible[vízəbl]: L. *visibilis, videre*, to see. (눈에)보이는, 가시(可視)의.

visible spectrum: 가시선 분광[법,학]. The portion of the electromagnetic spectrum that is visible to (can be detected by) the human eye. Electromagnetic radiation in this range of wavelengths is called visible light or simply light. A typical human eye will respond to wavelengths from about 390 to 750 nm. A light-adapted eye generally has its maximum sensitivity at around 555 nm, in the green region of the optical spectrum (see: luminosity function). The spectrum does not, however, contain all the colors that the human eyes and brain can distinguish. Unsaturated colors such as pink, or purple variations such as magenta(짙은 분홍색 아닐린 염료), are absent, for example, because they can only be made by a mix of multiple wavelengths.

vision[víʒən]: L. *videre, vision*, to see. 시력, 시각.

vitamin[váitəmin / vít-]: L. *vitamin, vita*, life + amine, because vitamins were originally thought to contain an amino acid. 비타민. An organic compound required as a vital nutrient in tiny amounts by an organism. In other words, an organic chemical compound (or related set of compounds) is called a vitamin when it cannot be synthesized in sufficient quantities by an organism, and must be obtained from the diet. Thus, the term is conditional both on the circumstances and on the particular organism.

vivo[ví:vou]: L. *vivus*, living, in an organism. 생체, 살아있는. VS vitro-[vitrə]: in a test tube.

volatile[válətl, -til / vɔ́lətail]: L. *volatilis, volare*, to fly. 휘발성의, 증발하기 쉬운.

volume[válju:m / vɔl-]: L. *valumen, valumin,* a roll, from *volvere*, to roll. 부피, 용적.

water[wɔ́:tər, wát-]: G. *hudōr*, water. 물, 음료수. A chemical substance with the chemical formula H_2O. A water molecule contains one oxygen and two hydrogen atoms connected by covalent bonds. Water is a liquid at ambient conditions, but it often co-exists on Earth with its solid state, ice, and gaseous state (water vapor or steam). Water also exists in a liquid crystal state near hydrophilic surfaces. Under nomenclature used to name chemical compounds, Dihydrogen monoxide is the scientific name for water, though it is almost never used.

wave[weiv]: Middle English. *wawe*, (sea) wave. 파(波), 파동. A disturbance (an oscillation) that travels through space and time, accompanied by the transfer of energy. Waves travel and the wave motion transfers energy from one point to another, often with no permanent displacement of the particles of the medium— that is, with little or no associated mass transport.

They consist, instead, of oscillations or vibrations around almost fixed locations. One type of wave is a mechanical wave, which propagates through a medium in which the substance of this medium is deformed. For example, sound waves propagate via air molecules bumping into their neighbors. This transfers some energy to these neighbors, which will cause a cascade of collisions between neighbouring molecules. When air molecules collide with their neighbors, they also bounce away from them (restoring force). This keeps the molecules from continuing to travel in the direction of the wave. Another type of wave can travel through a vacuum, e.g. electromagnetic radiation (including visible light, ultraviolet radiation, infrared radiation, gamma rays, X-rays, and radio waves). This type of wave consists of periodic oscillations in electrical and magnetic fields. Waves are described by a wave equation which sets out how the disturbance proceeds over time. The mathematical form of this equation varies depending on the type of wave.

wave function: 파동 함수. A probability amplitude in quantum mechanics describing the quantum state of a particle and how it behaves. Typically, its values are complex numbers and, for a single particle, it is a function of space and time. The laws of quantum mechanics (the Schrödinger equation) describe how the wave function evolves over time. The wave function behaves qualitatively like other waves, like water waves or waves on a string, because the Schrödinger equation is mathematically a type of wave equation. This explains the name "wave function", and gives rise to wave-particle duality. The most common symbols for a wave function are ψ or Ψ (lower-case and capital psi). Although ψ is a complex number, $|\psi|^2$ is real, and corresponds to the probability density of finding a particle in a given place at a given time, if the particle's position is measured.

wavelength: 파장. The spatial period of the wave — the distance over which the wave's shape repeats. It is usually determined by considering the distance between consecutive corresponding points of the same phase, such as crests, troughs, or zero crossings, and is a characteristic of both traveling waves and standing waves, as well as other spatial wave patterns. Wavelength is commonly designated by the Greek letter lambda (λ). The concept can also be applied to periodic waves of non-sinusoidal shape. The term wavelength is also sometimes applied to modulated waves, and to the sinusoidal envelopes of modulated waves or waves formed by interference of several sinusoids(곡선). The SI unit of wavelength is the meter. Assuming a sinusoidal wave moving at a fixed wave speed, wavelength is inversely proportional to frequency: waves with higher frequencies have shorter wavelengths, and lower frequencies have longer wavelengths. Examples of wave-like phenomena are sound waves, light, and water waves. A sound wave is a periodic variation in air pressure, while in light and other electromagnetic radiation the strength of the electric and the magnetic field vary. Water waves are periodic variations in the height of a body of water.

In a crystal lattice vibration, atomic positions vary periodically in both lattice position and time. Wavelength is a measure of the distance between repetitions of a shape feature such as peaks, valleys, or zero-crossings, not a measure of how far any given particle moves. For example, in waves over deep water a particle in the water moves in a circle of the same diameter as the wave height, unrelated to wavelength.

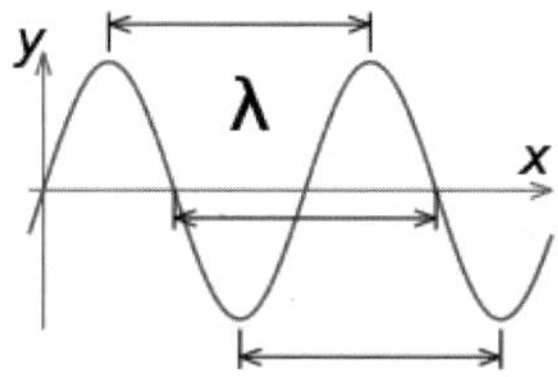

wavenumber: 파수(波數). A property of a wave, its spatial frequency, that is proportional to the reciprocal of the wavelength. It is also the magnitude of the wave vector. Its usual symbols are ν, $\bar{\nu}$, σ or k, the first three used for one definition, the last for another. The wavenumber has dimensions of reciprocal length, so its SI unit is m^{-1} and cgs unit cm^{-1} (in this context formerly called the *kayser*, after Heinrich Kayser). It can be defined as either $\tilde{\nu} = 1/\lambda$, the number of wavelengths per unit distance, where λ is the wavelength, sometimes termed the spectroscopic wavenumber, or $k = 2\pi / \lambda$, the number of wavelengths per 2π units of distance, sometimes termed the angular or circular wavenumber, but more often simply *wavenumber*. For electromagnetic radiation in vacuum, wavenumber is proportional to frequency and to photon energy. Because of this, wavenumbers are used as a unit of energy in spectroscopy. In the SI units, wavenumber is expressed in units of reciprocal meters (m^{-1}), but in spectroscopy it is usual to give wavenumbers in reciprocal centimeters (cm^{-1}). The angular wavenumber is expressed in radians per meter ($rad \cdot m^{-1}$).

xenon[zí:nɑn, zé-/zénɔn]: G. *xenos, xene, xenon,* foreign, strange, the hitherto unknown inert gas. Xe. 지논[희(稀)가스류의 원소]

zwitterion[tsvitəráiən]: German. *zwitterion*, *zwitter*, a hybrid + *ion*, ion. 양성(兩性) 이온(양전기와 음전기의 양쪽을 띈 이온).

사전

English for You. 영한사전. YBM-sisa 2000.

English for You. 한영사전. YBM-sisa 2000.

Oxford Dictionary of English (2nd ed). Oxford: Oxford University Press 2003.

Oxford Advanced Learner's Dictionary (7th ed). Oxford: Oxford University Press 2005.

The Concise Oxford Dictionary of English Etymology. Hoad TF (ed). Oxford: Oxford University Press, 1996.

화학 술어집 (4th ed). 대한화학회. 1993.

유기화합물 명명법 (2nd ed). 청문각. 2000.

의학어원론. 정상우. 군자출판사. 2009.

Wikipedia, The Free Encyclopedia.

유기화학 서적

Organic Chemistry (6th ed). Paula Yurkanis Bruice. Pearson. 2011.

Organic Chemistry (7th ed). McMurry. Tomson(Brooks/Cole). 2011.

March's Advanced Organic Chemistry 6th ed. Michael B. Smith and Jerry March. Wiley 2007.

Index

B

C

D

E

F

G

H

M

N

O

P

Q

R

S

T

저자_ **이왕근**

서울대학교 화학교육과
서울대학교 대학원 화학과
미국 Texas Tech 대학교 대학원 화학 및 생화학과(이학박사 Ph.D.)
전남대학교 사범대학 화학교육과 교수(유기화학)
현재 전남대학교 명예교수

유기화학의 어원과 용어

인쇄 2014년 12월 20일
발행 2014년 12월 30일

편저자 이왕근
발행인 지병문
발행처 전남대학교출판부

등록 1981. 5. 21. 제53호
주소 500-757 광주광역시 북구 용봉로 77
전화 (062) 530-0571~2 **마케팅** 530-0573
팩스 (062) 530-0579
홈페이지 http://www.cnup.co.kr
이메일 cnup0571@hanmail.net

값 17,000원

ISBN 978-89-6849-164-1 (93430)

이 도서의 국립중앙도서관 출판예정도서목록(CIP)은
서지정보유통지원시스템 홈페이지(http://seoji.nl.go.kr)와
국가자료공동목록시스템(http://www.nl.go.kr/kolisnet)에서 이용하실 수 있습니다.
(CIP제어번호 : CIP2014036596)